Math Mammoth
Grade 8 Answer Keys

for the complete curriculum
(Light Blue Series)

Includes answer keys to:

- Worktext part A
- Worktext part B
- Tests
- Cumulative Reviews

By Maria Miller

Contents

Math Mammoth Grade 8-A
Answer Key

By Maria Miller

Contents

Chapter 1: Exponents and Scientific Notation

Powers and the Order of Operations, pp. 13-14

Page 13

1. a. 64   b. 16   c. 125
   d. 0.008   e. 1   f. 10,000

2. a. Neither; they are equal.   b. 2^5

3.

a.	b.	c.
$10^1 = 10$	$2^1 = 2$	$0.1^1 = 0.1$
$10^2 = 100$	$2^2 = 4$	$0.1^2 = 0.01$
$10^3 = 1,000$	$2^3 = 8$	$0.1^3 = 0.001$
$10^4 = 10,000$	$2^4 = 16$	$0.1^4 = 0.0001$
$10^5 = 100,000$	$2^5 = 32$	$0.1^5 = 0.00001$
$10^6 = 1,000,000$	$2^6 = 64$	$0.1^6 = 0.000001$
$10^7 = 10,000,000$	$2^7 = 128$	$0.1^7 = 0.0000001$

Page 14

4.

a. $4,000 - 500$ $= \underline{3,500}$	b. $4(5^2 - 2^3)$ $= 4(25 - 8)$ $= \underline{68}$	c. $\dfrac{3}{1^8} + \dfrac{5}{3^2}$ $= \dfrac{3}{1} + \dfrac{5}{9} = 3\dfrac{5}{9}$
d. $7 \cdot 10^3 - 5(800 - 10^2)$ $= 7,000 - 5(700)$ $= 7,000 - 3,500 = \underline{3,500}$	e. $500 - \dfrac{3 \cdot 8}{2^3} + 2 \cdot 8^2$ $= 500 - \dfrac{24}{8} + 2 \cdot 64$ $= 500 - 3 + 128$ $= \underline{625}$	f. $\dfrac{2 \cdot 17 + 2^4}{7 \cdot 7 - 3^2} + 20$ $= \dfrac{34 + 16}{49 - 9} + 20$ $= \dfrac{50}{40} + 20$ $= \underline{21 \frac{1}{4}}$

5.

a. $0.5^2 - 0.2^2 - 0.1^2$ $= 0.25 - 0.04 - 0.01 = \underline{0.2}$	b. $3(0.1^2 - 0.2^3)$ $= 3(0.01 - 0.008)$ $= 3(0.002) = \underline{0.006}$	c. $0.6^2 + 2(1 - 0.3^2)$ $= 0.36 + 2(1 - 0.09)$ $= 0.36 + 2(0.91) = \underline{2.18}$

6. a. $4,096 \cdot 4 = \underline{16,384}$
   b. It is 4^8.
   c. No, because $16 + 64$ is not equal to 1,024.
   d. Yes: $4^2 \cdot 4^3 = 16 \cdot 64 = 1,024 = 4^5$.

7. a. Answers will vary; check the student's answer. For example, $3^4 = 81$ is greater than $7^2 = 49$.

   b. Answers will vary; check the student's answer. For example, $5^5 = 3,125$ is greater than $10^3 = 1,000$.

   c. This is not possible. All powers of 1 equal one.

8. a. It is $729 \cdot 3 \cdot 3 = 729 \cdot 9 = 6,561$.   b. It is $256 \cdot 2 \cdot 2 \cdot 2 = 256 \cdot 8 = 2,048$.

9. a. 2   b. 3   c. 4   d. Answers will vary; any whole number will work.

   e. 4   f. 5   g. 4   h. 7

10. a. 1/36   b. 27/1000   c. 16/81   d. 27/64

11. a. $14x^5$ b. $36x^5y^3$

   c. $60a^3b^3$ d. $0.6p^2r^4$

12. a. $10 \cdot 2^4 \cdot 3^2 = 10 \cdot 16 \cdot 9 = \underline{1,440}$.

   b. $14 \cdot 2^3 \cdot 0^5 = \underline{0}$.

13. Six times: $(1/2)^6 = 1/64$.

Puzzle corner. It is $\underline{9}$. In the expression $\dfrac{9 \cdot 9 \cdot 9 \cdot 9 \cdot 9 \cdot 9}{9 \cdot 9 \cdot 9 \cdot 9 \cdot 9}$, you can cancel out almost all the nines, leaving only

one nine in the numerator.

Powers with Negative Bases, pp. 16-17

1. a. 14   b. −9   c. −56   d. −6

   e. −4   f. −81   g. 28   h. 9

2. Answers will vary; check the student's answer. For example:

   A product of a positive and a negative integer can be written as a repeated addition where we add the negative number repeatedly, as many times as the positive integer indicates. For example: $3 \cdot (-4) = (-4) + (-4) + (-4)$. When you add a negative number repeatedly, the final sum will be negative, so a positive times a negative integer will be negative.

3. a. −14   b. 20   c. −30

   d. 12   e. −20   f. 60

   g. 240   h. −27/20 = −1 7/20   i. 9

4. a. −2.5   b. −8   c. 3

5. a. negative   b. positive   c. negative   d. positive

6.

> When there is an <u>odd</u> number of factors, the product is negative.
> For example, $(-1)^7$ has seven factors, and thus its value is negative 1.
>
> When there is an <u>even</u> number of factors, the product is positive.
> For example, $(-2)^6$ has six factors, and thus its value is (positive) 64.

7. a. −8   b. 1   c. −600

   d. −1   e. 256   f. 2,400

8. a. 1/25   b. −1/8   c. 16/81

   d. −7/8   e. −1/288   f. −8/15

Powers with Negative Bases, cont.

9.

a. $7 \cdot (-1{,}000) \cdot 100$ $= \underline{-700{,}000}$	b. $(4-6)^3 - (1+2)^3$ $= (-2)^3 - 3^3$ $= -8 - 27 = \underline{-35}$	c. $2(1-5)^2 + 3(-1+2)^4$ $= 2(-4)^2 + 3(1)^4$ $= 2(16) + 3 = \underline{35}$
d. $\dfrac{(-1)^4}{(-1)^7} + \dfrac{5}{(-2)^3}$ $= \dfrac{1}{-1} + \dfrac{5}{-8}$ $= -1 - \dfrac{5}{8} = -1\dfrac{5}{8}$	e. $\dfrac{-8}{(-2)^4} + \dfrac{(-1)^5}{(-3)^2}$ $= \dfrac{-8}{16} + \dfrac{-1}{9}$ $= -\dfrac{1}{2} - \dfrac{1}{9} = -\dfrac{11}{18}$	f. $\dfrac{6 \cdot 8}{(-4)^3} + 2 \cdot (-7)^2$ $= \dfrac{48}{-64} + 2 \cdot 49$ $= -\dfrac{3}{4} + 98 = 97\dfrac{1}{4}$

10.

a. -25 25	b. $-1{,}000$ $-1{,}000$	c. 92 108	d. 74 -54
e. 2 4	f. -1 7	g. -8 -98	h. -46 12

11. a. This is exponentiation, not multiplication. $2^5 = 32$.
   b. The negative sign is applied last and is not raised to any power. $-2^4 = -16$.
   c. When a negative integer is raised to an even power, the answer is positive.
     $(-1)^4 = 1$.

12. a. 3   b. 3   c. 0.04
   d. 8   e. $-1/2$ f. $-2/3$

13. See the answer on the right.

	1. 5		2. −9	
3. −1	0	0	0	
0				4. −1
2		5. −2	1	6
6. 4	5	0		2
		0		

Laws of Exponents, Part 1, p. 19

1. a. $2^4 2^3 = 2 \cdot 2 \cdot 2 \cdot 2 \cdot 2 \cdot 2 \cdot 2 = 2^7$ b. 3^6 c. a^{12}

2. a. No. The exponents don't get multiplied, but added: $2^4 2^2 = 2^6$.
   b. No. The base (2) does not change when powers (the 2^3 and 2^3) are multiplied: $2^3 2^3 = 2^6$.
   c. Yes.

3. $\dfrac{4^7}{4^5} = \dfrac{4 \cdot 4 \cdot 4 \cdot 4 \cdot 4 \cdot 4 \cdot 4}{4 \cdot 4 \cdot 4 \cdot 4 \cdot 4} = 4^2$

4. $\dfrac{5^5}{5^2} = \dfrac{5 \cdot 5 \cdot 5 \cdot 5 \cdot 5}{5 \cdot 5} = 5^3$

Page 19

5. x^4

6. No. The expression $\dfrac{2^5}{2^4}$ simplifies to 2, and thus $\dfrac{2^5}{2^4} \cdot 2 = 2 \cdot 2 = 4$.

   If, instead, the equation was $\dfrac{2^4}{2^5} \cdot 2 = 1$, it would be true.

7. a. No.   b. Yes.
   c. No.   d. Yes.

Page 20

8.

a. 4^{14}	b. $(-7)^{19}$	c. $\left(-\dfrac{2}{3}\right)^7$	
d. 5^{11}	e. 0.3^7	f. a^{16}	g. x^{m+7}

9. a. 10^7 b. 2^8 c. 3^9 d. 4^5

10. a. $x = 10$ b. $x = 2$ c. $x = 4$ d. $x = 2$

11. a. $8^9 10^7$ b. $3^7 4^3 5^8$ or $3^7 2^6 5^8$ c. $2^{11} 3^7$
    d. $2^4 3^4 5^4$ e. $2^{11} 5^2 7$ f. $2^5 3^7$

12. a. $x^{10} y^{12}$ b. $2a^{10} b^7$ c. $30 r^5 s^6 t^3$

Page 21

13. $\dfrac{a^6}{a^3} = \dfrac{a \cdot a \cdot a \cdot a \cdot a \cdot a}{a \cdot a \cdot a} = a^3$

14. a. 8^6 b. $(-7)^7$ c. 3^3 d. 0.4^5
    e. y^{11} f. $7x$ g. 4^2 h. 5^5

15. a. $x = 3$ b. $y = 9$ c. $w = 12$ d. $x = 5$

16. To find how many times one number is greater than another, we divide the numbers. When dividing powers of the same base, the exponents are subtracted, not divided. So the answer is actually 2^4 times, or 16 times.

17. a. $\dfrac{8^6 \cdot 5^3}{8^3 \cdot 5^2} = 8^3 \cdot 5$ b. $\dfrac{x^6 y^8}{y^3 x^3} = x^3 y^5$ c. $\dfrac{3 x^4 y^6}{5 x y^2} = \dfrac{3}{5} x^3 y^4$

18. Expressions (c) and (f).

Zero and Negative Exponents, p. 22

Page 22

Teaching box:

Zero and Negative Exponents, cont.

Page 22

1. The value for a^0 must be 1 because $a^m \cdot 1 = a^m$.

2. When $m = 4$ and $n = 4$, the quotient law becomes $\dfrac{a^4}{a^4} = a^{4-4} = a^0$. Since $\dfrac{a^4}{a^4} = 1$, this shows us that a^0 must equal 1.

3. a. Applying the quotient law, we get $\dfrac{3^2}{3^3} = 3^{2-3} = 3^{-1}$. Since $\dfrac{3^2}{3^3} = \dfrac{3 \cdot 3}{3 \cdot 3 \cdot 3} = \dfrac{1}{3}$, 3^{-1} must equal 1/3.

   b. Applying the quotient law, we get $\dfrac{5^2}{5^3} = 5^{2-3} = 5^{-1}$. Since $\dfrac{5^2}{5^3} = \dfrac{5 \cdot 5}{5 \cdot 5 \cdot 5} = \dfrac{1}{5}$, 5^{-1} must equal 1/5.

4.

2^3	8	÷ 2
2^2	4	÷ 2
2^1	2	÷ 2
2^0	1	÷ 2
2^{-1}	1/2	÷ 2
2^{-2}	1/4	÷ 2
2^{-3}	1/8	÷ 2

5^3	125	÷ 5
5^2	25	÷ 5
5^1	5	÷ 5
5^0	1	÷ 5
5^{-1}	1/5	÷ 5
5^{-2}	1/25	÷ 5
5^{-3}	1/125	÷ 5

10^3	1000	÷ 10
10^2	100	÷ 10
10^1	10	÷ 10
10^0	1	÷ 10
10^{-1}	1/10	÷ 10
10^{-2}	1/100	÷ 10
10^{-3}	1/1000	÷ 10

Page 23

5. a. 1/36   b. 1/16   c. 1/125   d. 7/10   e. 10/27

6. a. $2 \cdot 5^{-1}$ b. $5 \cdot 2^{-3}$ c. $11 \cdot 3^{-2}$ d. $37 \cdot 10^{-3}$

7. Yes, it does: $a^{-0} = \dfrac{1}{a^0} = \dfrac{1}{1} = 1$.

8.

a. $5 \cdot 2^8 2^{-5} = 5 \cdot 2^3 = 40$	b. $5 \cdot 2^{-2} 2^{-3} \cdot 3 = 15 \cdot 2^{-5} = \dfrac{15}{2^5} = \dfrac{15}{32}$	c. $x^{-3} x^2 = x^{-1} = \dfrac{1}{x}$
d. $y \cdot y^{-2} y^5 = y^4$	e. $3a^{-4} a^7 a^{-5} = 3a^{-2} = \dfrac{3}{a^2}$	f. $2b^5 \cdot a^{-3} a^{-2} = \dfrac{2b^5}{a^5}$

9. a. $x = -3$ b. $x = -6$ c. $x = 7$ d. $b = -5$
   e. $z = -4$ f. $a = -7$ g. $y = 0$ h. $x = 6$

Page 24

10. The former is −4 raised to the power of negative 2, and equals $\dfrac{1}{(-4)^2} = \dfrac{1}{16}$.

   The latter is the opposite of 4^{-2}, and equals $-\dfrac{1}{4^2} = -\dfrac{1}{16}$.

11. a. 1/8   b. −1/8   c. −1/8   d. −8

12. a. 1/25   b. −1/27   c. −1/8
    d. −53/100   e. −7/64   f. −50

13. a. $x = -4$ b. $y = -5$ c. $w = -7$

14. a. 6^2 b. 2^{-7} c. 9^{-7} d. $(-4)^{-8}$

15. a. 25/10 = 2 1/2   b. 1/12   c. 12
    d. 90/80 = 1 1/8   e. 6/40 = 3/20

16. (b), (c), (d), and (f)

13

More on Negative Exponents, pp. 25-26

Page 25

1. a. 81   b. 27   c. $5 \cdot 16 = \underline{80}$
   d. x^5 e. $7a^4$ f. x^n

2. a. 7^3 b. $2x^6$ c. $\dfrac{1}{2y^4}$ d. $\dfrac{4w^7}{s^2}$

3.

a.	b.	c.
$\dfrac{9 \cdot 2^{-4}}{5 \cdot 2^{-2}} = \dfrac{9 \cdot 2^{-2}}{5} = \dfrac{9}{5 \cdot 2^2} = \dfrac{9}{20}$	$\dfrac{3 \cdot 10^{-1}}{5 \cdot 10^{-3}} = \dfrac{3 \cdot 10^3}{5 \cdot 10^1} = \dfrac{3 \cdot 10^2}{5} = \dfrac{300}{5} = 60$	$\dfrac{10 \cdot 5^{-1}}{3 \cdot 5^2} = \dfrac{10}{3 \cdot 5^3} = \dfrac{2}{3 \cdot 5^2} = \dfrac{2}{75}$

Page 26

4. a. $\dfrac{7^5}{2^3}$ b. $\dfrac{2 \cdot 5^6}{3^{11}}$

   c. $\dfrac{2 \cdot 3 \cdot 10^5}{7}$ d. $\dfrac{1}{7^7 \cdot 2^8}$

   e. $\dfrac{x^3}{y}$ f. $2a^6 b^4$

   g. $\dfrac{2a^5}{3b^5}$ h. $\dfrac{x^2}{3y}$

5. a. 262,144   b. −16,384
   c. 1,024   d. 1/1024
   e. 65,536   f. −1/64
   g. 1/16   h. 1

Laws of Exponents, Part 2, p. 27

Page 27

1. a. $(9^5)^2 = 9^5 \cdot 9^5 = 9^{10}$ b. $(a^4)^3 = a^4 \cdot a^4 \cdot a^4 = a^{12}$

   c. $(x^{-1})^3 = \left(\dfrac{1}{x}\right)^3 = \dfrac{1}{x} \cdot \dfrac{1}{x} \cdot \dfrac{1}{x} = \dfrac{1}{x^3}$

2.

a. $(8^2)^4 = 8^8$	d. $x^6 x^{-3} = x^3$	g. $((-2)^{-3})^5 = (-2)^{-15}$	j. $s^{-1} s^{-7} = s^{-8}$
b. $8^2 8^4 = 8^6$	e. $(x^6)^{-3} = x^{-18}$	h. $(-2)^{-3}(-2)^5 = (-2)^2$	k. $(s^{-1})^{-7} = s^7$
c. $\dfrac{8^2}{8^4} = 8^{-2}$	f. $\dfrac{x^6}{x^{-3}} = x^9$	i. $\dfrac{(-2)^{-3}}{(-2)^5} = (-2)^{-8}$	l. $\dfrac{s^{-1}}{s^{-7}} = s^6$

3. a. $x = 2$ b. $x = 4$ c. $x = -5$ d. $x = -3$

4.

a.	b.	c.	d.
$27^6 = (3^3)^6 = 3^{18}$	$100^6 = (10^2)^6 = 10^{12}$	$125^3 = (5^3)^3 = 5^9$	$4^{12} = (2^2)^{12} = 2^{24}$

Page 28

5.

a.	b.	c.	d.
$(-7)^2 = 49$	$2 \cdot 5^3 = 250$	$3 \cdot 4^{-2} = 3/16$	$\left(\frac{2}{3}\right)^3 = \frac{8}{27}$
$-7^2 = -49$	$(2 \cdot 5)^3 = 1{,}000$	$(3 \cdot 4)^{-2} = 1/144$	$\frac{2^3}{3} = \frac{8}{3}$

6. See the solution on the right.

	1. 1	3	2. 5
	2		4
3. −2	8	4. 8	
1		1	
5. 6	6. 4		7. −1
	8. 2	4	3

7. a. $(a + b)^2 = 11^2 = 121$. $a^2 + b^2 = 4 + 81 = 85$.
   b. In the model, a^2 corresponds to section (i), and b^2 to section (iv).
   In contrast, $(a + b)^2$ corresponds to all four sections together (i, ii, iii, iv).

Page 29

8. $(xy)^3 = xy \cdot xy \cdot xy = x^3 \cdot y^3$

9. a. $16x^2$ b. $8a^3$ c. $2x^6$
   d. a^4b^4 e. $\frac{1}{9y^2}$ f. $\frac{1}{8x^3}$

10. Neither of them is correct. Using the definition of exponent as repeated multiplication, $(5a^3)^2 = (5a^3)(5a^3) = 25a^6$. Apparently, Robert added the exponents, ignoring the fact that 5 also is squared. Xavier multiplied the exponents correctly, but instead of raising 5 to the second power, he added 5 + 5 = 10.

11. The expression $(5\text{ m})^3$ refers to the quantity "5 meters" being cubed, whereas 5 m^3 simply means 5 cubic meters (only the unit "m" is cubed, not the number 5). The expression $(5\text{ m})^3 = 125\text{ m}^3$ gives us the volume of a cube with an edge of 5 m.

12.

$\frac{4x^2}{y}$	$7x^6$	$4x^2$	$-4x^2$
$(-2x)^2$	$\frac{(2x)^2}{y}$	$\left(\frac{x}{y}\right)^2$	$7x^9$
$\frac{x^2}{y^2}$	$\frac{x^2}{y}$	$7x^{3^2}$	$\frac{2x^2}{y}$

Page 30

1.

a. $(4y)^3 = 64y^3$	g. $\left(\dfrac{x}{y^2}\right)^3 = \dfrac{x^3}{y^6}$
b. $(11x)^2 = 121x^2$	h. $\left(\dfrac{7g}{8h}\right)^2 = \dfrac{49g^2}{64h^2}$
c. $(3ab)^3 = 27a^3b^3$	i. $\left(\dfrac{ab}{3c^5}\right)^4 = \dfrac{a^4b^4}{81c^{20}}$
d. $(-2w)^3 = -8w^3$	
e. $(-x^2)^5 = -x^{10}$	j. $\left(\dfrac{-2b}{b^2}\right)^5 = \dfrac{-32b^5}{b^{10}}$
f. $(2x)^4 \cdot (-3x)^2 = 16x^4 \cdot 9x^2 = 144x^6$	k. $\left(\dfrac{-x^2}{5x}\right)^3 = \dfrac{-x^6}{125x^3}$

Page 31

2. There are often several ways to work the process and to arrive to the same end result. Only one way is shown below.

a. $(5x^2)^{-1} = \dfrac{1}{5x^2}$	e. $(10x^{-3})^3 = \left(\dfrac{10}{x^3}\right)^3 = \dfrac{1000}{x^9}$	i. $(7e^2d^{-4})^2 = 49e^4d^{-8} = \dfrac{49e^4}{d^8}$
b. $(4y^3)^{-2} = \dfrac{1}{(4y^3)^2} = \dfrac{1}{16y^6}$	f. $(ab^2)^{-2} = \dfrac{1}{(ab^2)^2} = \dfrac{1}{a^2b^4}$	j. $(-2x)^3 \cdot (3x)^{-2} = -8x^3 \cdot \dfrac{1}{9x^2} = \dfrac{-8x^3}{9x^2} = \dfrac{-8x}{9}$
c. $(3a^4)^{-3} = \dfrac{1}{(3a^4)^3} = \dfrac{1}{27a^{12}}$	g. $(5x^2y)^{-3} = \dfrac{1}{(5x^2y)^3} = \dfrac{1}{125x^6y^3}$	k. $(-3y^2)^{-4} \cdot y^5 = \dfrac{y^5}{(-3y^2)^4} = \dfrac{y^5}{81y^8} = \dfrac{1}{81y^3}$
d. $(7b^{-4})^2 = \left(\dfrac{7}{b^4}\right)^2 = \dfrac{49}{b^8}$	h. $(-3w^{-2})^5 = \left(\dfrac{-3}{w^2}\right)^5 = \dfrac{-243}{w^{10}}$	l. $(-3c)^3 \cdot (5c)^{-2} = \dfrac{(-3c)^3}{(5c)^2} = \dfrac{-27c^3}{25c^2} = \dfrac{-27c}{25}$

3.

a. This is correct.	b. This is incorrect. The correct version is below: $\left(\dfrac{4ab}{-3b^3}\right)^3 = \dfrac{64a^3b^3}{-27b^9} = -\dfrac{64a^3}{27b^6}$

4.

a. $(4^4)^3 = 4^{12} = (4^2)^6 = 16^6$	b. $81^4 = (9^2)^4 = (3^4)^4 = 3^{16} = (3^8)^2$	c. $(32 \cdot 5^x)^{-3} = (2^5 \cdot 5^5)^{-3} = (10^5)^{-3} = \dfrac{1}{10^{15}}$

5.

$\left(\dfrac{6x}{y}\right)^2$	$8a^3b^{-6}$	$\dfrac{6x^2}{y^2}$	$36x^2y^2$
$36x^2y^{-2}$	$\dfrac{(6x)^2}{y^{-2}}$	$\left(\dfrac{2a}{b^2}\right)^3$	$8a^6b^6$
$\dfrac{6a^3}{b^5}$	$\dfrac{2a^2}{b^6}$	$-6x^2y^2$	$\dfrac{8a^3}{b^6}$

16

Laws of Exponents, Part 3, cont.

Page 31

Puzzle corner. a. 3   b. 10 and 8.
A number raised to the exponent of ½ is the square root of the number. Using similar reasoning, one can show that other fractional exponents correspond to roots, also.

Scientific Notation: Large Numbers, pp. 32-34

Page 32

1.

Scientific Notation	Decimal notation	Scientific Notation	Decimal notation
$6 \cdot 10^5$	600,000	$8.904 \cdot 10^3$	8,904
$2.5 \cdot 10^5$	250,000	$1.5594 \cdot 10^8$	155,940,000
$2.03 \cdot 10^6$	2,030,000	$3.6002 \cdot 10^{11}$	360,020,000,000

2. a. No, he is not correct. $3.58 \cdot 10^9 = 3,580,000,000$.
   b. When a is 1, 2, 3, 4, 5, 6, 7, 8, or 9.

Page 33

3. a. $1.3 \cdot 10^4$ b. $2.04 \cdot 10^5$ c. $4.506 \cdot 10^6$
   d. $4.528 \cdot 10^7$ e. $9.7005 \cdot 10^9$ f. $4.051 \cdot 10^{11}$

4. a. <   b. >   c. <
   d. >   e. <   f. <

5. Charlie is correct, because changing 210 to 2.1 makes it hundred times smaller, and so 10^5 becomes 100 times larger, or 10^7.

6.

7. a. $2.6 \cdot 10^7$ b. $9 \cdot 10^4$ c. $3.58 \cdot 10^6$
   d. $2.08 \cdot 10^6$ e. $2 \cdot 10^6$ f. $1.01 \cdot 10^7$

Page 34

8. a. =   b. <   c. <

9. a. The cost is $0.8 \cdot 20,000,000 \cdot \$20,000$
      $= 0.8 \cdot 2 \cdot 10^7 \cdot \$2 \cdot 10^4 = 3.2 \cdot 10^{11}$ dollars.
      This is such a high cost that this program may not happen!

   b. Writing the number out in decimal notation ($320,000,000,000) is one possibility. Another is to state it as 0.32 trillion dollars, or about one-third of a trillion dollars.

10. The cost would be $3/4 \cdot 8,000,000,000 \cdot \$20,000$
      $= 6 \cdot 10^9 \cdot \$2 \cdot 10^4 = 12 \cdot 10^{13}$ dollars
      $= 1.2 \cdot 10^{14}$ dollars.

11. See the answer on the right.

a. 8	7	b. 4	0	0	0		c. 7
9		2					4
d. 3	0	0			e. 5		0
0		f. 5	0	3	2	0	0
		0			0		0
					0		0
g. 2	9	9	0	0	0	0	0

Scientific Notation: Small Numbers, pp. 35-36

1.

Scientific Notation	Decimal notation	Scientific Notation	Decimal notation
$3 \cdot 10^{-5}$	0.00003	$2.388 \cdot 10^{-7}$	0.0000002388
$8 \cdot 10^{-4}$	0.0008	$8.2 \cdot 10^{-4}$	0.00082
$2.03 \cdot 10^{-6}$	0.00000203	$3.08 \cdot 10^{-9}$	0.00000000308
$6.108 \cdot 10^{-8}$	0.00000006108	$4.539 \cdot 10^{-7}$	0.0000004539

2. Yes.

3. a. >   b. <   c. >
   d. <   e. <   f. >

4. $9 \cdot 10^{-7} < 10^{-6} < 0.00002 < 0.0003 < 5.6 \cdot 10^{7} < 6 \cdot 10^{7} < 10^{8}$

5. a. $7 \cdot 10^{27}$

   b. There are more oxygen atoms.

6. $3.26964 \cdot 10^{-22}$ grams

7. a. $8.9 \cdot 10^{-4}$ b. $4.79 \cdot 10^{-4}$

   c. $3 \cdot 10^{-5}$ d. $2.08 \cdot 10^{-7}$

   e. $4.5 \cdot 10^{-10}$ f. $2 \cdot 10^{-8}$

8.

Scientific Notation	Decimal notation	Scientific Notation	Decimal notation
$-4 \cdot 10^{5}$	−400,000	$-7 \cdot 10^{-3}$	−0.007
$-5.9 \cdot 10^{7}$	−59,000,000	$-2.81 \cdot 10^{-7}$	−0.000000281
$-1.506 \cdot 10^{6}$	−1,506,000	$-9.8 \cdot 10^{-6}$	−0.0000098
$-1.0082 \cdot 10^{9}$	−1,008,200,000	$-5.03 \cdot 10^{-8}$	−0.0000000503

9. USA: $-2.682 \cdot 10^{13}$ dollars. India: $-1.261 \cdot 10^{12}$ dollars. The USA has more debt.

10. $-4 \cdot 10^{4} < -4 \cdot 10^{-4} < 4 \cdot 10^{-4} < 0.004 < 4 \cdot 10^{4}$

Significant Digits, p. 37

1. 0.5060 kg

2. a. 3   b. 3   c. 4   d. 1
   e. 2   f. 5   g. 2   h. 3
   i. 5   j. 1   k. 3   l. 4

3. a. 2.5   b. 1,040   c. 29,500
   d. 1,736   e. 0.30   f. 1.0329
   g. 500   h. 4,390,000 i. 656,800,000

Significant Digits, cont.

Page 38

4. a. 19 m^2 b. 7 m^2
   c. $28,000 \text{ ft}^2$ d. $120,000 \text{ kg}$
   e. $\$3,500$ f. $\$5,000$
   g. 880 km/h h. 22 mi/gal

5. $4,000 \cdot \$4.59 \cdot 365 \approx \$6,700,000$

6. The amount of power they used (in kilowatt-hours) is $0.95 \text{ kW} \cdot 8.0 \text{ hr} \cdot 30 = 228 \text{ kWh}$. This will cost $228 \text{ kWh} \cdot 14¢/\text{kWh} = \$31.92 \approx \underline{\$32}$.

7. a. $24.5 \text{ m} \cdot 13.8 \text{ m} \approx 338 \text{ m}^2$
   b. $24.56 \text{ m} \cdot 13.89 \text{ m} \approx 341.1 \text{ m}^2$

Page 39

8. a. 56 in __140__ cm d. 350 gal __1,300__ L
   b. 240 cm __94__ in e. 375 L __99.1__ gal
   c. 46 m __150__ ft f. 125 ft __38.1__ m

9. a. $5.6 \text{ kg} + 2.04 \text{ kg} - 0.078 \text{ kg} \approx \underline{7.6}$ kg b. $7.6 \text{ m} + 0.752 \text{ m} + 2.09 \text{ m} \approx \underline{10.4}$ m

   c. $14 \text{ lb} + 7.8 \text{ lb} + 55 \text{ lb} \approx \underline{77}$ lb d. $506 \text{ mi} + 78 \text{ mi} + 5.9 \text{ mi} \approx \underline{590}$ mi

10. a. Since 6.2 cm has two significant digits and the scale ratio has five, we give the final answer to two significant digits: $6.2 \text{ cm} \cdot 50,000 = 310,000 \text{ cm} = \underline{3.1 \text{ km}}$

    b. Since 12.5 cm has three significant digits and the scale ratio has six, we give the final answer to three significant digits: $12.5 \text{ cm} \cdot 200,000 = 2,500,000 \text{ cm} = \underline{25.0 \text{ km}}$

11. First we calculate the distance in inches: $3.0 \text{ in} \cdot 10,000 = 30,000 \text{ in}$. Now we convert this to miles and round to two significant digits: $30,000 \text{ in} \cdot (1 \text{ ft})/(12 \text{ in}) \cdot (1 \text{ mi})/(5,280 \text{ ft}) = 0.473\overline{48} \text{ mi} \approx \underline{0.47 \text{ mi.}}$

12. The dimensions 5.0 cm and 3.5 cm are given to two significant digits. Let's calculate those dimensions in reality:

    $5.0 \text{ cm} \cdot 8,000 = 40,000 \text{ cm} = 400 \text{ m} = 0.40 \text{ km}$
    $3.5 \text{ cm} \cdot 8,000 = 28,000 \text{ cm} = 280 \text{ m} = 0.28 \text{ km}$

    The area needs be given to two significant digits, since both numbers we multiply have two significant digits:

    $A = 0.40 \text{ km} \cdot 0.28 \text{ km} = 0.112 \text{ km}^2 \approx \underline{0.11 \text{ km}^2}$.

Using Scientific Notation in Calculations, Part 1, p. 40

Page 40

1. The sun is $\dfrac{2 \cdot 10^{30}}{6 \cdot 10^{24}} = \dfrac{1}{3} \cdot 10^6$, or about 333,000 times more massive than the earth.

2. a. It is $\dfrac{6 \cdot 10^{-20}}{3 \cdot 10^{-30}} = 2 \cdot 10^{10}$ times bigger.

   b. It is $\dfrac{2 \cdot 10^4}{8 \cdot 10^{-4}} = 0.25 \cdot 10^8 = 2.5 \cdot 10^7$ times bigger.

3. a. $150,000,000 \text{ km} = 1.5 \cdot 10^8 \text{ km}$
   b. From the formula governing distance, velocity, and time, $d = vt$, we can solve that $t = d/v$.

   So, the time it takes is distance divided by velocity, or $t = \dfrac{1.5 \cdot 10^8 \text{ km}}{3 \cdot 10^5 \text{ km/s}} = 0.5 \cdot 10^3 \text{ sec} = 500 \text{ sec} = \underline{8\ 1/3 \text{ minutes}}$.

Using Scientific Notation in Calculations, Part 1, cont.

Page 41

4. a. It signifies the number $1.5 \cdot 10^{26}$.

   b. Answers will vary; check the student's answer. For example: $5 \cdot 10^{10}$ and $0.3 \cdot 10^{16}$.

5. Typically, calculators show it the same way as very large numbers, just using a negative number for the exponent. For example, the calculator may show 2.4E-14 or 7.93e-25.

6. One light year is $299{,}792{,}458$ m $\cdot 60 \cdot 60 \cdot 24 \cdot 365 = 9{,}454{,}254{,}955{,}488{,}000 \approx 9.454 \cdot 10^{15}$ m $= 9.454 \cdot 10^{12}$ km.

7. The average mass of one bacterium is 0.2 kg $\div (3.8 \cdot 10^{13}) \approx 0.05263 \cdot 10^{-13}$ kg $= 0.05263 \cdot 10^{-10}$ g $\approx 5.3 \cdot 10^{-12}$ g.

8. The speed of the garden snail is $\dfrac{8.27 \cdot 10^6 \text{ in/hour}}{4{,}600} \approx 1{,}797.8261$ in/hour. This would be more reasonable to give in

   some unit per minute so we will convert it: $1{,}797.8261 \dfrac{\text{in}}{\text{hour}} \cdot \dfrac{1 \text{ ft}}{12 \text{ in}} \cdot \dfrac{1 \text{ hr}}{60 \text{ min}} \approx 2.496981$ ft/min. Since the speed of

   the eagle had 3 significant digits, and the 4,600 has two, we will give our answer with two significant digits, as 2.5 feet/minute.

Page 42

9. a. A trillion gold atoms have a mass of $10^{12} \cdot 3.2696 \cdot 10^{-22}$ grams $= 3.2696 \cdot 10^{-10}$ grams.

   b. 326.96 picograms or 0.32696 nanograms

10. The radius of a silicon atom is about 110 picometers $= 110 \cdot 10^{-12}$ m $= 1.1 \cdot 10^{-10}$ m.

    The radius of the nucleus of a silicon atom is about 3.6 femtometers $= 3.6 \cdot 10^{-15}$ m.

    The diameters are double that, or $2.2 \cdot 10^{-10}$ m and $7.2 \cdot 10^{-15}$ m.

    The diameter of the entire atom is $\dfrac{2.2 \cdot 10^{-10} \text{ m}}{7.2 \cdot 10^{-15} \text{ m}} \approx 0.31 \cdot 10^5 = \underline{31{,}000 \text{ times}}$ the diameter of the nucleus.

    (We use two significant digits in the answer, since 110 picometers and 3.6 femtometers both have 2 significant digits.)

Using Scientific Notation in Calculations, Part 2, p. 43

Page 43

1. a. $(2 \cdot 10^6) \cdot (3 \cdot 10^4) = 6 \cdot 10^{10} = 60{,}000{,}000{,}000$

   b. $2 \cdot 10^6 + 3 \cdot 10^4 = 2{,}000{,}000 + 30{,}000 = 2{,}030{,}000$

   c. $8 \cdot 10^3 + 7 \cdot 10^5 = 8{,}000 + 700{,}000 = 708{,}000$

   d. $(8 \cdot 10^3) \cdot (7 \cdot 10^5) = 56 \cdot 10^8 = 5{,}600{,}000{,}000$

2.

a. $4.8 \cdot 10^8 + 5 \cdot 10^7$	b. $9.3 \cdot 10^6 + 8 \cdot 10^7$	c. $5 \cdot 10^7 - 7 \cdot 10^5$	d. $8.4 \cdot 10^9 - 4.7 \cdot 10^8$
$= 48 \cdot 10^7 + 5 \cdot 10^7$	$= 9.3 \cdot 10^6 + 80 \cdot 10^6$	$= 5 \cdot 10^7 - 0.07 \cdot 10^7$	$= 84 \cdot 10^8 - 4.7 \cdot 10^8$
$= 53 \cdot 10^7$	$= 89.3 \cdot 10^6$	$= 4.93 \cdot 10^7$	$= 79.3 \cdot 10^8$
$= 5.3 \cdot 10^8$	$= 8.93 \cdot 10^7$		$= 7.93 \cdot 10^9$

3. Neither answer is correct. Using decimal notation, we get $5 \cdot 10^{-3} + 2 \cdot 10^{-4} = 0.005 + 0.0002 = 0.0052$, or $5.2 \cdot 10^{-3}$.

Page 43

4.

a. $8 \cdot 10^{-2} + 6 \cdot 10^{-3}$ $= 8 \cdot 10^{-2} + 0.6 \cdot 10^{-2} = \underline{8.6 \cdot 10^{-2}}$ Or, using decimal notation: $0.08 + 0.006 = 0.086 = \underline{8.6 \cdot 10^{-2}}$	b. $3 \cdot 10^{-6} + 5 \cdot 10^{-5}$ $= 0.3 \cdot 10^{-5} + 5 \cdot 10^{-5} = \underline{5.3 \cdot 10^{-5}}$
c. $2 \cdot 10^{-4} - 7 \cdot 10^{-6}$ $= 200 \cdot 10^{-6} - 7 \cdot 10^{-6} = 193 \cdot 10^{-6} = \underline{1.93 \cdot 10^{-4}}$	d. $5.4 \cdot 10^{-3} - 7 \cdot 10^{-4} =$ $= 5.4 \cdot 10^{-3} - 0.7 \cdot 10^{-3} = \underline{4.7 \cdot 10^{-3}}$

Page 44

5. a. The ratio of the volume of the sun to the volume of Jupiter is $\dfrac{1.4093 \cdot 10^{18} \text{ km}^3}{1.4313 \cdot 10^{15} \text{ km}^3} \approx 0.98463 \cdot 10^3 \approx 984.63$.

   So, <u>about 985 Jupiters would fit in the sun.</u>

   b. We convert one of the numbers so that both will have the same power of ten, and then subtract the decimals only:
   $1.0832 \cdot 10^{12} \text{ km}^3 - 1.6318 \cdot 10^{11} \text{ km}^3 = (10.832 - 1.6318) \cdot 10^{11} \text{ km}^3 \approx 9.200 \cdot 10^{11} \text{ km}^3$.
   (Note that since 10.832 had three decimal digits and 1.6318 had four, the answer is given to the accuracy of three decimals.)

   c. The total volume is $1.0832 \cdot 10^{12} \text{ km}^3 + 2.1968 \cdot 10^{10} \text{ km}^3 = 108.32 \cdot 10^{10} \text{ km}^3 + 2.1968 \cdot 10^{10} \text{ km}^3$
   $= (108.32 + 2.1968) \cdot 10^{10} \text{ km}^3 \approx 110.52 \cdot 10^{10} \text{ km}^3 = 1.1052 \cdot 10^{12} \text{ km}^3$.

6. Check the student's results. Answers will vary. You may take 10-20 breaths per minute. Using 12 breaths per minute, we can estimate the total number of breaths in a 70-year lifetime to be: $12 \cdot 60 \cdot 24 \cdot 365 \cdot 70 = 441{,}504{,}000$. This means that <u>option (b)</u>, or $5 \cdot 10^8$, is the closest estimate.

7. a. $\dfrac{4.22 \cdot 10^{27}}{7 \cdot 10^{27}} \approx 0.60285 \approx 60\%$.

   b. It has about $\dfrac{1.61 \cdot 10^{27}}{3.9 \cdot 10^{25}} \approx 0.4128 \cdot 10^2 \approx 41$ times more oxygen than nitrogen atoms.

   c. It has $1.61 \cdot 10^{27} - 8.03 \cdot 10^{26} = 1.61 \cdot 10^{27} - 0.803 \cdot 10^{27} = (1.61 - 0.803) \cdot 10^{27} \approx 0.81 \cdot 10^{27}$ more oxygen than carbon atoms.

Page 45

8. a. The total mass is $1 \text{ mg} \cdot 10^{15} = 10^{15} \text{ mg}$. Milligrams is not a sensible unit so we will use kilograms.
   Now, $1{,}000 \text{ mg} = 1 \text{ g}$, and $1{,}000 \text{ g} = 1 \text{ kg}$, so $1 \text{ kg} = 1{,}000{,}000 \text{ mg} = 10^6 \text{ mg}$. We can use a ratio to convert the
   amount to kilograms: $10^{15} \text{ mg} \cdot \dfrac{1 \text{ kg}}{10^6 \text{ mg}} = 10^9 \text{ kg}$.

   b. $4{,}800 \cdot 10^6 \cdot 60 \text{ kg} = 288{,}000 \cdot 10^6 \text{ kg} \approx 2.9 \cdot 10^{11} \text{ kg}$
   c. The people living in Asia. But both masses are quite large!

9. a. $2{,}750 \text{ km}^3 = 2{,}750 \cdot (1{,}000 \text{ m})^3 = 2{,}750 \cdot 1{,}000^3 \text{ m}^3 = 2{,}750 \cdot (10^3)^3 \text{ m}^3 = 2{,}750 \cdot 10^9 \text{ m}^3 = 2.75 \cdot 10^{12} \text{ m}^3$

   b. Since 1 cubic meter is 1,000 liters, the volume of water in Lake Victoria is $2{,}750 \cdot 10^{12} \text{ L}$. We will now divide that
   by 20 to get the number of bucketfuls of water. It has about $\dfrac{2{,}750 \cdot 10^{12} \text{ L}}{20 \text{ L}} = 137.5 \cdot 10^{12} \approx 1.4 \cdot 10^{14}$ bucketfuls
   of water.

Chapter 1 Review, pp. 46-47

1.

a. $(-2)^4 = 16$	b. $-2^4 = -16$	c. $8^{-2} = 1/64$	d. $5^2 \cdot 5^8 \cdot 5^{-7} = 5^3 = 125$
e. $11 \cdot 10^{-2} = 11/100$	f. $10^3 + 10^4 = 11{,}000$	g. $\left(\dfrac{2}{-3}\right)^3 = -\dfrac{8}{27}$	h. $\dfrac{12^7}{12^5} = 12^2 = 144$

2.

e. $(a^{-1})^4 = \dfrac{1}{a^4}$	b. $(2x)^3 = 8x^3$	c. $(5x)^{-2} = \dfrac{1}{25x^2}$	d. $-2s^5 t^7 t^3 \cdot 4s^8 = -8s^{13} t^{10}$
e. $\dfrac{9a^7}{30a^5} = \dfrac{3a^2}{10}$ or $\dfrac{3}{10}a^2$	f. $\dfrac{x^3}{x^{-2}} = x^5$	g. $\left(\dfrac{3x}{-4}\right)^3 = -\dfrac{27x^3}{64}$	h. $\left(\dfrac{2a^2}{b}\right)^5 = \dfrac{32a^{10}}{b^5}$

3.

a. $8^x 8^5 = 8^{24}$ $x = 19$	b. $(7^8)^{-3} = \dfrac{1}{7^y}$ $y = 24$	c. $\left(\dfrac{2}{3}\right)^x = \dfrac{16}{81}$ $x = 4$	d. $\dfrac{(-3)^x}{(-3)^5} = -27$ $x = 8$	e. $(3^z)^2 = 9^4$ $z = 4$

4. (a) and (b) are true statements.

5. The error is on the next to the last line. It should say $2^2 \cdot 5^2 \cdot 2^{\textbf{3}} \cdot 3^3$.
   The final prime factorization then becomes $2^5 \cdot 3^3 \cdot 5^2$.

6. $x = 0$

7. a. 66 km/h (answer given to two significant digits)
   b. 93 (answer rounded to the nearest meter, since the least accurate addend, 89 m, was accurate to the nearest meter.
   c. \$14,300,000 or $1.43 \cdot 10^7$ dollars (answer given to three significant digits)

8. a. 3.5 km $\cdot$ 104 $\approx$ 360 km. The answer is given to two significant digits, since 3.5 has two, and 104 has three. If you think of 104 as coming from 52 $\cdot$ 2, the two is exact. It's not possible it could be 2.1 or 1.9. The two is actually exact, as in 2.000 or however many zeros you'd like to include.

   b. 624.6 mi $\div$ 24.3 gal $\approx$ <u>25.7 mi/gal</u>

   c. 2.3 m $\cdot$ 11.9 m $\approx$ <u>27 m^2</u>

9.

Planet	Average distance from sun (km)	In scientific notation (km)
Mercury	58,000,000	$5.8 \cdot 10^7$ km
Jupiter	778,570,000	$7.7857 \cdot 10^8$ km
Neptune	4,495,000,000	$4.495 \cdot 10^9$ km

10. a. $2.40 \cdot 10^{-4}$ seconds and $4.3 \cdot 10^{-3}$ seconds.
    b. We subtract: $4.3 \cdot 10^{-3}$ s $- 2.40 \cdot 10^{-4}$ s $= 43 \cdot 10^{-4}$ s $- 2.4 \cdot 10^{-4}$ s $= 40.6 \cdot 10^{-4}$ s.
    This needs rounded to $41 \cdot 10^{-4}$ s since 43 is accurate to the nearest one and 2.4 is accurate to the nearest tenth.
    Lastly, given in milliseconds, the answer is <u>4.1 milliseconds</u>.

11. The cheetah is about $\dfrac{2.77 \cdot 10^1 \text{ m/s}}{1.3 \cdot 10^{-2} \text{ m/s}} \approx 2.1 \cdot 10^3 = 2{,}100$ times faster than the snail.

Chapter 2: Geometry

Geometric Transformations and Congruence, Part 1, pp. 51-54

Page 52

1. a. reflection b. rotation c. translation

2. a. No. The owl on the right (the blue owl) is "slimmer"
(less wide) than the owl on the left. You can rotate
the red owl on top of the blue owl to see this, or simply
use a ruler to measure the width of both owls.
    b. No.
    c. Yes.

Page 53

3. a.

   b. The side lengths measure the same.
   c. The angles measure the same.

4. a.

   b. (i) $\overline{XY}$ and $\overline{X'Y'}$ are congruent.

Page 54

5. a. & b.

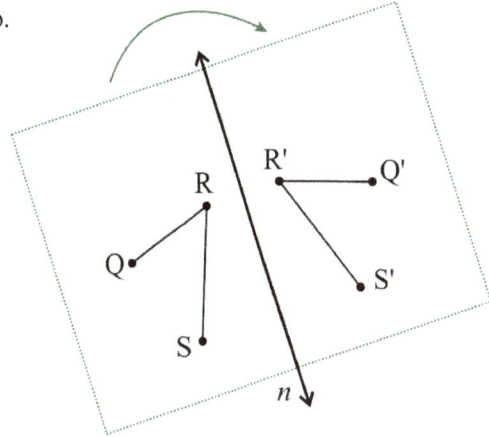

   c. The line segments are congruent (have the same length).
   d. The angles ∠QRS and ∠Q'R'S' are congruent
   (about 51°).

6. They will stay as parallel lines under each of those
transformations.

Geometric Transformations and Congruence, Part 2, p. 55

Page 55

1.

Since reflection is a congruent transformation, lengths
are preserved. The perimeter is
50 cm + 33 cm + 24 cm + 18 cm = 125 cm.

2. Yes, it is. It is congruent because both rotation and
reflection are rigid transformations: they preserve the
shape and size of the figure.

3. a.

   b. Perimeter, area, measure of angle DEF, and angle sum
   are all preserved. Position is the only one that is not.

23

Geometric Transformations and Congruence, Part 2, cont.

Page 56

4. Both Jeannie and Matthew are correct. Kim is not.

5. a.

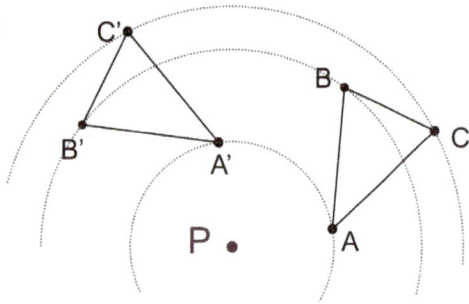

The above image shows the circular arcs involved in the rotation for the illustration's sake. The student's drawing won't show those arcs.

It is also possible to do this using a compass and a protractor. As an enrichment activity, you can encourage your student(s) to figure out how, and try do it.

   b. The angle measures don't change.
   c. The area does not change either.

Page 56

6. Translation. Starting from one small fish image, copies of it were translated (moved) to produce the larger image.

7. a. Yes; reflection.
   b. No.
   c. No.
   d. Yes; rotation.

Page 57

8. Umbrella B.

9. Yes. Figure B is indeed a reflection of Figure A across the line *s*.

10. a.

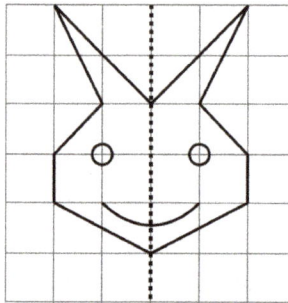

   b. & c. Answers will vary. Check the student's work.

Translations in the Coordinate Grid, pp. 58-59

Page 58

1. a.

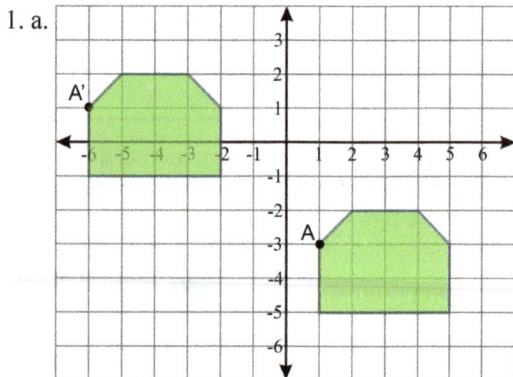

   b. A(1, −3) and A'(−6, 1)
   c. It becomes S'(−4, 0).
   d. The area of the figure is 11 square units and so is the area of its image.

2. a. (4, −3)
   b. Q(−3, 2)

Page 59

3. Jane is incorrect. Points A and B indeed map to A' and B' under that translation, but point C does not map to C'. Point C is (−2, 1) and in the translation of six units to the right and four units up, it would map to (4, 5), but in the image, point C' has the coordinates (3.5, 5). So, triangle A'B'C' is NOT congruent to triangle ABC.

4. A translation of figure F five units to the left and 3 units up maps F onto F'. Since translation preserves congruence, the figures are congruent.
   Note: students should demonstrate understanding that they are mapping F onto F', and not the other way around.

5. a. It is 2 · 20 + 2 · 35 = 40 + 70 = 110 units.
   b. Translation preserves lengths, so the side lengths of the translated rectangle are the same as of the original.

6. a. It is a trapezoid.
   b. Line segments $\overline{A'D'}$ and $\overline{B'C'}$.
   c. It is also 105°.
   d. Answers will vary. These are preserved: area, perimeter, angle sum, angle measure of any of the individual angles A, B, C, or D, any of the side lengths. Position is not preserved.

Translations in the Coordinate Grid, cont.

Page 60

7. a. No.

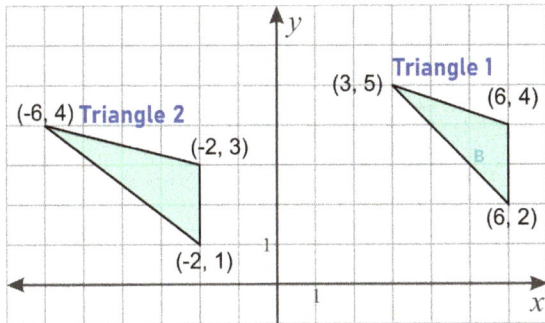

b. Move the vertex at (−6, 4) to (−5, 4) instead. Or, move the other two vertices to (−3, 1) and (−3, 3).

c. It is easiest to use the vertical side as the base. The area of triangle 1 is $2 \cdot 3 \div 2 = 3$ square units. The area of triangle 2 is $2 \cdot 4 \div 2 = 4$ square units. The areas are not the same.

8. a. It is translated 6 units to the right and 5 units down.

b. No.

c. Translation is a congruent transformation so it preserves parallel lines (or line segments). Since angle ABC was translated to become angle A'B'C', the line segment $\overline{BC}$ is parallel to $\overline{B'C'}$. It follows that the lines containing those line segments ($\overleftrightarrow{BC}$ and $\overleftrightarrow{B'C'}$) are parallel also, and thus never meet.

9. The x-coordinate of C" is $-4 + 6 - 3 = -1$.
   Its y-coordinate is $1 - 2 - 3 = -4$. So, C" is (−1, −4).

Puzzle corner. There are an infinite number of solutions.

Point C can be located anywhere on the vertical line $x = 7$ (the line that passes through point (7, 0)) or on the vertical line $x = -5$ (the line that passes through point (−5, 0)). That way, the altitude of the triangle will be exactly 6 units, and the area will be 12 square units.

Reflections in the Coordinate Grid, pp. 61-62

Page 61

1. a.

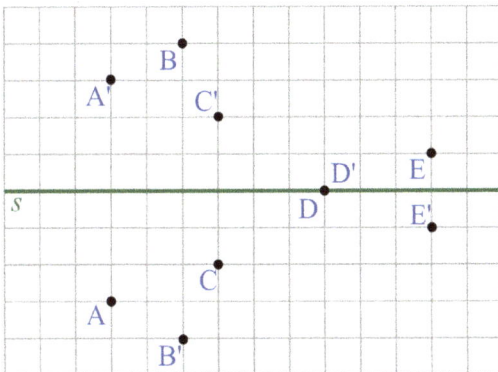

b.

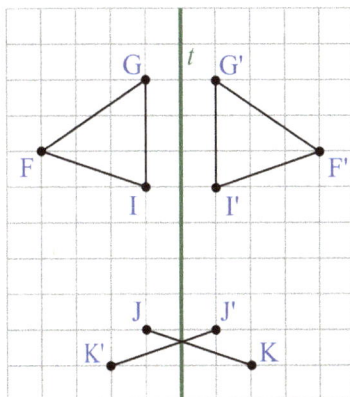

2. a. b. c. d.

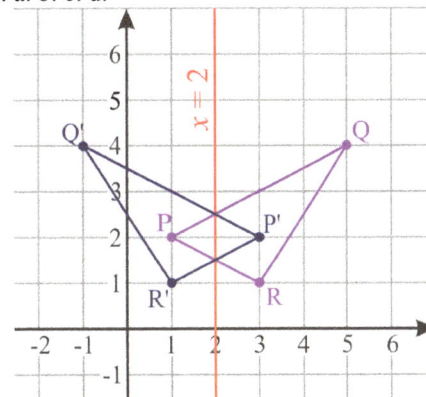

Page 62

3. a. Figures 1 and 2 are not congruent. They don't have the same shape.

b. Change point R' to be at (−4, 3) instead of (−5, 3).

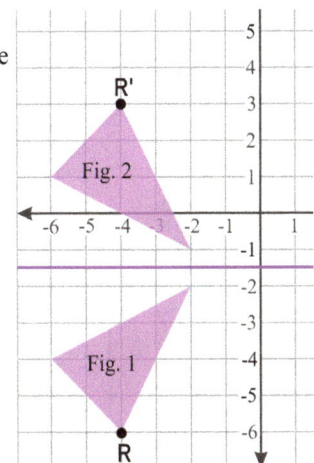

Reflections in the Coordinate Grid, cont.

Page 62

4. H (−2, 3) → H' (−2, −3)
   I (1, −1) → I' (1 , 1)
   J (3 , 5) → J' (3, −5)
   K (−5 , −4) → K' (−5, 4)

   Compare the coordinates of each point and its image. What do you notice?

   The *x*-coordinate stays the same.
   The *y*-coordinate becomes the opposite.

   What do you suppose happens to the coordinates of points that are reflected in the *y*-axis?

   Their *y*-coordinate stays the same, but the *x*-coordinate becomes the opposite.

5. The coordinates are M'(−3, −1), N'(−1, −4), O'(3, −4), P'(5, −1), and Q'(0, 1).

Page 63

6. The preserved attributes are: a. perimeter, d. area, and e. measure of angles.

7. a.

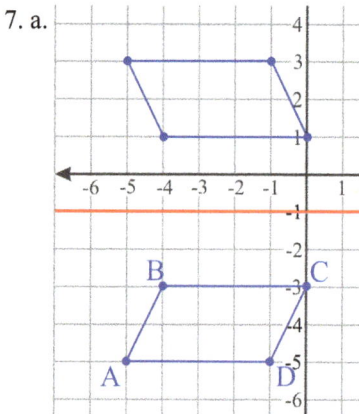

7. b.

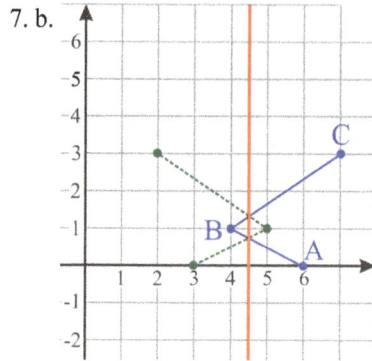

8.

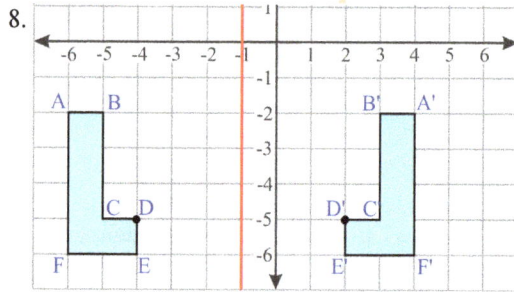

Puzzle corner. It is the horizontal line that passes through point (0, −20). (Algebraically, we denote that line with the equation $y = -20$ because for all the points on that line, their *y*-coordinate is −20.)

Translations and Reflections, p. 64

Page 64

1. a.& b. It does matter. The image on the right shows how the two sets of transformations don't result in the same final position for our triangle.

2. a.

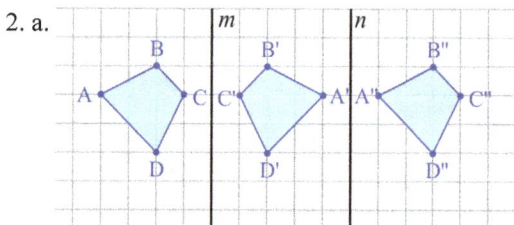

   b. A translation (10 units to the right).

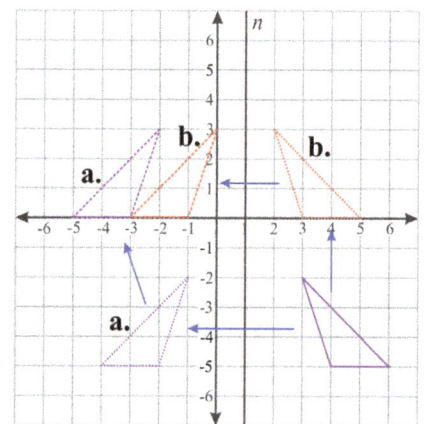

26

Translations and Reflections, cont.

3. Answers will vary. There are multitudes of possible answers. Check the student's answer. For example:

   a. A reflection in the vertical line at $x = -1$ followed by a translation 3 units up.
   Or, a reflection in the vertical line at $x = -2$ followed by a translation 2 units to the right and 3 units up.

   b. A translation 3 units up, followed by a reflection in the vertical line at $x = -1$.
   Or, a translation 2 units to the right and 3 units up, followed by a reflection in the y-axis.

4. Answers will vary. There are multitudes of possible answers. Check the student's answer.

   a. For example: a reflection in the x-axis, followed by a translation 8 units to the right and 2 units down.
   Or, a translation 8 units to the right, followed by a reflection in the horizontal line at $y = -1$.

   b. For example: a reflection in the line $x = 2.5$ followed by a translation 6 units up.
   Or, a translation 6 units up, followed by a reflection in the vertical line at $x = 3$, followed by a translation 1 unit to the left.

5. The vertices of the original triangle are $(0, -2)$, $(1, 0)$, and $(4, -1)$.

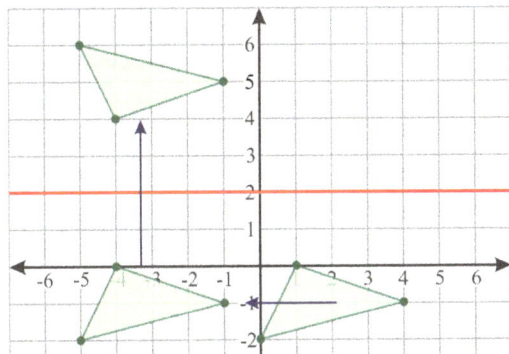

6. a. Yes — a reflection at the vertical line at $x = 1$.

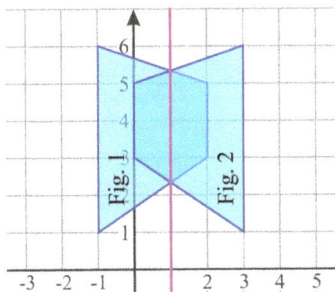

   b. No. Figure 2 is a translation of Figure 1, and you cannot do a translation using a single reflection. However, a sequence of *two* reflections in two parallel lines produces a translation.

7. Answers will vary. There are multitudes of possible answers. Check the student's answer. For example:

   Reflect the figure in the line $y = -2$, and then translate it two units to the right and one unit down.

   Or: reflect the figure in the line $y = -2.5$, and then translate it two units to the right.

8. When a figure that has a vertical symmetry line is reflected in a vertical line (that doesn't touch the figure), this is equivalent to a translation. The movement of that translation is perpendicular to the line of reflection.

   Here is one example. The drop is reflected in the vertical line, and this is identical to a translation.

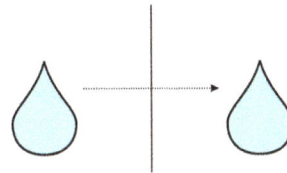

   Similarly, when a figure that has a horizontal symmetry line is reflected in a horizontal line that doesn't touch the figure, it is equivalent to a translation where the movement is perpendicular to the line of reflection.

Puzzle corner. There are many possible solutions. For example:

   a. A 180-degree rotation around $(-1, -1.5)$ followed by a reflection in the x-axis.

   Or, a 180-degree rotation around $(-1, -0.5)$ followed by a reflection in the horizontal line at $x = 1$.

   Or, a 180-degree rotation around $(-1, 0.5)$ followed by a reflection in the horizontal line at $x = 2$.

   Etc.

   b. A reflection in the horizontal line at $y = -3$, followed by a 180-degree rotation around $(-1, -1.5)$.

   Or: a reflection in the horizontal line at $y = -2$, followed by a 180-degree rotation around $(-1, -0.5)$.

   Or: a reflection in the horizontal line at $y = -1.5$, followed by a 180-degree rotation around $(-1, 0)$.

   Etc.

Page 67

1. a. b.

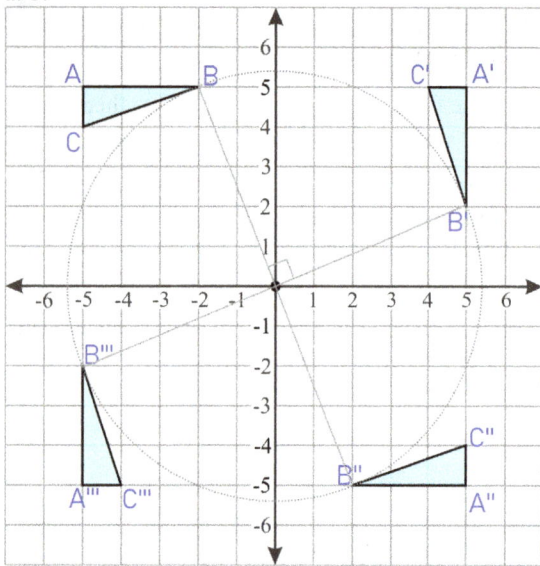

In a 180-degree rotation around the origin, a point P is mapped onto point P' with coordinates that are _opposites_ of the coordinates of P.

For example, point (−2, 1) is mapped onto point (2, −1).

Note: Besides "opposites", the terms "negations" and "additive inverses" are also correct.

3. a. Angle 2.

b.

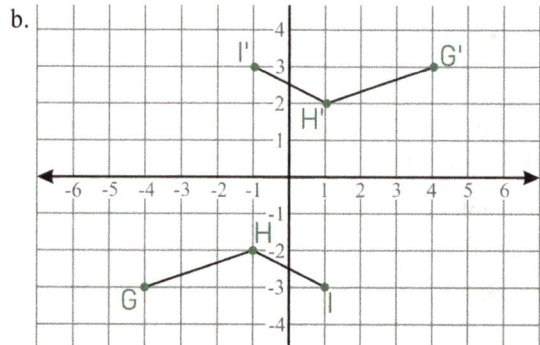

4. The original points are (−5, 6), (−3, 2), (−2, 0), and (0, 4).

Page 68

2. a. b.

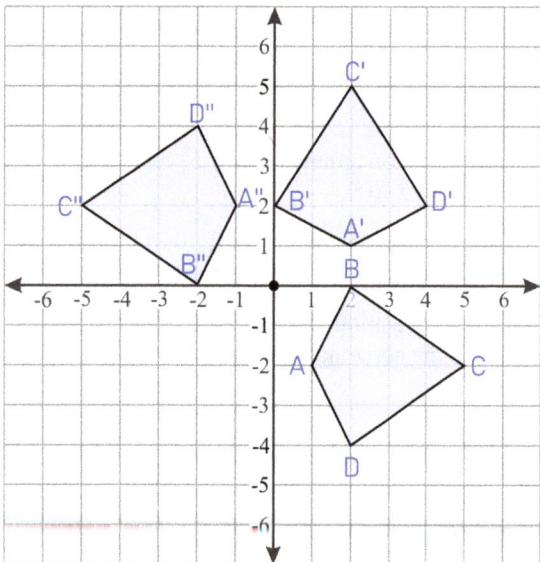

c. Kite A"B"C"D" is the image of kite ABCD under a _180_-degree rotation around the origin.

Now compare the coordinates of kite ABCD and kite A"B"C"D". What do you notice?

Both the x and y-coordinates are the opposites of the corresponding coordinates in the original figure. Here are the coordinates:

A(1, −2) and A"(−1, 2)
B(2, 0) and B"(−2, 0)
C(5, −2) and C"(−5, 2)
D(2, −4) and D"(−2, 4)

Page 69

5. a. b. c. d.

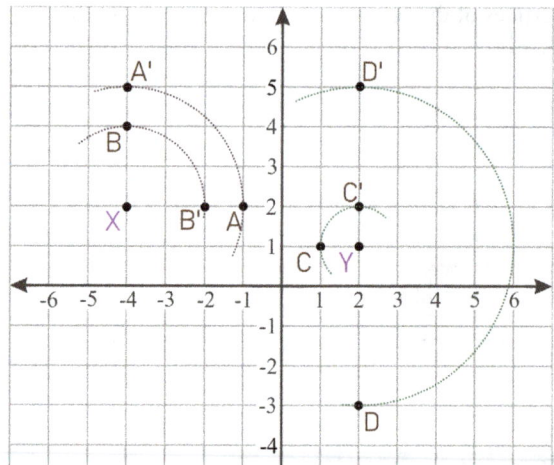

The coordinates of the rotated points are:

A'(−4, 5), B'(−2, 2), C'(2, 2), and D'(2, 5).

6. a. b. c. d.

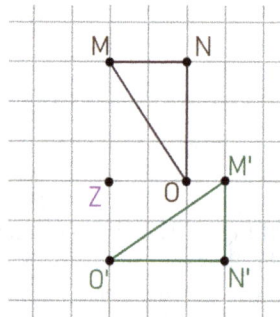

Rotations in the Coordinate Grid, cont.

Page 69

Puzzle corner. A rotation 90 degrees counterclockwise around point (5.5, 0.5) will do.

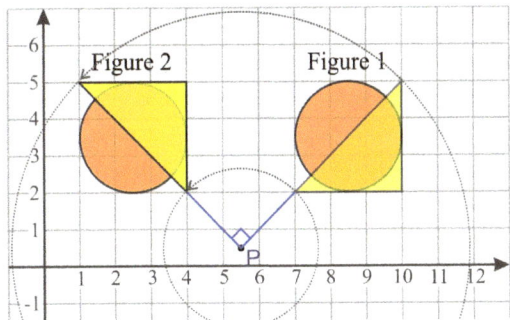

Page 70

7. a. Rotate Figure 1 around point A 180 degrees (clockwise or counterclockwise; it won't matter which.)

   b. Answers will vary. Check the student's answer. For example: Rotate the pentagon 180 degrees around point A. Then translate it 4 units to the left and 4 units down.

   Or, reflect the pentagon in the x-axis. Then reflect it in the y-axis.

8. a. b. c. d.

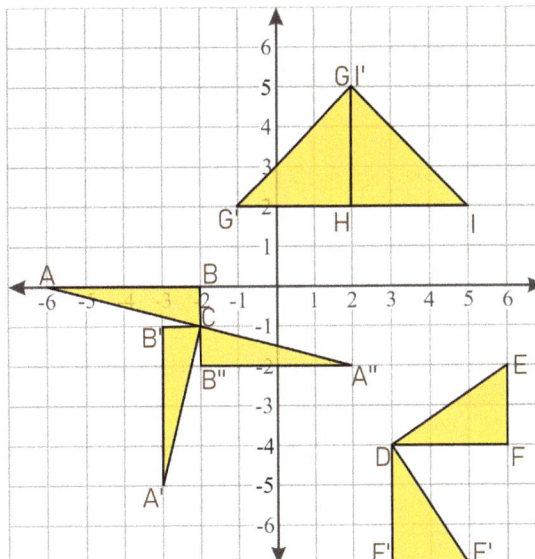

Sequences of Transformations, pp. 71-72

Page 71

1. a. b. Answers will vary. Check the student's answer. For example:

   Reflect the figure in the line $y = 2$, then translate it six units to the left.

   Or, first reflect it in the x-axis, then translate it six units to the left and four units up.

   Or, first translate it six units to the left, then reflect it in the line $y = 2$.

2. Answers will vary. Check the student's answer. For example:

   a. Rotate the parallelogram 90° counterclockwise around the origin. Then translate it one unit to the right.

   b. Rotate it 90° counterclockwise around point A. Then translate it two units up and one unit to the right.

      Or, first rotate it 90° clockwise around point D, and then translate it seven units up and two to the right.

Page 72

3. The edits below show the mistakes in the proof and the corrections for them.

   A rotation 90 degrees ~~clockwise~~ **counterclockwise** *around point E, followed by a translation five units down and* ~~three~~ **two** *to the left transforms Figure 1 to Figure 2.*

   Since both rotations and translations preserve congruence, the two figures are congruent.

4. Answers will vary. Check the student's answer. For example:

   When you rotate triangle ABC 90° counterclockwise around the origin, and then translate the resulting triangle two units to the right and one unit down, you will get Figure 2. Since both rotations and translations are congruent transformations, the two figures are congruent.

   Or:
   A rotation 90 degrees counterclockwise around point A, followed by a translation three units to the left and two units down will transform triangle ABC to Figure 2. Since both rotations and translations preserve congruence, the two figures are congruent.

29

Page 72

5. a.

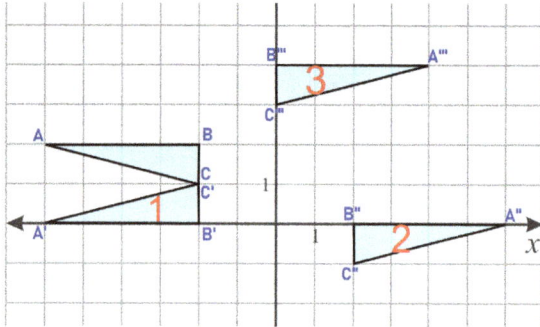

Puzzle corner.

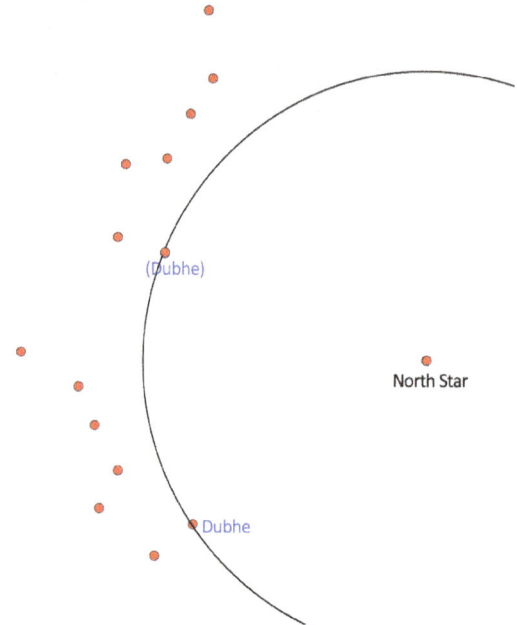

   b. Answers will vary. Check the student's answer.
      For example:

      Reflect the triangle in the line $x = -2$, and then translate
      it two units to the right and two units up.

      Or, first translate it two units up and two units to the
      right, then reflect it in the y-axis.

Page 73

6. a. Answers will vary. Check the student's answer.
      For example:

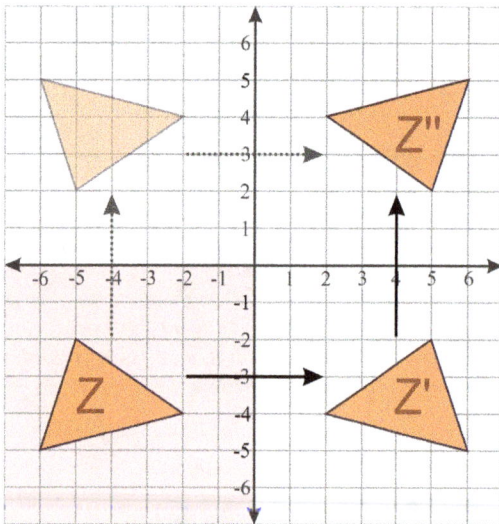

   b. Yes.
   c. A rotation 180 degrees around the origin.

Sequences of Transformations, Part 2, pp. 74-75

Page 74

1. After the reflection, the points are A'(−1, 2), B'(−5, 3), and C'(−4, 1). After the translation, the points are <u>A"(1, −4), B"(−3, −3), and C"(−2,−5)</u>.

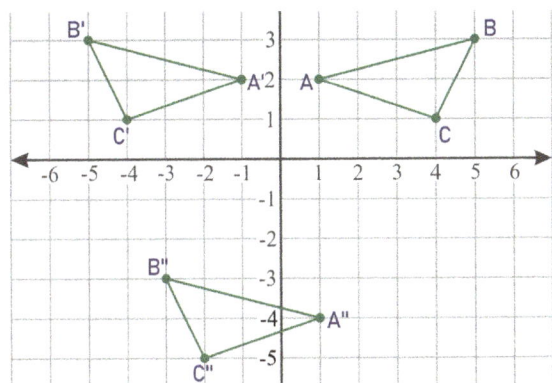

2. After the rotation, the points are A'(2, −4) and B'(0, −2). After the translation, they are <u>A"(−3, 3) and B"(−5, 5)</u>.

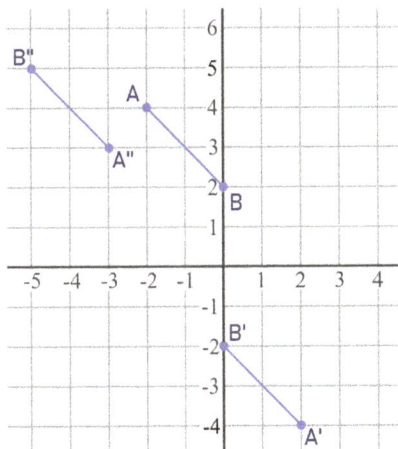

3. The original points were at <u>(−5, −3), (−2, −5), (−1, −3), and (−2, −2)</u>.

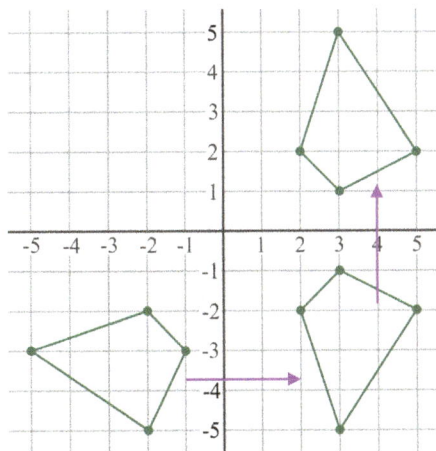

4. In Ashley's answer, the *x*-coordinates are 1 unit too much. She said the points are (5, −2), (1, −3), and (2, −1). In reality, they are (4, −2), (0, −3), and (1, −1).

The coordinates of the original points are A(1, 2), B(5, 3), and C(4, 1). After the first rotation, the points become A'(−2, 1), B'(−3, 5), and C'(−1, 4). After the translation, the points become A"(−2, −4), B"(−3, 0), and C"(−1, −1). After the second rotation, the points become A‴(4, −2), B‴(0, −3), and C‴(1, −1).

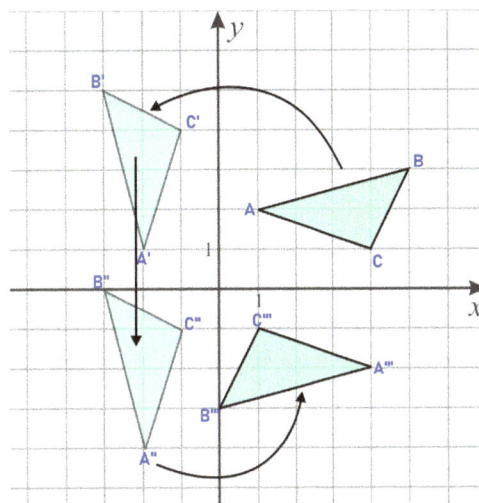

Page 75

5. Greg is correct and Jenny is not. In the way Jenny suggests moving the quadrilateral, point A would map with point B', point B with A', D with C', and C with D'.

If the points were not labeled, they would both be correct.

6. After the reflection, the points become H'(−5, 0), I'(−4, −2), J'(−2, −2), and K'(−4, 1). After the rotation, they become <u>H"(5, 0), I"(4, 2), J"(2, 2), and K"(4, −1)</u>.

7. The translation was one unit to the left and five units up. The reflection was in the line $x = −1$.

Original figure	Translation	Reflection
P(−5, −2)	P'(−6, 3)	P"(4, 3)
Q(−3, −2)	Q'(−4, 3)	Q"(2, 3)
R(−4, 1)	R'(<u>−5</u> , <u>6</u>)	R"(<u>3</u> , <u>6</u>)

8. Figure 3.

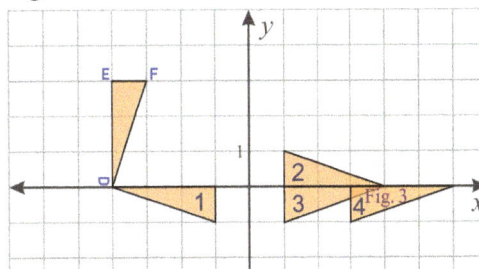

31

Dilations, pp. 76-78

Page 76

Teaching box. What about if it is 1? Then the size of the dilated figure is the same as the original; there is no change.

1. Figures 2, 3, and 6.

2. a. $\overline{AB}$ corresponds to $\overline{DE}$. $\overline{AC}$ corresponds to $\overline{DF}$. $\overline{BC}$ corresponds to $\overline{EF}$.
   The side lengths of triangle DEF are double the side lengths of triangle ABC.

   b. Angle A corresponds with angle D, Angle B with E, and angle C with F.
   The corresponding angles are congruent (their angle measures are the same).

3. Check the student's sketches.
   a. The side lengths should be half the side lengths of the given rectangle. If you are using the print version, or have printed from the digital version at 100% scaling, the sides of the original rectangle should measure 4.2 cm and 1.7 cm, and the sides of the scaled rectangle 2.1 and 0.8 cm.

   b. The side lengths should be three times the side lengths of the given triangle. If you are using the print version, or have printed from the digital version at 100% scaling, the two perpendicular sides of the original triangle should measure 1.7 cm and 1.1 cm, and the sides of the scaled triangle 5.1 and 3.3 cm.

Page 77

4. a. b.

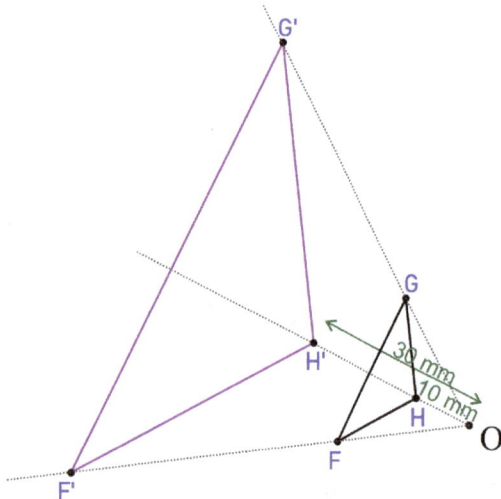

   c. It has the same angle measure as angle GHF.

Page 78

5. a.

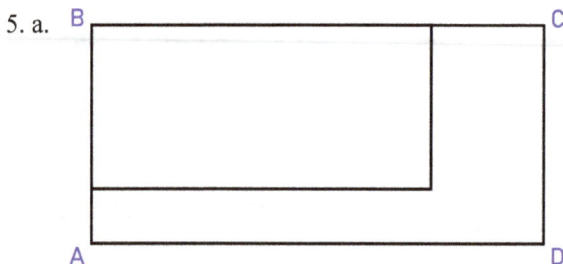

   b. They stay or remain as parallel lines.

6. Nothing happens; the image of the figure is the figure itself.

7. The true statements are: a, c, e, and f.

1.

a. scale factor 2	b. scale factor 1/3

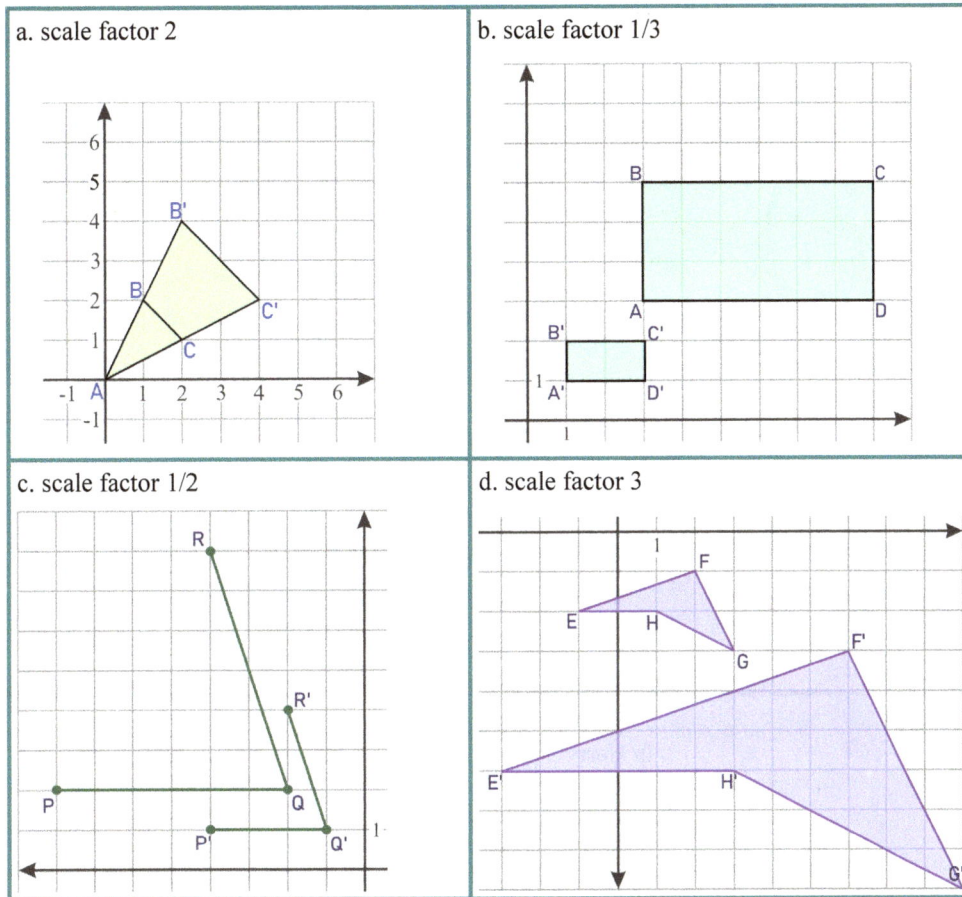

c. scale factor 1/2	d. scale factor 3

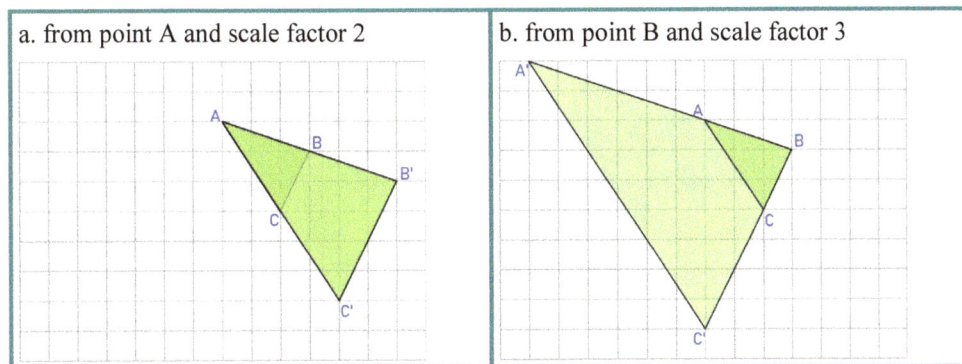

2.

a. from point A and scale factor 2	b. from point B and scale factor 3

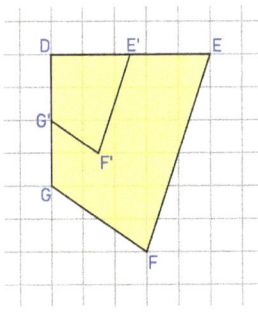

3. a.   b.

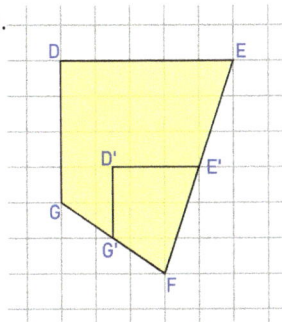

Dilations in the Coordinate Grid, cont.

Page 80

4. a. After the dilation, the coordinates are
   $(-2, 2)$, $(0, 4)$, $(6, 2)$, and $(0, 0)$.

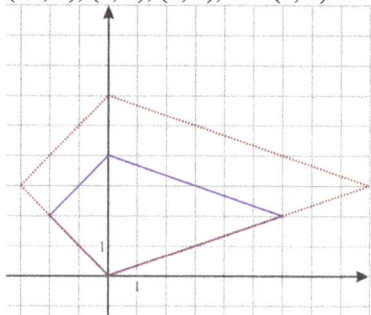

b. The center of dilation is at $(0, 3)$ After the dilation,
   the coordinates are $(-2, 3)$, $(0, 5)$, $(6, 3)$, and $(0, 1)$.

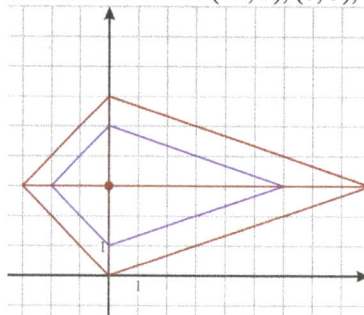

Page 81

5. a. Center point _B_ Scale factor: _4_   b. Center point: _origin_ Scale factor: _3_

6. a. Center point: _origin_ Scale factor: _1/2_   b. Center point: _G_ Scale factor: _1/3_

7. The center of dilation is different. For Figure 1, the center of dilation is point B. For Figure 2, the center of dilation is origin.

Similar Figures, Part 1, p. 82

Page 82

1. a. A dilation and a translation.
   b. A reflection and a dilation.
   c. A rotation and a dilation.

2. Henry's proof is correct, and Harry's isn't.

In Henry's proof, △ABC is first dilated with origin as center and with the scale factor 2. The image below shows the resulting triangle after the dilation (in purple). This is then reflected in the horizontal line at $y = 1$, which results in △A'B'C'.

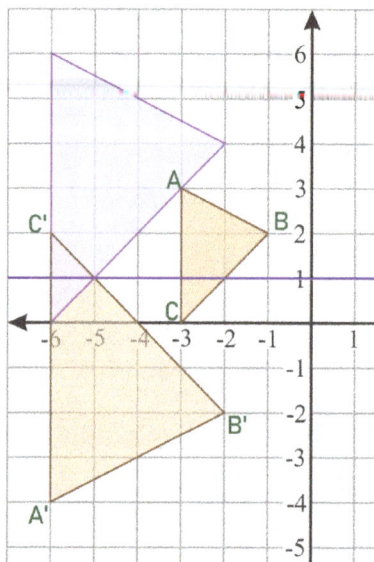

In Harry's proof, △ABC is first reflected in the x-axis, then the resulting triangle is dilated from point B'. The purple triangle in the picture below is the result of reflecting triangle ABC in the x-axis. However, the image of point B is not B', but the point $(-1, -2)$, and if you dilate the purple triangle from that point, you will not get triangle A'B'C'. After that dilation, you would need to, additionally, translate the resulting triangle 1 unit to the left, and then you would have triangle A'B'C'.

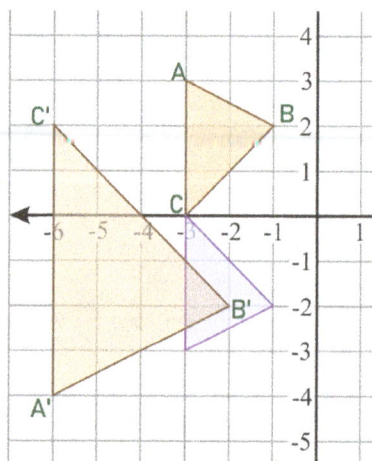

Similar Figures, Part 1, cont.

Page 82

3. Answers will vary; check the student's answer.

For example: First, translate triangle ABC one unit to the right and one unit down. Then reflect it in the line $x = -3$. Lastly, dilate it from the image of point A, with scale factor 1/2.

Or, first dilate it from point A and with scale factor 1/2. Then reflect it in the line $x = -4$. Lastly, translate it one unit to the right and one unit down.

Page 83

4. The coordinates of point D''' are (4, 0).

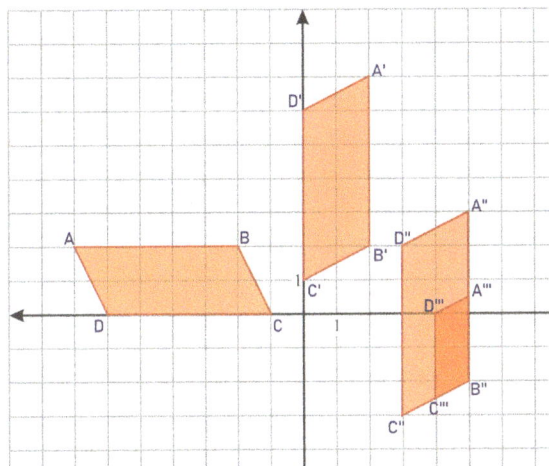

5. Transformation 1 is dilation with origin as center and scale factor 1/4.
Transformation 2 is a translation of two units to the right and four units down.

6. a. Yes. Two congruent figures are similar. The definition of similarity requires there to be a sequence of transformations consisting of translations, reflections, rotations, and dilations that maps one figure to the other. If figures are congruent, this means there is a sequence of *congruent* transformations — translations, reflections, rotations — that maps one figure to the other. Clearly this sequence of transformations fulfills the sequence of transformations required for similarity.

b. No. One figure could be a dilation of the other, say with scale factor 2. Then they would not be congruent, but they would be similar.

7. The dilation is from point F, since it does not change. The reflection is in the x-axis.

Original figure	Dilation	Reflection
E(−1, −1)	E'(−5, −2)	E"(−5, 2)
F(3, 0)	F'(3, 0)	F"(3, 0)
G(3, −2)	G'(3, −4)	G"(3 , 4)
H(2, −2)	H'(1 , −4)	H"(1 , 4)

Page 84

8. a. Yes. There are many possible sequences of transformations that will map the one figure to the other. Check the student's answer. For example:

Rotate the figure 180 degrees around point H. Then dilate it from point H with scale factor 1/2. Lastly, translate the figure seven units to the right and one unit up.

b. No. While it almost looks like you can get the smaller figure with a dilation by scale factor 1/2, when you look at the arrowhead part carefully, you can see it doesn't work. The original arrowhead is 6 units wide (from A to C) and 2 units tall. The smaller arrowhead is 3 units wide and 2 units tall. The smaller one would need to be 3 units wide 1 unit tall, to be a dilation of the bigger.

9. Statement (2) is correct, the others are not.

Puzzle Corner. Transformation 1 is a reflection in the line $x = -2$, and Transformation 2 is a 90-degree rotation clockwise around point W'.

Original figure	Reflection	Rotation
W(−6, 0)	W'(2, 0)	W"(2, 0)
X(−3, 1)	X'(−1, 1)	X"(3, 3)
Y(−3, −3)	Y'(−1, −3)	Y"(−1 , 3)
Z(−6, −2)	Z'(2 , −2)	Z"(0 , 0)

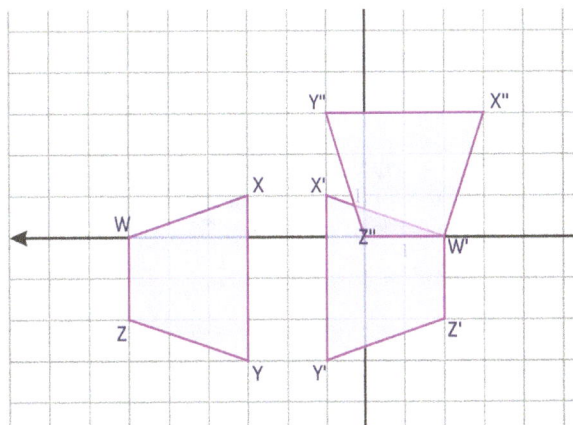

35

Page 86

1. a. Point C
   b. 1.5
   c. 2:3
   d. 87 inches
   e. 50 inches

2. a. The corresponding angles are congruent.
   b. $6/15 = 2/5 = \underline{0.4}$
   c. 5:2
   d. 12.4 cm · 0.4 = 4.96 cm ≈ <u>5.0 cm</u>

3. The 0.90-m side corresponds with the 0.63-m side.
   The 2.20-m side corresponds with the side marked with x.
   The 1.90-m side corresponds with the side marked with y.

   To find x, we can use the proportion $x/(2.2\text{ m}) = (0.63\text{ m})/(0.9\text{ m})$, and get $x = 0.63/0.9 \cdot 2.2\text{ m} = \underline{1.54\text{ m}}$.

   To find y, we can use the proportion $y/(1.9\text{ m}) = (0.63\text{ m})/(0.9\text{ m})$, and get $x = 0.63/0.9 \cdot 1.9\text{ m} = \underline{1.33\text{ m}}$.

4. You can use a proportion: width/(94 mm) = 245/270, from which width = 245/270 · 94 mm ≈ <u>85 mm</u>.

Similar Figures: More Practice, pp. 87-89

Page 87

1. a. Answers will vary. Check the student's answer. For example: Rotate Figure 1 counterclockwise 90° around point Q. Then dilate it from point Q with scale factor 1.5. Lastly translate the figure two units to the right and three units down.
   b. 2:3
   c. 1.5
   d. Perimeter = 2 + 2 + 2 + 0.5 · π · 2 ≈ 9.14 units.
   e. Perimeter = 1.5 · 9.14 units = 13.71 units.

2. a. PQ is 3 units, and looking at the sides QR and Q'R, we can see that scale ratio is 5:3. This means the scale factor is 3/5 = 0.6. So, P'Q' = 0.6 · 3 units = <u>1.8 units</u>.
   b. P'R = 0.6 · 5.83 units ≈ <u>3.50 units</u>.

Page 88

3. For each of these unknowns, you can either write a proportion (in several different ways) or use a scale ratio or a scale factor. So, there are many ways to do the calculations. Here we only give one example way to calculate the answers.

   Looking at the two triangles at the top, we can see from the side lengths of 7 and 35 that the scale factor between them is 5. So, w = 50/5 = <u>10 units</u>.

   Looking at the two triangles at the top, we can see from the side lengths of 7 and 35 that the scale factor between them is 5. So, x = 5 · 13 = <u>65 units</u>.

   Looking at the two triangles on the left, we can see from the side lengths of 13 and 52 that the scale factor between them is 4. So, y = 4 · 7 = <u>28 units</u>.

   To calculate z, we will write a proportion based on the triangle on the top left and the triangle with side z: $z/10 = 63/7$ from which $z = 63/7 \cdot 10 = \underline{90\text{ units}}$.

4. a. There are many ways to reason this out, but essentially, all methods will boil down to these calculations:
   $x = 11/15.7 \cdot 20\text{ cm} \approx \underline{14.0\text{ cm}}$
   $y = 11/15.7 \cdot 25\text{ cm} \approx \underline{17.5\text{ cm}}$.

   b. The value of z cannot be determined from the given information. It could be determined if we knew the angle measures, or were given the length of the corresponding side in the triangle on the right.

5. Statement (c). Statement (a) is almost true; it would be true if we changed point B''' to point A'''.

Page 89

6. (i) and (iv). Their sides are proportional while their angle measures are congruent.

7. Its area quadruples (becomes four-fold). Originally, the area is 15 square units. After the dilation, the area will be 6 units · 10 units = 60 square units.

8. a. The altitude will be 12 units and the base 22 units.
   b. It is four times as much. The area of the original triangle is 33 square units. The area of the dilated triangle is 12 units · 22 units /2 = 132 square units.

Puzzle corner. Answers will vary; check the student's answer. The student should find that the area varies as the *square* of the scale factor. In other words, if you scale a rectangle with a scale factor r, then the area of the scaled rectangle is r^2 times the area of the original. For example:

Original rectangle			Scale Factor	Dilated rectangle		
Width	Height	Area		Width	Height	Area
2	3	6	3	6	9	54
2	5	10	2	4	10	40
1	5	5	5	5	25	125
3	6	18	4	12	24	288
6	2	12	3	18	6	108
3	1	3	4	12	4	48

Review: Angle Relationships, pp. 90-92

Page 91

1. $x = 46°$

2. a. One.
   b. $\angle\alpha = 54°$ $\angle\beta = 126°$

      $\angle\gamma = 54°$ $\angle\delta = 126°$

3. $x = 180° - 108° = 72°$.
   $y = 180° - 152° = 28°$.

4. a. Any two neighboring angles in a parallelogram are <u>supplementary</u>.

      The opposite angles in a parallelogram are <u>congruent</u>.

   b. 360 degrees.

Page 92

5. $\angle\alpha = 180° - 45° = $ <u>135</u> ° $\angle\beta = $ <u>45</u> ° (vertical angle)

   $\angle\gamma = 180° - 30° - 45° = $ <u>105°</u>.

6. The two marked angles are vertical angles, thus congruent. We can write the equation $9x - 30 = 3x + 66$. Solving that, we get $6x = 96$, from which <u>$x = 16$</u>.

7. The other two angles are complementary (since their sum is 90°).

8. $28° + x + 3x + 2x + 2x = 360°$; $8x = 332°$; <u>$x = 41.5°$</u>.

9. a. It is a rhombus (all four sides are congruent).

   b. Right scalene triangles.

   c. $180° - 26° - 26° = $ <u>128°</u>.

Corresponding Angles, pp. 93-94

Page 93

1. a. The exact way the student marks the angles may vary.

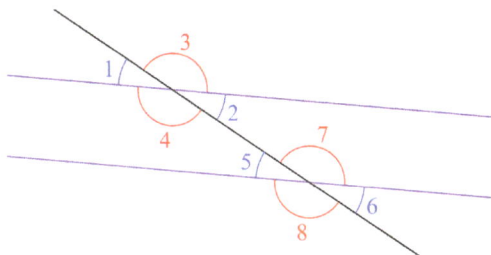

   b. Each of the eight angles measures (approximately) either 28° or 152°. Angles 1, 2, 5, and 6 are congruent, and so are angles 3, 4, 7, and 8.

2. What is the same is that again, we have a line that intersects two other lines. Again, there are some vertical angles formed. But the vertical angles formed at the intersection of the transversal and the line on top are not congruent with the angles formed at the intersection of the transversal with the other (lower) line.

Page 94

3. a. $\angle 6$
   b. $\angle 4$
   c. $\angle 5$ is also 127° because angles 5 and 7 are vertical angles.
   d. $\angle 6$ is 53° because angles 6 and 7 are supplementary (or because angles 5 and 6 are supplementary).
   e. Angles 1, 3, and 5 are congruent to angle 7.

4. A translation two units to the right and four units up maps angle ABC to angle α, so they are congruent.

   To be more precise, we could word the proof this way:

   A translation two units to the right and four units up maps point B to the point where line m intersects the line containing A and B. Since lines m and n are parallel, angle ABC will map to angle α in this same translation, so the two angles are congruent.

More Angle Relationships with Parallel Lines, pp. 95-96

Page 95

1. Use a 180-degree rotation around point M. That rotation will map $\angle 1$ to $\angle 2$, proving that they are congruent.

2. Angles 5 and 7 are <u>vertical</u> angles.
   Angles 3 and 5 are <u>alternate interior</u> angles.
   Angles 1 and 7 are <u>alternate exterior</u> angles.
   Angles 2 and 6 are <u>corresponding</u> angles.

Page 96

3. a. They are alternate exterior angles.
   b. Use a 180-degree rotation around point M. That rotation will map $\angle 1$ to $\angle 2$, proving that they are congruent.

4. a. It is also 53°, because angles 3 and 5 are alternate interior angles, thus congruent.

   b. It is 127°, because angles 8 and 5 are supplementary.

Page 96

5. a. A parallelogram.

   b. Since lines L_1 and L_2 are parallel, angles 1 and 7 are alternate interior angles, and thus are congruent. Now, since lines L_3 and L_4 are parallel, angles 7 and 3 are corresponding angles, and thus are congruent.

   Since angle 1 is congruent to angle 7, and angle 7 is congruent to angle 3, angle 1 is also congruent to angle 3.

The Angle Sum of a Triangle, pp. 97-99

Page 97

1. Drawings will vary; check the student's work. The angle sum should be 180° or close. If it is not, the measurement of some angle has probably been inaccurate.

2. a. b. c.

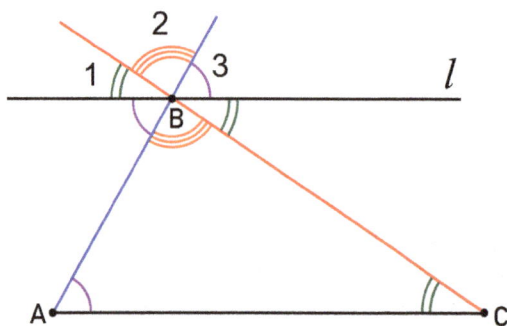

   d. Angles 1, 2, 3 are adjacent and form together a straight line, or a 180° angle. Angle 1 is congruent to ∠C (they are corresponding angles), angle 2 is congruent to ∠B (vertical angles), and angle 3 is congruent to ∠A (again, they are corresponding angles).

Page 98

3. In the diagram, we see triangle ABC, and its angles α, β, and γ. The line l is drawn so that it is parallel to AC.

   Angles α and α′ are congruent because they are <u>corresponding</u> angles.

   Angles β and β′ are <u>congruent</u> because they are vertical angles.

   And angles γ and γ′ are congruent because they are, again, <u>corresponding</u> angles.

   Therefore, the sum ∠α′ + ∠β′ + ∠γ′ is equal to the sum ∠α + ∠β + ∠γ.

   Since the three angles α′, β′, and γ′ are adjacent and form a <u>straight</u> angle, the sum of their angle measures is <u>180°</u>. This means that ∠α + ∠β + ∠γ = <u>180°</u>, too.

4. a. ? = 180° − 82° − 55° = <u>43°</u>
   b. β = 90° − 24.6° = <u>65.4°</u>
   c. x = 180° − 47° − 59° = <u>74°</u>.
   (The third angle of the triangle and the angle marked with x are vertical angles, thus congruent.)
   d. ? = 180° − 72° − 67° = <u>41°</u>.
   (The third angle of the triangle and the angle marked with ? are corresponding angles, thus congruent.)

Page 99

5. a. (180° − 24°)/2 = <u>78°</u>
   b. 180° − 2 · 73° = <u>34°</u>

6. Yes. An equilateral triangle has three congruent sides, so it fulfills the definition of an isosceles triangle (of having at least two congruent sides).

7. 60°

8. No. An obtuse angle is more than 90°. The angle sum of two of them would be more than 180°. But the angle sum in a triangle is always 180°, and thus you cannot "fit" two obtuse angles into a triangle.

9. Start out by drawing the 50°-angle. Then measure the two 2 ½-inch sides. The third side is approximately 2 1/8 inches. The base angles measure 65°. The image below is not to scale.

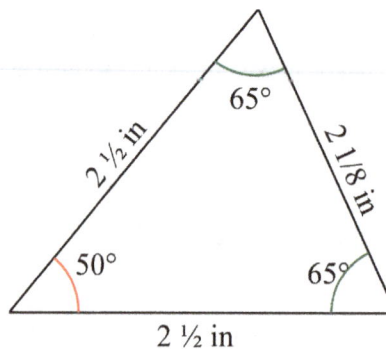

Exterior Angles of a Triangle, pp. 100-102

1. a. $x = 151°$; $y = 40°$
   b. $x = 119°$; $y = 119°$

2. Angle H is supplementary to a right angle so H is 90°.
   Now we can calculate G; $G = 180° − 90° − 57° = 33°$.
   G is supplementary to z, so $z = 180° − 33° = \underline{147°}$.

3.

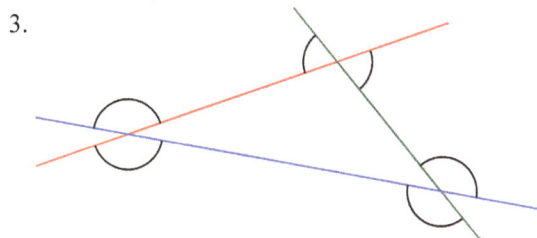

Page 101

4. Angles M and N are supplementary, so $M = 180° − 140° = 40°$.

   The angles in a triangle add up to 180°, so
   $K = 180° − 40° − 104° = 36°$. And since x and K are
   vertical angles, they are congruent, and thus $x = 36°$.

5. a.
$$\begin{aligned} x + 3x + 7 + x + 23 &= 180 \\ 5x + 30 &= 180 \\ 5x &= 150 \\ x &= 30 \end{aligned}$$

   b.
$$\begin{aligned} 2y − 51 + y + 32 + y − 29 &= 180 \\ 4y − 48 &= 180 \\ 4y &= 228 \\ y &= 57 \end{aligned}$$

   c.
$$\begin{aligned} w − 15 + 2w + 53 + (180 − 4w) &= 180 \\ −w + 218 &= 180 \\ −w &= −38 \\ w &= 38 \end{aligned}$$

Page 102

6. a. In triangle 1, $A = 180° − 91° − 48° = 41°$.

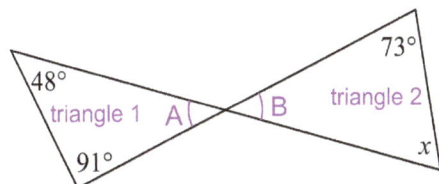

   Since A and B are vertical angles, $B = 41°$.
   Lastly, in triangle 2, $x = 180° − 41° − 73° = \underline{66°}$.

6. b. See the illustration below.

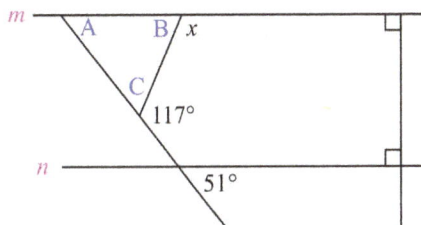

   Angle $C = 180° − 117° = 63°$ (supplementary angles).

   Lines m and n are parallel, so A and the 51° angle are
   corresponding angles. Thus $A = 51°$.

   We can calculate B from the fact that angles in
   a triangle add up 180°: $B = 180° − 51° − 63° = 66°$.

   And lastly, x is supplementary to B, so $x = 180° − 66° = \underline{114°}$.

7. The idea is to draw a line that is parallel to one of the sides
   of the triangle and that goes through the point that is not
   on that side. In this image, line l is drawn so it is parallel
   to $\overline{AC}$.

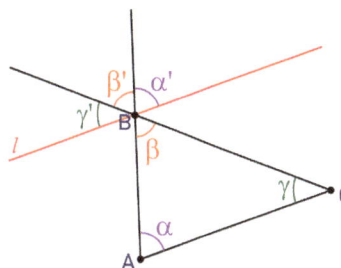

   Then, we continue $\overline{AB}$ and $\overline{BC}$. The line containing $\overline{AB}$ is a
   transversal, and so is the line containing $\overline{BC}$, and thus,
   corresponding angles are formed.

   The proof should include these thoughts:

   Angles α and α' are corresponding angles, thus congruent.
   Angles β and β' are vertical angles, thus congruent.
   Angles γ and γ' are corresponding angles, thus congruent.

   The three angles α', β', and γ' form a straight line together,
   or a 180-degree angle. So, their sum is 180°. Since they are
   congruent to the three angles of the triangle, the angle
   sum of the triangle is also 180°.

8. The angle sum in a triangle is 180°, from which we get that
   $a = 180° − 61° − 65° = 54°$ and $b = 180° − 84° − 35° = 61°$.

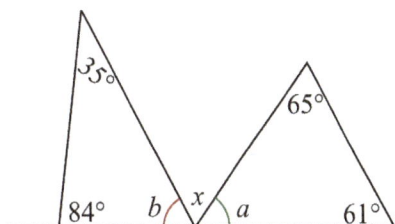

   Now, angles b, x, and a, form a straight line, so
   $x = 180° − a − b = 180° − 54° − 61° = \underline{65°}$.

Page 102

Puzzle corner:

a. It is 360°.

b. The basic idea of the proof is that, since any quadrilateral can be divided into two triangles, the sum of its angles is just twice the sum of the angles of a triangle, or $2 \cdot 180° = 360°$. Here is the proof written out with more detail:

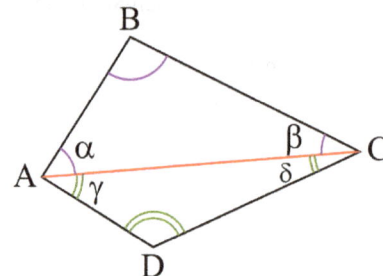

We divide the quadrilateral into two triangles, triangle ABC and triangle ACD.

Notice that the two angles ∠A and ∠C of the quadrilateral are composed of angles of the triangles: $A = \alpha + \gamma$ and $\angle C = \beta + \delta$. The sum of the angles of the quadrilateral ABCD is
$$\angle A + \angle B + \angle C + \angle D = (\alpha + \gamma) + \angle B + (\beta + \delta) + \angle D.$$

Since addition is commutative and associative, we can rewrite the last expression as $(\alpha + \angle B + \beta) + (\gamma + \delta + \angle D)$, where $(\alpha + \angle B + \beta)$ is the sum of the angles of triangle ABC, and $(\gamma + \delta + \angle D)$ is the sum of the angles of triangle ACD. Since the sum of the angles of each of the triangles is 180°, the sum of the angles of the quadrilateral is just $180° + 180° = 2 \cdot 180° = 360°$.

Angles in Similar Triangles, Part 1, pp. 103-104

Page 103

1. Yes, they are, because their angles are congruent (with angle measures 31°, 73°, and 76°).

2. Yes, you can know that the triangles are similar. This is because if two of the angles are congruent, the third angles are also. And if all three angles in two triangles are congruent, they are similar triangles.

3. Student drawings will vary; please check them. For example:

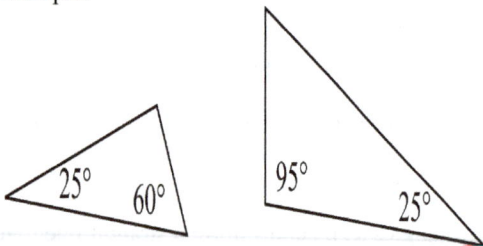

a.

b. The ratios of the corresponding sides are:
$17/21 \approx 0.810$; $31/37 \approx 0.838$; $39/47 \approx 0.830$.

These are close to each other. Recall that measurements to the nearest millimeter are not that accurate. In reality, if we could measure the lengths, say, to the hundredth of a millimeter, the ratios would be much closer to each other.

c. Using the average of the three ratios that were already calculated for part (b), the one scale ratio is 0.826.

For the other, we write the ratios the other way:
$21/17 \approx 1.235$; $37/31 \approx 1.194$; $47/39 \approx 1.205$.
The scale ratio as the average of those is 1.211.

d. The product of the two scale ratios is
$0.826 \cdot 1.211 = 1.000286$

Page 104

4. Yes, they are similar. We know that because the three angles in the first triangle are **115°**, **24°**, and $180° - 115° - 24° = \mathbf{41°}$, and the three angles in the second triangle are **115°**, **41°**, and $180° - 115° - 41° = \mathbf{24°}$. So, all three angles in these two triangles are congruent.

5. a. Yes, it is. Since $\overline{AC}$ is parallel to $\overline{A'C'}$, angle C' and angle y are corresponding angles, thus congruent. Similarly, angles A and A' are corresponding angles and thus congruent. So, triangles ABC and A'BC' have three congruent angles, which makes them similar.

   b. It is $180° - 58° - 83° = \underline{39°}$. This is because y is congruent to ∠C', and we can find the value of C' by subtracting 83 and 58 from 180 (the angle sum of a triangle is 180°).

6. Since $\overline{AB}$ is parallel to $\overline{CD}$, the 124° and the angle formed by 40° and x are alternate interior angles and are congruent.

   So, $124° = 40° + x$, from which $\underline{x = 84°}$.

7. It is not possible to find the value of x (without further information).

   We can figure out the third angle in triangle ABD, but that will not help us. If we knew that AD was parallel to BC, then we could figure out x (and x would be 62°), but without knowing that, or knowing some of the other angles, it is not possible.

Angles in Similar Triangles, Part 2, pp. 105-106

Page 105

1. a. The two triangles formed have congruent angles. First, the two angles at K are vertical angles so are congruent. Then, both triangles have a right angle. Since two of their angles are congruent, the third angles are also. And thus they are similar triangles.

   b. From the proportion $x/(67 \text{ ft}) = 156/104$, we get $x = 156/104 \cdot 67 \text{ ft} = \underline{100.5 \text{ ft}}$.

2. They are not. The bigger triangle has angles of 55°, 90°, and 35°. The smaller triangle has angles of 62°, 90°, and 28°. Since the angles are not congruent, the triangles are not similar.

3. We cannot tell whether they are similar, with the given information.

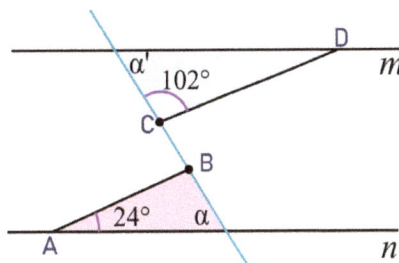

We can know that α and α' are congruent, being alternate interior angles. But we cannot conclude anything else, without further knowledge.

If we knew that AB was parallel to CD, then we could conclude that the two triangles are similar.

Page 106

4. Let h be the height of the tree. Then, $h/(600 \text{ cm}) = 145/207$, from which $h = 145/207 \cdot 600 \text{ cm} \approx \underline{420 \text{ cm or 4.2 m}}$.

5. Answers will vary. Check the student's work.

6. Triangle ABC and triangle CDE are similar, and so we can write a proportion: $AB/(6 \text{ m}) = (6.5 \text{ m}) / (8 \text{ m})$, from which $AB = 6.5/8 \cdot 6 \text{ m} = 4.875 \text{ m} \approx \underline{4.9 \text{ m}}$.

Volume of Prisms and Cylinders, pp. 107-108

Page 107

1. Yes, it does. In the formula $V = A_b h$, the area of the base is its width times the depth. So, that makes this formula the same as multiplying the width, the depth, and the height.

Page 108

2. $V = \pi \cdot (4 \text{ cm})^2 \cdot 12 \text{ cm} \approx \underline{600 \text{ cm}^3}$.

3. a. It is a triangular prism.
   b. $V = 1 \text{ m} \cdot 0.6 \text{ m} / 2 \cdot 1.4 \text{ m} = \underline{0.42 \text{ m}^3}$.

4. a. The volume of the first tank is
   $V_1 = \pi \cdot (2.5 \text{ ft})^2 \cdot 6 \text{ ft} \approx \underline{117.8 \text{ ft}^3}$.

   The volume of the second and third is
   $V_2 = \pi \cdot (4 \text{ ft})^2 \cdot 10 \text{ ft} \approx \underline{502.7 \text{ ft}^3}$.

   In total, the volume of the three is
   $117.8 \text{ ft}^3 + 502.7 \text{ ft}^3 + 502.7 \text{ ft}^3 = \underline{1{,}123.2 \text{ ft}^3}$.

   b. In gallons, the tanks contain
   $1{,}123.2 \text{ ft}^3 \div (0.16054 \text{ ft}^3/\text{gal}) \approx 6995.39$
   gallons. This amount provides water for the family for 6995.39 gal $\div$ (120 gal/day) $\approx \underline{58 \text{ days}}$.

5. Answers will vary; check the student's work.

Volume of Pyramids and Cones, pp. 109-110

Page 109

1. a. A triangular pyramid
   b. A circular cone
   c. A pentagonal pyramid
   d. A (circular) cylinder

2. a. A triangular pyramid
   b. A (circular) cylinder
   c. A circular cone
   d. A hexagonal pyramid

Page 110

3. a. $V = 21 \text{ cm}^2 \cdot 18 \text{ cm} / 3 = 126 \text{ cm}^3$
   b. $V = 2.0 \text{ ft} \cdot 1.5 \text{ ft} \cdot 2.0 \text{ ft} / 3 = 2.0 \text{ ft}^3$
   c. $V = \pi \cdot (2 \text{ cm})^2 \cdot 7 \text{ cm} / 3 \approx 29 \text{ cm}^3$
   d. $V = \pi \cdot (2.125 \text{ in})^2 \cdot 4.75 \text{ in} / 3 \approx 22.5 \text{ in}^3$
   e. $V = 2 \cdot 10 \text{ cm} \cdot 10 \text{ cm} \cdot 9 \text{ cm} / 3 = 600 \text{ cm}^3$
   f. $V = \pi \cdot (1.5 \text{ in})^2 \cdot 1.9 \text{ in} + \pi \cdot (1.5 \text{ in})^2 \cdot 3.3 \text{ in} / 3$
   $\approx 21.2 \text{ in}^3$

Volume of Pyramids and Cones, cont

4. a. 9,000 cm^3 b. 30 cm

5. a. 1/3 ≈ 33.3%   b. 50%

6. The volume of the taller hat is $\pi \cdot (5 \text{ cm})^2 \cdot 25 \text{ cm} / 3 \approx 654.50 \text{ cm}^3$.
   The volume of the shorter is $\pi \cdot (10 \text{ cm})^2 \cdot 12.5 \text{ cm} / 3 \approx 1309.00 \text{ cm}^3$.
   The shorter hat's volume is twice the volume of the taller hat.

Volume of Spheres, pp. 112-113

1. $V = (4/3)\pi(1.5 \text{ in})^3 \approx 14 \text{ in}^3$

2. $V = (1/2)(4/3)\pi(11.5 \text{ m})^3 \approx 3{,}190 \text{ m}^3$

3. Answers will vary; check the student's work.

4. The volume of the first is 1/8 of the volume of the second. You can calculate the volumes of the balls, and get the right answer, but it works out neater if you don't actually calculate the volumes, but only write the expressions for them.

   The volume of the first ball is
   $V_1 = (4/3)\pi(1 \text{ in})^3 = (4/3)\pi \text{ in}^3$.

   The volume of the second ball is
   $V_2 = (4/3)\pi(2 \text{ in})^3 = (4/3)\pi \cdot 8 \text{ in}^3$.

   The fraction that the volume of the first is of the

   volume of the second is $\dfrac{(4/3)\pi \text{ in}^3}{(4/3)\pi \cdot 8 \text{ in}^3}$.

   In it, 4/3 and π cancel out, as does in^3, and we are left with 1/8.

5. The volume of the three tennis balls is
   $3 \cdot (4/3)\pi(3.0 \text{ cm})^3 = 4\pi \cdot 27 \text{ cm}^3$.

   The volume of the cylinder is $\pi \cdot (3.0 \text{ cm})^2 \cdot 18 \text{ cm}$
   $= \pi \cdot 9 \text{ cm}^2 \cdot 18 \text{ cm}$.

   Therefore, the tennis balls take up $\dfrac{4\pi \cdot 27 \text{ cm}^3}{\pi \cdot 9 \text{ cm}^2 \cdot 18 \text{ cm}}$

   of the space of the cylinder.

   In this expression, π cancels out, as do the units cm^3, cm^2, and cm. Also, 27 and 9 simplify to 3 and 1.
   So, we are left with only $\dfrac{4 \cdot 3.0}{18} = \dfrac{12}{18} = \dfrac{2}{3}$.

   The tennis balls take up 2/3 of the volume of the cylinder.

6. The volume of the earth is about $(4/3)\pi(6{,}370 \text{ km})^3$
   $\approx 1{,}082{,}696{,}932{,}430 \text{ km}^3 \approx 1{,}083{,}000{,}000{,}000 \text{ km}^3$.

7. The volume of one marble is $(4/3)\pi(8 \text{ mm})^3$. The density, 2.6 g/cm^3, is given in grams per cubic centimeter, which means we need to use centimeters as our unit in the formula for volume. So, we will use $(4/3)\pi(0.8 \text{ cm})^3$ for the volume of one marble. The total volume will be 50 times that. Then, the volume will need to be multiplied by the density, in order to get the total mass (M):

   $M = 50 \cdot (4/3)\pi(0.8 \text{ cm})^3 \cdot 2.6 \text{ g/cm}^3 \approx 278.81 \text{ g} \approx \underline{280 \text{ g}}$.

Volume Problems, p. 114

2. The volume of the cylinder is $V_1 = \pi(24 \text{ in})^2 \cdot 44 \text{ in} \approx 79{,}620.52$ cubic inches.

   The volume of the cone is $V_2 = \pi(24 \text{ in})^2 \cdot 25 \text{ in} /3 \approx 15{,}079.64$ cubic inches.

   The total volume of the tank is $V_1 + V_2 \approx 94{,}700.16 \text{ in}^3$.

   The diagram in the exercise shows us that 1 cubic foot = 1 ft $\cdot$ 1 ft $\cdot$ 1 ft = 12 in $\cdot$ 12 in $\cdot$ 12 in = 1,728 in^3.

   And so, in cubic feet, the total volume is 94,700.16 in^3 ÷ 1,728 in^3/ ft^3 = 54.80$\overline{3}$ cubic feet.

   To fill this at the rate of 2 cubic feet per minute will take 54.80$\overline{3}$ ÷ 2 ≈ 27.4 minutes or about 27 minutes 24 seconds.

Volume Problems, cont.

Page 114

3. a. The volume of the frustum is the volume of the entire cone minus the volume of the small, imaginary cone at the top.

$$V = \pi(32 \text{ cm})^2 \cdot 70 \text{ cm} \div 3 \ - \ \pi(17 \text{ cm})^2 \cdot 39 \text{ cm} \div 3$$

$$\approx 75,063.1 \text{ cm}^3 - 11,803.0 \text{ cm}^3 = 63,260.1 \text{ cm}^3$$

$$\approx \underline{63,300 \text{ cm}^3}.$$

   b. 63.3 liters

Page 115

4. The volume of the circular cylinder is $\pi(57.5 \text{ ft})^2 \cdot 36 \text{ ft} \approx 373,928.07 \text{ ft}^3$.

The volume of the half-sphere is $(4/3)\pi(57.5 \text{ ft})^3 \div 2 \approx 398,164.14 \text{ ft}^3$.

The total volume is $373,928.07 \text{ ft}^3 + 398,164.14 \text{ ft}^3$ $= 772,092.21 \text{ ft}^3 \approx \underline{772,000 \text{ ft}^3}$.

5. a. The bottom face consists of six identical triangles, so its area is $6 \cdot 3.2 \text{ cm} \cdot 2.8 \text{ cm} \div 2 = 26.88 \text{ cm}^2$.

The volume of the jar is then $8 \text{ cm} \cdot 26.88 \text{ cm}^2$ $= 215.04 \text{ cm}^3 \approx \underline{215 \text{ cm}^3}$.

   b. The volume of the cylindrical jar is $\pi(3.2 \text{ cm})^2 \cdot 8 \text{ cm}$ $= 257.36 \text{ cm}^3 \approx \underline{257 \text{ cm}^3}$.

   c. The percentage $= 215.04/257.36 = 0.83556 \approx \underline{83.6\%}$. Note that we included more digits in the two volumes than our final answers for (b) and (c), to keep it more accurate.

If you use the rounded (final) answers from (b) and (c), your answer to this percentage question will be different.

Mixed Review Chapter 2, pp. 116-117

Page 116

1. a. -16 b. 1/125   c. 57/1000   d. 3
   e. 1/64   f. 1/25   g. $5 \cdot 2^3 + 4 = 44$

2. a. 10,000   b. 2/343   c. 1

3.

a. $8b^{15}$	b. $\dfrac{1}{27y^3}$	c. $\dfrac{6}{x^2}$	d. $-10a^{10}b^9$
e. $\dfrac{1}{x}$	f. $\dfrac{2}{a^3}$	g. $\dfrac{-8x^3}{27}$	h. $\dfrac{x^2}{z^4}$

4. a. $>$ b. $<$ c. $<$

5. No. The quantity $5x$ is in parentheses, and this means even the number 5 is raised to the power of -2. Compare:

$$(5x)^{-2} = \frac{1}{(5x)^2} = \frac{1}{25x^2} = 25x^{-2}.$$

6. a. 3   b. 2   c. 2   d. 3
   e. 1   f. 4   g. 3   h. 4

7. a. $4.03 \cdot 10^{-1}$ b. $6.6 \cdot 10^7$
   c. $2 \cdot 10^{-3}$ d. $2.91 \cdot 10^{-2}$

Page 117

8.

a. $2 \cdot 10^{-3} + 5 \cdot 10^{-2}$	b. $7 \cdot 10^{-5} + 0.03$
$= 0.002 + 0.05 = \underline{0.052}$	$= 0.00007 + 0.03 = \underline{0.03007}$
c. $3.2 \cdot 10^{-1} - 0.07$	d. $5.4 \cdot 10^4 - 2,000 + 8 \cdot 10^3$
$= 0.32 - 0.07 = \underline{0.25}$	$= 54,000 - 2,000 + 8,000$
	$= \underline{60,000}$

9. a. 1.2 m^2 b. 25 mi/gal
   c. 7,930 mi   d. 58,000 lb
   e. \$350/person   f. 72,000 kg

10. $\dfrac{6.09 \cdot 10^{12} \text{ km}^2}{12 \cdot 10^3} = 5.1 \cdot 10^8 \text{ km}^2$

11. a. Since 550,000 is accurate to the thousands, it has three significant digits, and our answer also needs to be given with three significant digits. They spent \$1,780 per student.

   b. 950 cm^2. Notice the answer needs to be given to two significant digits.

12. $\dfrac{9.3 \cdot 10^7 \text{ mi}}{1.38 \cdot 10^{-1} \text{ hr}} = 6.7 \cdot 10^8 \text{ mph}$

— which is the speed of light! The light of the sun reaches the earth in about 0.138 hours which is about 8 minutes and 17 seconds.

Page 118

1. Answers will vary. Check the student's answer.
For example: First rotate the figure around the origin 180 degrees. Then translate it one unit to the right.

2.

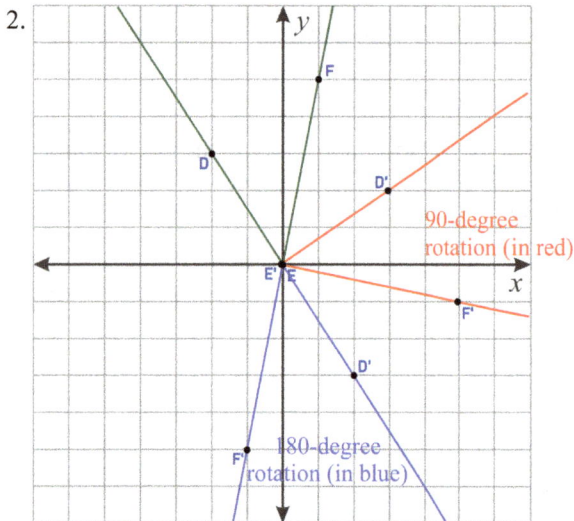

3. The coordinates of its vertices before these transformations were $(-5, 3)$, $(-2, 5)$, $(-1, 4)$, and $(-4, 1)$.

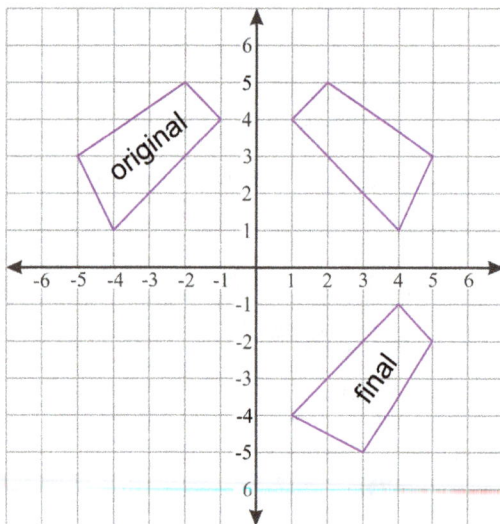

4. The true statements are (a), (c), and (d).

Page 119

5. a.

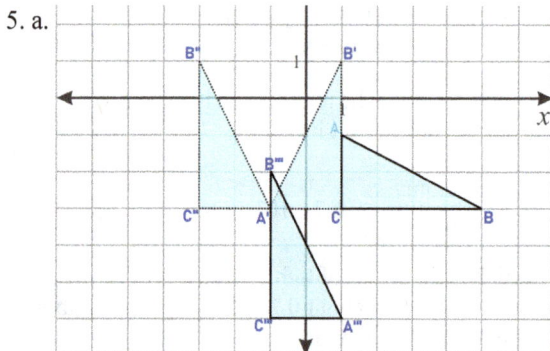

5. b. Answers will vary. Check the student's answer.
For example: First, reflect the triangle in the horizontal line $y = -3$. Then rotate it 90° counterclockwise around point C'. Lastly translate it three units down and two units to the left.

6. a.

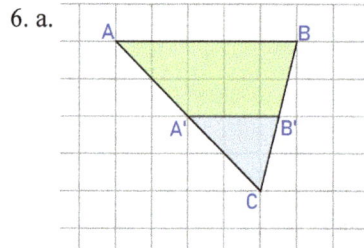

b.

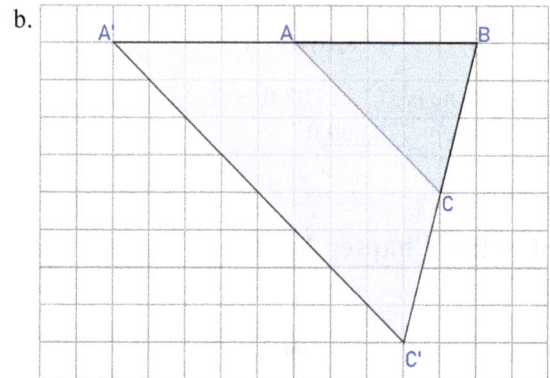

7. The coordinates of point F''' are $(-4, -1)$.

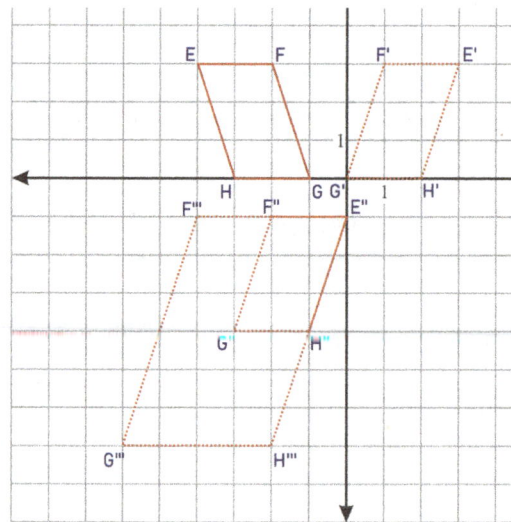

Page 120

8. Answers will vary. Check the student's answer.
For example: First, rotate the triangle 90 degrees clockwise around point E. Then dilate it with scale factor 1/3 from point E. Lastly, translate it four units to the right.

Chapter 2 Review, cont.

9. The dilation is from point S, since S' is the same as S. To figure out the scale factor, look at the distances from P to S, and then from P' to S'. From P to S is one unit horizontally and two units vertically. From P' to S' is two units horizontally and four units vertically, so the distances doubled. Thus, the scale factor is 2.

Since S' = S", the rotation is around point S' (or S). And since R' is six units horizontally from S', and R" is six units vertically from S", the rotation is 90° clockwise around S'.

Original figure	Dilation	Rotation
P(−5, 3)	P'(−6, 5)	P"(_0_ , _3_)
Q(0, 3)	Q'(4, 5)	Q"(_0_ , _−7_)
R(−1, 1)	R'(_2_ , _1_)	R"(−4 , −5)
S(−4, 1)	S'(−4, 1)	S"(−4, 1)

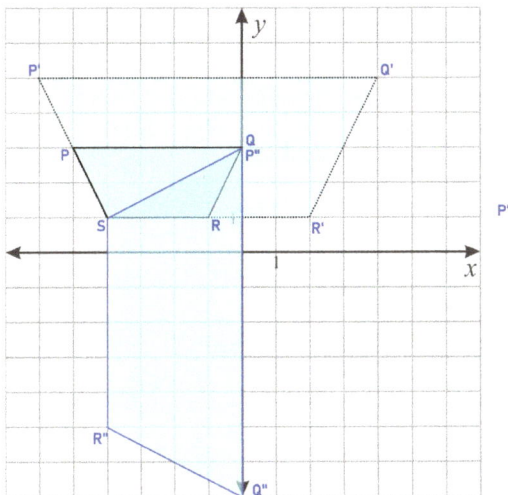

10. a. The three angles of the triangle are $x + 13$, $x + 18$, and $5x + 2$ (note there are vertical angles formed; that is why the unmarked angle of the triangle is $x + 13$). Those add up to 180°, so we can write the equation:

$$\begin{aligned} x + 13 + x + 18 + 5x + 2 &= 180 \\ 7x + 33 &= 180 \\ 7x &= 147 \\ x &= 21 \end{aligned}$$

b. Angle α is supplementary to the angle marked with $x + 13$. Since x is 21, $x + 13$ is 34°. So, α = 180° − 34° = _146°_.

11. In the larger triangle, the third angle, angle α', is 180° − 85° − 57° = 38°.

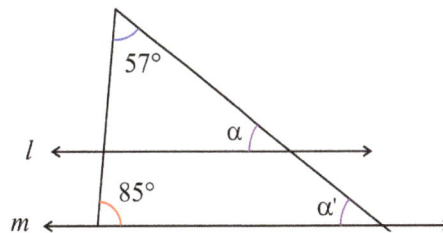

This is also the measure of the angle α, since α and α' are corresponding angles. So, α = _38°_.

12. Since m and n are parallel lines, angle A and the 64° angle are corresponding angles, thus congruent.

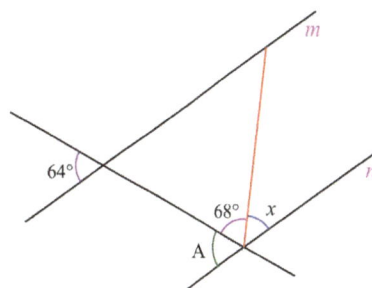

Angle A, the 68° angle, and x form a straight line so their sum is 180°. From that, we can get that $x = 180° − 68° − 64° = $ _48°_.

13. Its volume in cubic centimeters is $\pi \cdot (3 \text{ cm})^2 \cdot 17 \text{ cm} \approx 480.66 \text{ cm}^3 = 480.66$ ml. The shampoo bottle is about $473/480.66 \approx$ _98% full_.

Chapter 2 Review, cont.

14. The volume of the sphere is $(4/3) \cdot \pi \cdot (5 \text{ cm})^3$. The volume of a cone is $A_b \cdot h/3$, where A_b is the area of the base. In this case, A_b equals $\pi \cdot (5 \text{ cm})^2$.

Now we simply set the volume of the cone equal to the volume of the sphere, and solve for the height (h).

$$A_b \cdot h / 3 \;=\; (4/3) \cdot \pi \cdot (5 \text{ cm})^3$$
$$\pi \cdot (5 \text{ cm})^2 \cdot h / 3 \;=\; (4/3) \cdot \pi \cdot (5 \text{ cm})^3$$

Now, π cancels out from both sides:

$$(5 \text{ cm})^2 \cdot h / 3 = (4/3) \cdot (5 \text{ cm})^3$$
$$25 \text{ cm}^2 \cdot h / 3 = (4/3) \cdot 125 \text{ cm}^3$$

Dividing both sides by 25 cm^2, we get

$$h / 3 \;=\; (4/3) \cdot 5 \text{ cm}$$
$$h \;=\; 4 \cdot 5 \text{ cm} = 20 \text{ cm}$$

So, when the height is 20 cm, the cone will have an equal volume to the sphere with a radius of 5 cm.

Another way to solve this is to actually calculate the volume of the sphere and the area of the base of the cone, and use those calculated numbers in the equation.

The volume of the sphere is $(4/3) \cdot \pi \cdot (5 \text{ cm})^3 \approx 523.599 \text{ cm}^3$. The area of the base of the cone is $\pi \cdot (5 \text{ cm})^2 \approx 78.5398 \text{ cm}^2$.

Then, our equation looks like this:

$$A_b \cdot h / 3 \;=\; (4/3) \cdot \pi \cdot (5 \text{ cm})^3$$
$$78.5398 \text{ cm}^2 \cdot h / 3 \;=\; 523.599 \text{ cm}^3$$
$$78.5398 \text{ cm}^2 \cdot h \;=\; 3 \cdot 523.599 \text{ cm}^3$$
$$h \;=\; \frac{3 \cdot 523.599 \text{ cm}^3}{78.5398 \text{ cm}^2}$$
$$h \;\approx\; 20.0 \text{ cm}$$

15. The mound of ammonium nitrate consists of two parts: a cylinder and a cone.

The volume of the cylinder is $\pi \cdot (57.5 \text{ ft})^2 \cdot 36 \text{ ft}$.

The volume of the cone is $\pi \cdot (57.5 \text{ ft})^2 \cdot 57.5 \text{ ft} / 3$.

The total volume is
$$\pi \cdot (57.5 \text{ ft})^2 \cdot 36 \text{ ft} + \pi \cdot (57.5 \text{ ft})^2 \cdot 57.5 \text{ ft} / 3$$

$\approx$ 573,000 cubic feet.

Chapter 3: Linear Equations

Algebra Terms, p. 125

1. $6ab - 7a^2 + 8a + 25$

Review: Integer Addition and Subtraction, pp. 126-128

Page 126

1. a. 4   b. −10   c. −6   d. −43
   e. −19   f. 2   g. −2
   h. −10   i. 10

Page 127

2.

a. $8 - (-6)$ $\downarrow$ $8 + 6 = 14$	b. $-9 - (-14)$ $\downarrow$ $-9 + 14 = 5$	c. $-21 - 8$ $\downarrow$ $-21 + (-8) = -29$	d. $3 - 15$ $\downarrow$ $3 + (-15) = -12$

3. a. −7   b. −11   c. 7   d. 11
   e. −11   f. −3   g. 11   h. −3

4.

a. $7 + (-10)$ $7 - 10 = -3$	b. $-2 + (-1)$ $-2 - 1 = -3$	c. $14 + 3$ $14 - (-3) = 17$	d. $-8 + 5$ $-8 - (-5) = -3$

5. a. −1   b. −18   c. 1
   d. −16   e. 7   f. −21

Page 128

6.

a. $\frac{2}{7} + \left(-\frac{3}{4}\right)$ $= \frac{8}{28} + \frac{-21}{28} = -\frac{13}{28}$	b. $\frac{1}{8} + \left(-\frac{1}{2}\right) + \left(-\frac{3}{4}\right)$ $= \frac{1}{8} + \frac{-4}{8} + \frac{-6}{8} = \frac{-9}{8} = -1\frac{1}{8}$
c. $-\frac{5}{6} + \frac{2}{9}$ $= \frac{-15}{18} + \frac{4}{18} = -\frac{11}{18}$	d. $-\frac{2}{3} + \frac{2}{9} - \frac{1}{6}$ $= \frac{-12}{18} + \frac{4}{18} + \frac{-3}{18} = -\frac{11}{18}$
e. $-\frac{7}{8} + \left(-\frac{1}{10}\right)$ $= \frac{-35}{40} + \frac{-4}{40} = -\frac{39}{40}$	f. $\frac{1}{6} - \left(-\frac{7}{8}\right)$ $= \frac{4}{24} + \frac{21}{24} = \frac{25}{24} = 1\frac{1}{24}$

Equations Review, Part 1, pp. 129-131

<u>Page 129</u>

1. a. No, it isn't. Substituting 2 in place of x, we get the equation $\dfrac{3 \cdot 2^2 - 7}{5} = 2$.

   Simplifying the left side, this becomes $\dfrac{5}{5} = 2$, which is a false equation.

   b. Yes. Substituting −90 in place of x, we get $\dfrac{2}{3}(-90) + 11 = -49$. Further simplifying the left side, we get

   $(-60) + 11 = -49$. This is a true equation, so −90 is a root of the equation

2. Yes, you can simplify the left side of the equation before solving it. If we combine like terms on the left side, the equation becomes $11x - 8 = -5(x + 8)$. Now, let's substitute −3 in the place of x. See the simplification on the right:

   This is a false equation, so −3 is not a root.

   $$11(-3) - 8 = -5(-3 + 8)$$
   $$-33 - 8 = 15 - 40$$
   $$-41 = -25$$

3. Answers will vary. Check the student's answers. For example: $2x = -10$; $3x + 7 = -8$; $-6x + 5 = 35$.

4. No, it doesn't. If $2w + 6$ equaled $3w - 15$, then it would follow that $50 = 51$ (since $2w + 6 = 50$ and $3w - 15 = 51$).

<u>Page 130</u>

5. a. Substituting 14 in place of x would result in $10 - 14 = 24$.
   This is false, so 14 is not a root.

   b. Derek eliminated the minus sign from in front of x, which resulted in an incorrect solution. The correct root is $x = -14$. See the solution on the right.

$$10 - x = 24$$
$$\underline{-10 \qquad -10}$$
$$-x = 14$$
$$x = -14$$

6.

a. $\begin{aligned} -p &= 12 - 34 \\ -p &= -22 \\ p &= 22 \end{aligned}$	b. $\begin{aligned} 78 - x &= -8 \quad \|-78 \\ -x &= -86 \\ x &= 86 \end{aligned}$	c. $\begin{aligned} -2 - 7 &= -3z \\ -9 &= -3z \quad \|\div(-3) \\ 3 &= z \\ z &= 3 \end{aligned}$
d. $\begin{aligned} \dfrac{y}{-4} &= -22 \quad \|\cdot(-4) \\ y &= 88 \end{aligned}$	e. $\begin{aligned} 10x &= -40 - 5 \\ 10x &= -45 \quad \|\div 10 \\ x &= -9/2 \text{ or } -4\tfrac{1}{2} \end{aligned}$	f. $\begin{aligned} 2.1 - x &= -6.7 \quad \|-2.1 \\ -x &= -8.8 \\ x &= 8.8 \end{aligned}$

<u>Page 131</u>

7.

a. $\begin{aligned} 5x + 2 &= 67 \quad \|-2 \\ 5x &= 65 \quad \|\div 5 \\ x &= 13 \end{aligned}$	b. $\begin{aligned} -3y + 2 &= 71 \quad \|-2 \\ -3y &= 69 \quad \|\div(-3) \\ y &= -23 \end{aligned}$	c. $\begin{aligned} 25 - 3w &= 17 \quad \|-25 \\ -3w &= -8 \quad \|\div(-3) \\ w &= 8/3 \end{aligned}$
d. $\begin{aligned} -34 &= 2x - 11 \quad \|+11 \\ -23 &= 2x \quad \|\div 2 \\ -23/2 &= x \\ x &= -23/2 \end{aligned}$	e. $\begin{aligned} -98 &= -8z - 2 \quad \|+2 \\ -96 &= -8z \quad \|\div(-8) \\ 12 &= z \\ z &= 12 \end{aligned}$	f. $\begin{aligned} -8 - 4z &= 10 \quad \|+8 \\ -4z &= 18 \quad \|\div(-4) \\ z &= -9/2 \end{aligned}$

48

Equations Review, Part 1, cont.

8. The error is in the line $80x = 40$ where the minus sign is omitted. It should be $-80x = 40$, from which $x = -\frac{1}{2}$.
   In other words, when subtracting 14 from both sides, the student also erased the minus sign from in front of 80.

9.

a. $\begin{aligned} -5 + 15 &= -6w \\ 10 &= -6w \quad \div(-6) \\ -10/6 &= w \\ w &= -5/3 \end{aligned}$	b. $\begin{aligned} 6 &= \dfrac{d}{-1.1} \quad \cdot(-1.1) \\ -6.6 &= d \\ d &= -6.6 \end{aligned}$	c. $\begin{aligned} \dfrac{a}{5} &= -1.2 + (-3.1) \\ \dfrac{a}{5} &= -4.3 \quad \cdot 5 \\ a &= -21.5 \end{aligned}$
d. $\begin{aligned} 56 - 5x &= 28 \\ -5x &= -28 \quad \div(-5) \\ x &= 28/5 \end{aligned}$	e. $\begin{aligned} -35 &= -4q + 2 \quad -2 \\ -37 &= -4q \\ -4q &= -37 \quad \div(-4) \\ q &= 37/4 \end{aligned}$	f. $\begin{aligned} -150 + 30w &= 60 \quad +150 \\ 30w &= 210 \quad \div 30 \\ w &= 7 \end{aligned}$
g. $\begin{aligned} 13.5 - 2y &= 7 \quad -13.5 \\ -2y &= -6.5 \quad \div(-2) \\ y &= 3.25 \end{aligned}$	h. $\begin{aligned} 7.8 - 16.2 &= \dfrac{x}{7} \\ -8.4 &= \dfrac{x}{7} \quad \cdot 7 \\ -58.8 &= x \\ x &= -58.8 \end{aligned}$	i. $\begin{aligned} -55 &= -6w - 13 \quad +13 \\ -42 &= -6w \quad \div(-6) \\ 7 &= w \\ w &= 7 \end{aligned}$

The Distributive Property, pp. 133-134

1. The error is in forgetting to multiply all the terms inside the parentheses by the factor outside them.
   In the first example, the correct answer is $4a + 4b + 12$, in the second, $14x - 63$.

2. An optional step is shown in gray.

a. $-2(x + 9)$	b. $-2(x - 9)$	c. $-3(5x + 8)$	d. $-3(5x - 8)$	e. $-5(2x + 7)$	f. $-5(2x - 7)$
$-2x + (-2)9$	$-2x - (-2)9$	$-15x + (-3)8$	$-15x - (-3)8$	$-10x + (-5)7$	$-10x - (-5)7$
$-2x - 18$	$-2x + 18$	$-15x - 24$	$-15x + 24$	$-10x - 35$	$-10x + 35$

3.

a. $-2(x + y - 9)$	b. $-2(x - y + 9)$	c. $-3(9 + 2y)$	d. $-3(9 - 2y)$
$-2x - 2y + 18$	$-2x + 2y - 18$	$-27 - 6y$	$-27 + 6y$

4. a. $-y - 3$ b. $-y + 3$ c. $-6 + a - 2b$ d. $6 - a + 2b$

5. a. $-0.9a + 1.8b - 6.3$ b. $x - 4 + 5y$ c. $7w - 0.5$ d. $-8 + 10x$

   e. $\frac{1}{4}g - \frac{1}{2}h + 1$ f. $32v + 8w - 4$ g. $-4p + 8q$ h. $1 - 10z$

6. a. $56 + 14x$ b. $10y - 40 - 15x$ c. $-6x + 14$
   d. $20s - 4 + 13t$ e. $87w - 24$ f. $-9x + 27y$

The Distributive Property, cont.

7.

a. $14x - 10$	b. $-27x + 36$	c. $-12s + 18t - 6$	d. $-0.9x + 1.2y - 0.6$
$2(7x - 5)$	$-9(3x - 4)$ or $9(-3x + 4)$ or $-3(9x - 12)$ or $3(-9x + 12)$	$-6(2s - 3t + 1)$ or $6(-2s + 3t - 1)$	$-0.3(3x - 4y + 2)$ or $0.3(-3x + 4y - 2)$
e. $\frac{1}{4}w - \frac{1}{2}$ $\frac{1}{2}(\frac{1}{2}w - 1)$ or $\frac{1}{4}(w - 2)$	f. $\frac{3}{4}s + \frac{3}{8}$ $\frac{3}{4}(s + \frac{1}{2})$ or $\frac{3}{8}(2s + 1)$	g. $\frac{1}{5}y - x + \frac{6}{5}$ $\frac{1}{5}(y - 5x + 6)$	h. $-x - \frac{2}{3}y + \frac{1}{3}$ $-\frac{1}{3}(3x + 2y - 1)$ or $\frac{1}{3}(-3x - 2y + 1)$

8.

a. $-2(6 + \boxed{9x}) = -12 - 18x$	b. $\boxed{-3}(4y - 5) = -12y + 15$	c. $\boxed{-0.2}(4v - 6w + 5) = -0.8v + 1.2w - 1$

9. Jonathan is correct. The expression $-(-3x - 6)$ signifies the opposite of the expression $-3x - 6$. The sign of each term in it change, thus it becomes $3x + 6$.

10. Expressions (a) and (e).

11.

a. $\dfrac{8y + 48}{8} = \dfrac{\boxed{8y}}{8} + \dfrac{\boxed{48}}{8} = \boxed{y} + 6$	b. $\dfrac{2x - 12}{6} = \dfrac{\boxed{2x}}{6} - \dfrac{12}{6} = \dfrac{\boxed{x}}{3} - \boxed{2}$

12. a. $5x + 6y$ b. $0.5 - 0.8a + 3b$ c. $2x - y/4 + 3$

Puzzle corner. a. $5x - 2$ b. $-5y + 2$

Equations Review, Part 2, p. 136

1.

a. $5(x + 2) = 65$ $5x + 10 = 65$ $\mid -10$ $5x = 55$ $\mid \div 5$ $x = 11$	b. $3(y - 2) = 72$ $3y - 6 = 72$ $\mid +6$ $3y = 78$ $\mid \div 3$ $y = 26$
c. $-8(w - 11) = 18$ $-8w + 88 = 18$ $\mid -88$ $-8w = -70$ $\mid \div(-8)$ $w = 70/8$	d. $8 = -9(z + 5)$ $8 = -9z - 45$ $\mid +45$ $53 = -9z$ $\mid \div(-9)$ $z = -53/9$
e. $2.5 = -(6t - 1.5)$ $2.5 = -6t + 1.5$ $\mid -1.5$ $1 = -6t$ $\mid \div(-6)$ $t = -1/6$	f. $-10(-x + 7) = 27$ $10x - 70 = 27$ $\mid +70$ $10x = 97$ $\mid \div 10$ $x = 97/10$

Page 137

2.

Divide first:	Distribute the multiplication first:
$3(x-7) \ = \ 42 \quad \div 3$ $x - 7 \ = \ 14 \quad + 7$ $x \ = \ 21$	$3(x-7) \ = \ 42$ $3x - 21 \ = \ 42 \quad + 21$ $3x \ = \ 63 \quad \div 3$ $x \ = \ 21$

Page 138

3.

a. Way 1: $\ -20(q-5) \ = \ 80 \quad \div(-20)$ $q - 5 \ = \ -4 \quad + 5$ $q \ = \ 1$	Way 2: $\ -20(q-5) \ = \ 80$ $-20q + 100 \ = \ 80 \quad -100$ $-20q \ = \ -20 \quad \div(-20)$ $q \ = \ 1$
b. Way 1: $\ -5x - 8 \ = \ 27 \quad +8$ $-5x \ = \ 35 \quad \div(-5)$ $x \ = \ -7$	Way 2: $\ -5x - 8 \ = \ 27 \quad \div(-5)$ $x + 8/5 \ = \ -27/5 \quad -8/5$ $x \ = \ -35/5 = -7$

4.

a. $\quad -8(x+40) \ = \ 10 \quad \div(-8)$ $x + 40 \ = \ -1.25 \quad -40$ $x \ = \ -41.25$	b. $\quad \dfrac{4x+2}{7} = -6 \quad \cdot 7$ $4x + 2 = -42 \quad -2$ $4x = -44 \quad \div 4$ $x = -11$
c. $\quad 2.3 \ = \ -2x + 0.9 \quad -0.9$ $1.4 \ = \ -2x$ $-2x \ = \ 1.4 \quad \div(-2)$ $x \ = \ -0.7$	d. $\quad 90 \ = \ 20(x-3)$ $90 \ = \ 20x - 60 \quad +60$ $150 \ = \ 20x$ $20x \ = \ 150 \quad \div 20$ $x \ = \ 15/2$

Page 139

5. The solution processes may vary but the final answer does not.

a. $\quad -2(x+4) \ = \ 7$ $x + 4 \ = \ -7/2$ $x \ = \ -15/2$	b. $\quad 9 = 0.3(x-5)$ $30 = x - 5$ $35 = x$ $x = 35$	c. $\quad 500 \ = \ -20y - 50$ $550 \ = \ -20y$ $y \ = \ -55/2$
d. $\quad -19 - 4w \ = \ 3$ $-4w \ = \ 22$ $w \ = \ -11/2$	e. $\quad -0.5 = 0.2x - 1.4$ $0.9 = 0.2x$ $x = 4.5$	f. $\quad 5(-x-8) \ = \ -10$ $-x - 8 \ = \ -2$ $x + 8 \ = \ 2$ $x \ = \ -6$

Page 139

5.

g. $40(3 - q) = 65$	h. $200 = -7.5x - 40$	i. $9 - 2s = 78$
$3 - q = 65/40$	$240 = -7.5x$	$-2s = 69$
$-q = 65/40 - 3$	$x = -32$	$s = -69/2$
$q = 3 - 65/40$		
$q = 55/40 = 11/8$		

Puzzle corner. To solve these, substitute the given value of x into the equation, then solve for the unknown a or b.
a. Substituting $x = -5$ we get: $-5 + a = 9$, from which $a = 14$.
b. Substituting $x = -5$ we get: $3(-5) + b = 9$, from which $b = 24$.

Equations Review, Part 3, pp. 140-141

Page 140

1. a. The error is that in the second line, the term 7 did not get multiplied by 8. Here is the corrected solution:

$$\begin{array}{rcl|l} \frac{3}{8}y - 7 &=& 2 & \cdot\, 8 \\ 3y - \mathbf{56} &=& 16 & +\,\mathbf{56} \\ 3y &=& 72 & \div\, 3 \\ y &=& 24 & \end{array}$$

b. The error is that on the second line, the left side did not get multiplied by 5. Here is the corrected solution:

$$\begin{array}{rcl|l} 4(y + 2) &=& \frac{13}{5} & \cdot\, 5 \\ 5(4)(y + 2) &=& 13 & \\ 20(y + 2) &=& 13 & \\ 20y + 40 &=& 13 & -\,40 \\ 20y &=& -27 & \div\, 20 \\ y &=& -\frac{27}{20} & \end{array}$$

Page 141

2.

a. $\frac{1}{5}a + 7 = 3$ $\quad\vert\ \cdot\, 5$	b. $\frac{1}{5}(a + 7) = 3$ $\quad\vert\ \cdot\, 5$	c. $-\frac{2}{5}(a + 7) = 3$ $\quad\vert\ \cdot\, 5$
$a + 35 = 15$ $\quad\vert -35$	$a + 7 = 15$ $\quad\vert -7$	$-2(a + 7) = 15$
$a = -20$	$a = 8$	$-2a - 14 = 15$ $\quad\vert +14$
		$-2a = 29$ $\quad\vert \div(-2)$
		$a = -29/2$

3.

a. $2 = -\frac{9}{10}(4 - x)$ $\vert\ \cdot\, 10$	b. $2(1 - x) = \frac{5}{12}$ $\vert\ \cdot\, 12$	c. $2y - 5 = -\frac{4}{7}$ $\vert\ \cdot\, 7$
$20 = -9(4 - x)$	$24(1 - x) = 5$	$14y - 35 = -4$ $\quad\vert +35$
$20 = -36 + 9x$ $\quad\vert +36$	$24 - 24x = 5$ $\quad\vert -24$	$14y = 31$ $\quad\vert \div 14$
$56 = 9x$ $\quad\vert \div 9$	$-24x = -19$ $\quad\vert \div(-24)$	$y = 31/14$
$x = 56/9$	$x = 19/24$	

Page 141

4.

a. $0.4(x + 5) = -3.7$	b. $4.72w - 8.9 = 20$ $\mid + 8.9$	c. $98.5 = -3(y + 25.6)$
$0.4x + 2 = -3.7$ $\mid -2$	$4.72w = 28.9$ $\mid \div 4.72$	$98.5 = -3y - 76.8$ $\mid +76.8$
$0.4x = -5.7$ $\mid \div 0.4$	$w \approx 6.12$	$175.3 = -3y$ $\mid \div(-3)$
$x = -14.25$		$y \approx -58.43$

Page 142

5.

$$2(x + \frac{4}{5}) = -7$$
$$2x + \frac{8}{5} = -7 \qquad \mid -8/5$$
$$2x = -7 - \frac{8}{5}$$
$$2x = -\frac{43}{5} \qquad \mid \div 2$$
$$x = -\frac{43}{10}$$

6.

a. $-3(x + \frac{1}{6}) = 1$	b. $-3x + \frac{1}{6} = 1$ $\mid -1/6$	c. $-3x + 1 = -\frac{1}{6}$ $\mid -1$
$-3x - \frac{1}{2} = 1$ $\mid +\frac{1}{2}$	$-3x = \frac{5}{6}$ $\mid \div(-3)$	$-3x = -\frac{7}{6}$ $\mid \div(-3)$
$-3x = \frac{3}{2}$ $\mid \div(-3)$	$x = -\frac{5}{18}$	$x = \frac{7}{18}$
$x = -\frac{1}{2}$		

7. See the answer on the right.

$$2y - 7 = \frac{5}{9} \qquad \mid \cdot 9$$
$$9(2y - 7) = 5$$
$$18y - 63 = 5 \qquad \mid +63$$
$$18y = 68 \qquad \mid \div 18$$
$$y = 68/18 = 34/9$$

Page 143

8. a. Substituting $x = -4/3$ into the equation, we get:

$$6(-\frac{4}{3} - \frac{2}{3}) = -2$$
$$6(-2) = -2$$
$$-12 = -2$$

...which is a false equation. The error was done on the second line of the solution, when the right side was multiplied by 6.

8. b. See the solution below.

$$6(x - \frac{2}{3}) = -2$$
$$6x - \frac{12}{3} = -2$$
$$6x - 4 = -2$$
$$6x = 2$$
$$x = 1/3$$

Equations Review, Part 3, cont.

9.

T $\quad 3(x + \frac{2}{9}) = -3 \qquad \bigg| \div 3$ $\qquad x + \frac{2}{9} = -1 \qquad \bigg| -2/9$ $\qquad\qquad x = -11/9$	**R** $\quad 2 = \frac{1}{8}(7 - x) \qquad \bigg| \cdot 8$ $\qquad 16 = 7 - x \qquad \bigg| -7$ $\qquad -x = 9$ $\qquad\quad x = -9$	**A** $\quad -3x + 6 = \frac{3}{5} \qquad \bigg| \cdot 5$ $\qquad -15x + 30 = 3 \qquad \bigg| -30$ $\qquad -15x = -27 \qquad \bigg| \div 15$ $\qquad\qquad x = \frac{27}{15} = \frac{9}{5}$
H $\quad 0.2(6 - s) = 50 \qquad \bigg| \cdot 5$ $\qquad 6 - s = 250 \qquad \bigg| -6$ $\qquad -s = 244$ $\qquad\quad s = -244$	**E** $\quad 1.5 = 3(-T + 0.7)$ $\qquad 1.5 = -3T + 2.1 \qquad \bigg| -2.1$ $\qquad -0.6 = -3T \qquad \bigg| \div(-3)$ $\qquad\quad T = 0.2$	**W** $\quad 40 - 0.9x = 35.5 \qquad \bigg| -40$ $\qquad -0.9x = -4.5 \qquad \bigg| \div(-0.9)$ $\qquad\qquad x = 5$

Everyone always talks about it, but no one does anything about it. What is it?

The

5	0.2	9/5	−11/9	−244	0.2	−9
W	E	A	T	H	E	R

Combining Like Terms, pp. 144-145

1. a. $7x - 5$ b. $22a^2$

   c. $s + 8$ d. $-\frac{5}{4}x + 5$

2. Answers may vary in the fact that the terms in the student's response may not be in the same order as given here. The text has not taught about organizing polynomials by the alphabetical order of the variables.

a. $-3m - 15$	b. $8x$
c. $-25q - 16n$	d. $5x - 7y - 3$
e. $-10.5m^2 + 20$	f. $16a + 15c + 10d - 7$
g. $12x^2 y^2$	h. $2x^3 - 8x - 5x^3$ $\quad = -3x^3 - 8x$
i. $9 - \frac{1}{7}w - \frac{4}{5}w$ $= 9 - \frac{5}{35}w - \frac{28}{35}w$ $= 9 - \frac{33}{35}w$	j. $-\frac{1}{3}x - 2 + \frac{1}{8}x + 5$ $= -\frac{8}{24}x - 2 + \frac{3}{24}x + 5$ $= -\frac{5}{24}x + 3$
k. $\frac{3}{5}n^2 - \frac{2}{3}n^2 - 3m$ $= \frac{9}{15}n^2 - \frac{10}{15}n^2 - 3m$ $= -\frac{1}{15}n^2 - 3m$	l. $2x - \frac{1}{2}x^2 - \frac{7}{8}x + 3x^2$ $\quad = \frac{5}{2}x^2 + \frac{9}{8}x$

54

Combining Like Terms, cont.

Page 145

3. a. $2 - s$ b. $5 + x$ c. $-2x + 4$
   d. $2x^2 + y^2$ e. $4.9w$ f. w
   g. $-4x + 4$ h. $3.2y + 3.6$
   i. $-89x + 110$ j. $21z - 3$

Page 146

4.

a. $5x + 2x = 14$
          $7x = 14$
           $x = 2$
	b. $-7y + 3y - (-2y) = 50$
	            $-2y = 50$
	             $y = -25$

c. $-2y + (-5y) + 8y - 7 = 8$
            $y - 7 = 8$
              $y = 15$
	d. $20 - 36 = -8x + 9x - 6x$
	        $-16 = -5x$
	          $x = 16/5$

e. $1.5s - (-4.8s) - 1.3s = 3.5$
            $5s = 3.5$
             $s = 0.7$
	f. $3.2 = -(-2x) - x - 5x + 1.6$
	    $3.2 = -4x + 1.6$
	    $1.6 = -4x$
	       $x = -0.4$

g. $2t - 4.8t + 1.3t - 0.8t = 4.6$
          $-2.3t = 4.6$
            $t = -2$
	h. $11 = \frac{1}{2}x - \frac{1}{8}x + 5$
	
	     $11 = \frac{3}{8}x + 5$
	
	      $6 = \frac{3}{8}x$
	
	    $48 = 3x$
	
	      $x = 16$
	
	You could also start by first multiplying both sides of the equation by 8.

Word Problems, pp. 147-148

Page 147

1. $x + 3x + x + 3x = 28$; $8x = 28$; $x = 3\frac{1}{2}$.
   The length of the rectangle is <u>3.5 meters</u>, and its width is <u>10.5 meters</u>.

2. Let x be the width of the rectangle. Then, its length is $x + 12$. We get the equation $x + x + 12 + x + x + 12 = 136$.
   This simplifies to $4x + 24 = 136$, from which $x = 28$. So, the width of the rectangle is <u>28 cm</u> and its length is <u>40 cm</u>.
   (It is also possible that the student will choose x to be the length of the rectangle. The width will then be $x - 12$.
   The final answer will be the same.)

Page 148

3. $3x + 2x = 170$; $5x = 170$; $x = 34$. Henry washes <u>68 windows</u>.

4. The parts of the inheritance are x, $6x$, and $5x$. The equation is $x + 6x + 5x = 354,000$, from which $12x = 354,000$
   and $x = 29,500$. <u>The heirs got $29,500, $177,000, and $147,500.</u>

5. Let $3x$ and $5x$ be the two sides of the rectangle. The equation is: $3x + 5x + 3x + 5x = 416$; $16x = 416$; $x = 26$.
   The two sides are $3 \cdot 26$ cm = <u>78 cm</u> and $5 \cdot 26$ cm = <u>130 cm</u>.

Word Problems, cont.

6. The equation is: $x + x + 5 = 99$, from which we get: $2x + 5 = 99$; $x = 47$. Eric worked for 52 hours.

7. Let d be the number of ducks. Then, $d + 17$ is the number of chickens. Equation: $d + d + 17 = 135$; $d = 59$. She has $59 + 17 = \underline{76 \text{ chickens}}$.

8. a. Let d be the number of days Hans carpooled out of 10. When he carpools, he pays $2, and when he doesn't, he pays $6. We can write the equation $d \cdot \$2 + (10 - d) \cdot \$6 = \$36$, or written in the usual way: $2d + 6(10 - d) = 36$.

$$
\begin{aligned}
2d + 6(10 - d) &= 36 \\
2d + 60 - 6d &= 36 \\
60 - 4d &= 36 \\
-4d &= -24 \\
d &= 6
\end{aligned}
$$

He carpooled <u>on six days</u>.

8. b. Let d be the number of days Hans carpooled out of 22. We can write the equation $2d + 6(22 - d) = 96$.

$$
\begin{aligned}
2d + 6(22 - d) &= 96 \\
2d + 132 - 6d) &= 96 \\
132 - 4d &= 96 \\
-4d &= -36 \\
d &= 9
\end{aligned}
$$

He carpooled on nine days, so he did not carpool on <u>13 days</u>.

9. Let p be the price of the cheapest leash. Then, the other two cost $p + 5.4$ and $p + 11.6$. The equation is:

$$
\begin{aligned}
p + p + 5.4 + p + 11.6 &= 62.6 \\
3p + 17 &= 62.6 \\
3p &= 45.6 \\
p &= 15.2
\end{aligned}
$$

The most expensive leash cost $15.20 + $11.60 = \underline{\$26.80}$.

10. Let x, $x + 1$, and $x + 2$ be the three consecutive numbers. Then:

$$
\begin{aligned}
x + x + 1 + x + 2 &= 360 \\
3x + 3 &= 360 \\
3x &= 357 \\
x &= 119
\end{aligned}
$$

The numbers are 119, 120, and 121.

11. Let x, $x + 2$, and $x + 4$ be the three consecutive odd numbers. Then:

$$
\begin{aligned}
x + x + 2 + x + 4 &= 1{,}971 \\
3x + 6 &= 1{,}971 \\
3x &= 1{,}965 \\
x &= 655
\end{aligned}
$$

The numbers are 655, 657, and 659.

Word Problems, cont.

Page 150

12. Let x, $x + 5$, $x + 10$, and $x + 15$ be the four consecutive multiples of 5. Then:

$$x + x + 5 + x + 10 + x + 15 = 1{,}570$$
$$4x + 30 = 1{,}570$$
$$4x = 1{,}540$$
$$x = 385$$

The numbers are 385, 390, 395, and 400.

Puzzle corner. a. Make a table to organize your guesses and results. Observe whether your guesses are producing too low or too high a result (in this case the product), and adjust your guess accordingly. For an example, see the table on the right:

Numbers	Sum	Product
30, 5	35	150
26, 9	35	234
25, 10	35	250

From the table, one can see that the numbers need to be closer to each other for their product to get closer to 300.
So let's continue guessing. See the table on the right:

Here is our result: 20 and 15 work.

Numbers	Sum	Product
22, 13	35	286
21, 14	35	294
20, 15	**35**	**300**

b. Again, start with some numbers. Check the products to see whether you are going in the right direction with your guesses.

The numbers 160 and 60 work.

Numbers	Sum	Product
200, 20	220	4,000
190, 30	220	5,700
180, 40	220	7,200
170, 50	220	8,500
160, 60	220	9,600

A Variable on Both Sides, p. 151

Page 151

1.

First add 2s:

$10 - 2s = 4s + 9$	$+ 2s$
$10 = 6s + 9$	$- 9$
$1 = 6s$	$\div 6$
$s = 1/6$	

First subtract 4s :

$10 - 2s = 4s + 9$	$- 4s$
$10 - 6s = 9$	$- 10$
$-6s = -1$	$\div (-6)$
$s = 1/6$	

2. The solution processes may vary; check the student's solution. The final answer (the root) does not vary.

a.				b.				
	$3x + 2 = 2x - 7$	$- 2x$			$9y - 2 = 7y + 5$	$- 7y$		
	$x + 2 = -7$	$- 2$			$2y - 2 = 5$	$+ 2$		
	$x = -9$				$2y = 7$	$\div 2$		
					$y = 7/2$			

A Variable on Both Sides, cont.

Page 152

3.

$$
\begin{array}{ll}
7w + 8 = 2w - 5 & \quad -2w \\
5w + 8 = -5 & \quad -8 \\
5w = -13 & \quad \div 5 \\
w = -13/5 &
\end{array}
$$

4. The solution processes may vary; check the student's solution. The final answer (the root) does not vary.

a.
$$
\begin{array}{ll}
-2y - 6 = 20 + 6y & \quad +2y \\
-6 = 20 + 8y & \quad -20 \\
-26 = 8y & \quad \div 8 \\
y = -26/8 = -13/4 &
\end{array}
$$

b.
$$
\begin{array}{ll}
8x - 12 = -1 - 3x & \quad +3x \\
11x - 12 = -1 & \quad +12 \\
11x = 11 & \quad \div 11 \\
x = 1 &
\end{array}
$$

c.
$$
\begin{array}{ll}
6z - 5 = 9 - 2z & \quad +2z \\
8z - 5 = 9 & \quad +5 \\
8z = 14 & \quad \div 8 \\
z = 14/8 = 7/4 &
\end{array}
$$

5. a. Job 1: $19.50m + 150$ Job 2: $21m$

b. In Job 1, in 20 hours, he would earn $\$19.50 \cdot 20 + \$150 = \$540$. In Job 2, he would earn $\$21 \cdot 20 = \$420$.

c. Equation:
$$
\begin{array}{ll}
19.5m + 150 = 21m & \quad -19.5m \\
150 = 1.5m & \quad \div 1.5 \\
m = 100 &
\end{array}
$$

For 100 hours of work, both jobs would provide the same wages, that is, $\$2,100$.

Page 153

6. The solution processes may vary; check the student's solution. The final answer (the root) does not vary.

a.
$$
\begin{array}{ll}
3x + 7 - 5x = 6x + 1 - 4x & \\
-2x + 7 = 2x + 1 & \quad -7 \\
-2x = 2x - 6 & \quad -2x \\
-4x = -6 & \quad \div(-4) \\
x = 3/2 &
\end{array}
$$

b.
$$
\begin{array}{ll}
-7x - 5 + x + 4x = 8 - 2x - 5x & \\
-2x - 5 = 8 - 7x & \quad +7x \\
5x - 5 = 8 & \quad +5 \\
5x = 13 & \quad \div 5 \\
x = 13/5 &
\end{array}
$$

7. The solution processes may vary; check the student's solution. The final answer (the root) does not vary.

a.
$$
\begin{array}{ll}
6x + 3x + 1 = 9x - 2x - 7 & \\
9x + 1 = 7x - 7 & \quad -7x \\
2x + 1 = -7 & \quad -1 \\
2x = -8 & \quad \div 2 \\
x = -4 &
\end{array}
$$

b.
$$
\begin{array}{ll}
16y - 4y - 3 = -4y - y & \\
12y - 3 = -5y & \quad +5y \\
17y - 3 = 0 & \quad +3 \\
17y = 3 & \quad \div 17 \\
y = 3/17 &
\end{array}
$$

c.
$$
\begin{array}{ll}
-26x + 12x = -18x + 8x - 6 & \\
-14x = -10x - 6 & \quad +10x \\
-4x = -6 & \quad \div(-4) \\
x = 6/4 = 3/2 &
\end{array}
$$

Page 154

8. The solution processes may vary; check the student's solution. The final answer (the root) does not vary.

a.
$$
\begin{array}{ll}
0.9y + 1 - 1.4y = 4.6y - 4.8 + y & \\
-0.5y + 1 = 5.6y - 4.8 & \quad +0.5y \\
1 = 6.1y - 4.8 & \quad +4.8 \\
5.8 = 6.1y & \quad \div 6.1 \\
y = 5.8/6.1 \approx 0.95 &
\end{array}
$$

b.
$$
\begin{array}{ll}
4(0.7w + 0.9) = 1.6 - 0.8w & \\
2.8w + 3.6 = 1.6 - 0.8w & \quad +0.8w \\
3.6w + 3.6 = 1.6 & \quad -3.6 \\
3.6w = -2 & \quad \div 3.6 \\
w = -2/3.6 \approx -0.56 &
\end{array}
$$

A Variable on Both Sides, cont.

Page 154

9. The solution processes may vary; check the student's solution. The final answer (the root) does not vary.

a. $\begin{aligned} 6w - 6.5 &= 2w - 1 \\ 4w - 6.5 &= -1 \\ 4w &= 5.5 \\ w &= 1.375 \approx 1.38 \end{aligned}$	b. $\begin{aligned} 11 - 2q &= 7 - 5q \\ 11 + 3q &= 7 \\ 3q &= -4 \\ q &= -4/3 \end{aligned}$	c. $\begin{aligned} 5g - 5 &= -20 - 2g \\ 7g - 5 &= -20 \\ 7g &= -15 \\ g &= -15/7 \end{aligned}$
d. $\begin{aligned} 2.56x + 2 &= 5.1 - 4.89x \\ 7.45x + 2 &= 5.1 \\ 7.45x &= 3.1 \\ x &= 3.1/7.45 \approx 0.42 \end{aligned}$		e. $\begin{aligned} 2(14.85z + 0.8) + 2z &= 0.5(z - 3) \\ 29.7z + 1.6 + 2z &= 0.5z - 1.5 \\ 31.7z + 1.6 &= 0.5z - 1.5 \\ 31.2z + 1.6 &= -1.5 \\ 31.2z &= -3.1 \\ z &= -3.1/31.2 \approx -0.1 \end{aligned}$

Word Problems and More Practice, pp. 155-156

Page 155

1. a. $\begin{aligned} 11(9 + x) &= 253 \\ 99 + 11x &= 253 \qquad &| -99 \\ 11x &= 154 \qquad &| \div 11 \\ x &= 14 \end{aligned}$

b.

$\begin{aligned} 16(x + 11) &= 496 \\ 16x + 176 &= 496 \qquad &| -176 \\ 16x &= 320 \qquad &| \div 16 \\ x &= 20 \end{aligned}$

2. Let x be the shortest side of the triangle. Then, the second side is $x + 5$ and the third side is $x + 15$. The perimeter is $5x$. See the equation and its solution on the right:

$\begin{aligned} x + x + 5 + x + 15 &= 5x \\ 3x + 20 &= 5x \qquad &| -3x \\ 20 &= 2x \qquad &| \div 2 \\ x &= 10 \end{aligned}$

The sides of the triangle are 10, 15, and 25 units long.

3. Robert adds and subtracts "across the sides" of the equation.
In (a), he added $3y + 8y$ to get $11y$.
In (b), he subtracted $5x - 3x$ to get $2x$.

a. $\begin{aligned} 3y + 5 &= 8y + 7 \\ 5 &= 5y + 7 \\ -2 &= 5y \\ y &= -2/5 \end{aligned}$

b. $\begin{aligned} 5x - 7 &= 20 - 3x \\ 8x - 7 &= 20 \\ 8x &= 27 \\ x &= 27/8 \end{aligned}$

Page 156

4. Let p be the price of one workbook before the discount. Then the discounted price is $p - 3.5$. The equation is:

$\begin{aligned} 24(p - 3.5) &= 232.8 \\ 24p - 84 &= 232.8 \qquad &| +84 \\ 24p &= 316.8 \qquad &| \div 24 \\ p &= 13.2 \end{aligned}$

The price before the discount was \$13.20.

59

Word Problems and More Practice, cont.

Page 156

5. Let x be the discount amount. Then, his total was $20 \cdot 13.90 + 20 \cdot (13.90 - x)$, and this equals \$498. See the equation and its solution on the right:

The discount was \$2.90 (per bag).

$$
\begin{aligned}
20 \cdot 13.90 + 20 \cdot (13.90 - x) &= 498 \\
278 + 278 - 20x &= 498 \\
556 - 20x &= 498 \quad \big| -556 \\
-20x &= -58 \quad \big| \div (-20) \\
x &= 2.9
\end{aligned}
$$

6.

a.			b.			c.		
$-50w - 30$	$=$	$26w + 18$	$12x + 9$	$=$	$-3 - 5x - 8$	$2m - 1$	$=$	$9 - 2m - 7 - 8m$
-30	$=$	$76w + 18$	$12x + 9$	$=$	$-11 - 5x$	$2m - 1$	$=$	$2 - 10m$
-48	$=$	$76w$	$17x + 9$	$=$	-11	$12m - 1$	$=$	2
w	$=$	$-48/76 = -12/19$	$17x$	$=$	-20	$12m$	$=$	3
			x	$=$	$-20/17$	m	$=$	$\frac{1}{4}$

Page 157

7. The solution processes may vary; check the student's solution. The final answer (the root) does not vary.

a.
$$
\begin{aligned}
8 - 2m + 5 - 8m &= 20 - m + 5m - 2m \\
13 - 10m &= 2m + 20 \quad \big| + 10m \\
13 &= 12m + 20 \quad \big| - 20 \\
-7 &= 12m \quad \big| \div 12 \\
m &= -7/12
\end{aligned}
$$

b.
$$
\begin{aligned}
49.5 - 1.2s &= 2.4s - 22.5 \quad \big| + 1.2s \\
49.5 &= 3.6s - 22.5 \quad \big| + 22.5 \\
72 &= 3.6s \quad \big| \div 3.6 \\
s &= 20
\end{aligned}
$$

c.
$$
\begin{aligned}
7(50x - 5) &= 40x + 145 + 130x \\
350x - 35 &= 170x + 145 \quad \big| - 170x \\
180x - 35 &= 145 \quad \big| + 35 \\
180x &= 180 \quad \big| \div 180 \\
x &= 1
\end{aligned}
$$

d.
$$
\begin{aligned}
6(2 + 8y) - 12y + 4 &= 2(9 - 5y) - 2 \\
12 + 48y - 12y + 4 &= 18 - 10y - 2 \\
36y + 16 &= 16 - 10y \quad \big| + 10y \\
46y + 16 &= 16 \quad \big| - 16 \\
46y &= 0 \quad \big| \div 46 \\
y &= 0
\end{aligned}
$$

Puzzle corner. a. Substitute $x = 3$ in place of x to get ▢ $\cdot\, 3 - 8 = -2$. Now, treat this as an equation where the square is an unknown, and solve. Let's use c in place of the square. We have the equation $c \cdot 3 - 8 = -2$ or $3c - 8 = -2$. Now add 8 to both sides: $3c = 6$, from which $\underline{c = 2}$. So, the equation is $2x - 8 = -2$, and indeed $x = 3$ is its root.

b. Similarly, substitute $x = 1$ to the equation, and then solve the equation for b: $b \cdot 1 + 5 = -4$; $b + 5 = -4$; $\underline{b = -9}$.

Simplifying Linear Expressions, pp. 158-159

Page 158

1. a. Forgetting to change the last term inside parentheses to its opposite.
   b. $9x - 3(x - 4)$ becomes $9x - 3x + 12$ and $20 - (6 - 9y)$ becomes $20 - 6 + 9y$.

2. a. $22x - 27$ b. $-14x + 27$ c. $-3x - 3$ d. $-3x + 3$
   e. $-68y + 100$ (or $100 - 68y$)   f. $-5w - 4$ g. $-13s + 3$ h. $-20x + 10$

Page 159

3. a. $A + B = 12x - 9 + (3x - 8) = \underline{15x - 17}$
   b. $A - B = 12x - 9 - (3x - 8) = \underline{9x - 1}$
   c. $2A + B = 2(12x - 9) + (3x - 8) = 24x - 18 + 3x - 8 = \underline{27x - 26}$

Simplifying Linear Expressions, cont.

Page 159

4.

a. $5(9 - 2y) + 7(3 - 5y)$ $= 45 - 10y + 21 - 35y$ $= -45y + 66$	b. $5(6x - 12) + 7(x + 9)$ $= 30x - 60 + 7x + 63$ $= 37x + 3$
c. $5(1.2 - 0.8x) - 10(0.5x - 4.2)$ $= 6 - 4x - 5x + 42$ $= -9x + 48$	d. $2(0.9w - v) - 4(w + 0.5v)$ $= 1.8w - 2v - 4w - 2v$ $= -4v - 2.2w$
e. $\frac{1}{4}(60v - 16) - \frac{2}{3}(45v + 15)$ $= 15v - 4 - 30v - 10$ $= -15v - 14$	f. $-\frac{1}{2}(24x + 6) - 5(3x + \frac{1}{5})$ $= -12x - 3 - 15x - 1$ $= -27x - 4$
g. $36s - 9(3s + \frac{2}{3})$ $= 36s - 27s - 6$ $= 9s - 6$	h. $-\frac{1}{2}(30y + 12 - 6y) - 15y + 9$ $= -15y - 6 + 3y - 15y + 9$ $= -27y + 3$

Page 160

5. The solution processes may vary; check the student's solution. The final answer (the root) does not vary.

a. $\begin{aligned} 4(x - 2) &= 8(x + 1) + 20x \\ 4x - 8 &= 8x + 8 + 20x \\ 4x - 8 &= 28x + 8 \quad \vert -4x \\ -8 &= 24x + 8 \quad \vert -8 \\ -16 &= 24x \quad \vert \div 24 \\ x &= -16/24 = -2/3 \end{aligned}$	b. $\begin{aligned} 10y + 2(y - 6) &= 3y + 5(4 - 2y) \\ 10y + 2y - 12 &= 3y + 20 - 10y \\ 12y - 12 &= 20 - 7y \quad \vert +7y \\ 19y - 12 &= 20 \quad \vert +12 \\ 19y &= 32 \quad \vert \div 19 \\ y &= 32/19 \end{aligned}$
c. $\begin{aligned} -2(y - 12) &= 30 - 6(y + 4) \\ -2y + 24 &= 30 - 6y - 24 \\ -2y + 24 &= 6 - 6y \quad \vert +6y \\ 4y + 24 &= 6 \quad \vert -24 \\ 4y &= -18 \quad \vert \div 4 \\ y &= -18/4 = -9/2 \end{aligned}$	d. $\begin{aligned} 30(-5x - 7) + 20x &= 80x - 5(10x + 1) \\ -150x - 210 + 20x &= 80x - 50x - 5 \\ -130x - 210 &= 30x - 5 \quad \vert -30x \\ -160x - 210 &= -5 \quad \vert +210 \\ -160x &= 205 \quad \vert \div(-160) \\ x &= -205/160 = -41/32 \end{aligned}$

More Practice, pp. 161

Page 161

1. a. The error is on line 3. Instead of -24, it should have $+24$.

   b. See the simplification on the right.

(1) $\quad -(14a + 11) - \frac{3}{4}(12a - 32)$

(2) $\quad -14a - 11 - \frac{3}{4}(12a - 32)$

(3) $\quad -14a - 11 - 9a + 24$

(4) $\quad -23a + 13$

Page 161

2. a. Answers will vary. Check the student's answer. For example: $20(10x - 4y + 21)$.

   b. Yes. For example: $10(20x - 8y + 42)$ or $2(100x - 40y + 210)$ or $5(40x - 16y + 84)$.

3. The solution processes may vary; check the student's solution. The final answer (the root) does not vary.

a. $\begin{aligned} 1.5(x-2) &= 0.6(x+4)+x \\ 1.5x-3 &= 0.6x+2.4+x \\ 1.5x-3 &= 1.6x+2.4 \quad \vert \;-1.6x \\ -0.1x-3 &= 2.4 \quad\quad\;\; \vert \;+3 \\ -0.1x &= 5.4 \quad\quad\;\; \vert \;\cdot(-10) \\ x &= -54 \end{aligned}$	b. $\begin{aligned} 10y-2(y-6) &= 3y+5(4-2y) \\ 10y-2y+12 &= 3y+20-10y \\ 8y+12 &= 20-7y \quad \vert \;+7y \\ 15y+12 &= 20 \quad\quad\; \vert \;-12 \\ 15y &= 8 \quad\quad\;\; \vert \;\div 15 \\ y &= 8/15 \end{aligned}$
c. $\begin{aligned} 1{,}060-20(a-10) &= 50a \\ 1{,}060-20a+200 &= 50a \\ 1{,}260-20a &= 50a \quad \vert \;+20a \\ 1{,}260 &= 70a \quad \vert \;\div 70 \\ a &= 18 \end{aligned}$	d. $\begin{aligned} -\tfrac{1}{2}(36x+12) &= \tfrac{1}{3}(3x+24-12x) \\ -18x-6 &= x+8-4x \\ -18x-6 &= 8-3x \quad \vert \;+3x \\ -15x-6 &= 8 \quad\quad \vert \;+6 \\ -15x &= 14 \quad\quad \vert \;\div(-15) \\ x &= -14/15 \end{aligned}$

Page 162

4. Let x be the number of cheap geese he bought. Then, the number of the more expensive geese is $15 - x$.
His total cost was $29x + 58 \cdot (15 - x) = 725$. We can now solve this equation for x:

$$\begin{aligned} 29x + 58 \cdot (15-x) &= 725 \\ 29x + 870 - 58x &= 725 \\ 870 - 29x &= 725 \quad \vert \;-870 \\ -29x &= -145 \quad \vert \;\div(-29) \\ x &= 5 \end{aligned}$$

He bought 5 cheap geese and 10 more expensive ones.
Note: It's also perfectly possible to choose the unknown to be the number of the expensive geese.

5.

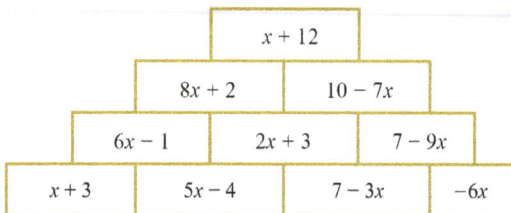

6. a. $2(5x + 4) - 1 = 10x + 7$

  b. $2(5x + 4 - 1) = 10x + 6$

  c. $(4x - 5) - (10x + 9) = -6x - 14$

  d. $1 - 2(0.3 + 4x) = 1 - 0.6 - 8x = 0.4 - 8x$

  e. $2[1 - (0.3 + 4x)] = 2[1 - 0.3 - 4x] = 2[0.7 - 4x] = 1.4 - 8x$

More Practice, cont.

<u>Page 163</u>

7. The solution processes may vary; check the student's solution. The final answer (the root) does not vary.

a.			
$-12 - 4(z - 3)$	$=$	$5z + 8$	
$-12 - 4z + 12$	$=$	$5z + 8$	
$-4z$	$=$	$5z + 8$	$+ 4z$
0	$=$	$9z + 8$	$- 8$
-8	$=$	$9z$	$\div 9$
z	$=$	$-8/9$	

b.			
$4.5w + 0.8(-2w + 4)$	$=$	$4 - 0.5(3 - w)$	
$4.5w - 1.6w + 3.2$	$=$	$4 - 1.5 + 0.5w$	
$2.9w + 3.2$	$=$	$2.5 + 0.5w$	$- 0.5w$
$2.4w + 3.2$	$=$	2.5	$- 3.2$
$2.4w$	$=$	-0.7	$\div 2.4$
w	$\approx$	-0.29	

8. Let x be the number of rainy days in each of the less rainy months. The rainier months had $x + 5$ rainy days each. In total, there were $7x + 5(x + 5) = 109$ rainy days in the year. Solving this, we get $12x + 25 = 109$, from which $x = 7$. The less rainy months had <u>7 rainy days each</u>.

9. Let x be the oldest son's share. Then, the middle son's share is $x/2$ and the youngest son's share is $3x$. We get the equation $x + x/2 + 3x = 1{,}215{,}000$, or $4.5x = 1{,}215{,}000$, from which $x = 270{,}000$. The oldest got \$270,000, the middle son got \$135,000, and the youngest \$810,000.

Age and Coin Word Problems, p. 164

<u>Page 164</u>

1.
$x + 10$	$=$	$2(x - 22)$
$x + 10$	$=$	$2x - 44$
10	$=$	$x - 44$
54	$=$	x

Right column annotations: $- x$, $+ 44$

Ann is 54 years old now.

2. Let x be Harry's age now. Five years ago, he was $x - 5$ years old, and that is 1/3 of $x + 45$. We get the equation $x - 5 = (1/3)(x + 45)$. See the solution on the right.

Harry is <u>30 years old now</u>.

$x - 5$	$=$	$\frac{1}{3}(x + 45)$	$\cdot 3$
$3x - 15$	$=$	$x + 45$	$- x$
$2x - 15$	$=$	45	$+ 15$
$2x$	$=$	60	$\div 2$
x	$=$	30	

3. Let x be Hannah's age now. In 18 years, she will be $x + 18$ years old, and that is three times $x - 10$.

Hannah is <u>24 years old now</u>.

$x + 18$	$=$	$3(x - 10)$	
$x + 18$	$=$	$3x - 30$	$- x$
18	$=$	$2x - 30$	$+ 30$
48	$=$	$2x$	$\div 2$
x	$=$	24	

4. Let x be my age now. The equation and its solution are on the right.

I am <u>33 years old now</u>.

$x - 15$	$=$	$\frac{1}{2}(x + 3)$	$\cdot 2$
$2x - 30$	$=$	$x + 3$	$- x$
$x - 30$	$=$	3	$+ 30$
x	$=$	33	

Page 165

5. The equation is $y + 2y + y - 15 = 153$.
   See the solution of the equation on the right.

   She has 42 twenties, 84 tens, and 27 fifties.

$$
\begin{aligned}
y + 2y + y - 15 &= 153 \\
4y - 15 &= 153 \\
4y &= 168 \\
y &= 42
\end{aligned}
$$

6. Let x be the number of dimes. Then, Mom has $x/2$ pennies, $2x$ nickels, and $x - 4$ quarters. See the solution of the equation on the right.

   Mom has 14 dimes, 7 pennies, 28 nickels, and 10 quarters.

$$
\begin{aligned}
x + x/2 + 2x + x - 4 &= 59 \\
4.5x - 4 &= 59 \\
4.5x &= 63 \\
x &= 14
\end{aligned}
$$

7. Let x be the amount that Joel earned. Then, Jill earned $x + 150$ and Jack earned $x - 70$. See the solution of the equation on the right.

   Joe earned \$400, Jill earned \$550, and Jack earned \$330.

$$
\begin{aligned}
x + x + 150 + x - 70 &= 1{,}280 \\
3x + 80 &= 1{,}280 \\
3x &= 1{,}200 \\
x &= 400
\end{aligned}
$$

Page 166

8.
$$
\begin{aligned}
0.25x + 0.1(x + 15) &= 5.35 &&\mid \cdot 100 \\
25x + 10(x + 15) &= 535 \\
25x + 10x + 150 &= 535 \\
35x + 150 &= 535 &&\mid -150 \\
35x &= 385 &&\mid \div 35 \\
x &= 11
\end{aligned}
$$

Ryan has <u>11 quarters</u>.

9. Let m be the number of quarters Sheila has. Then she has $m/3$ dimes. The value of her coins is $0.25m + 0.1 \cdot m/3 + 4 \cdot 0.5$, which simplifies to $0.25m + 0.1/3 \cdot m + 2$. Here is the equation and its solution:

$$
\begin{aligned}
0.25m + 0.1/3 \cdot m + 2 &= 12.20 &&\mid \cdot 3 \\
0.75m + 0.1m + 6 &= 36.60 &&\mid \cdot 100 \\
75m + 10m + 600 &= 3{,}660 \\
85m + 600 &= 3{,}660 &&\mid -600 \\
85m &= 3{,}060 &&\mid \div 85 \\
m &= 36
\end{aligned}
$$

Sheila has <u>36 quarters, 12 dimes, and four half-dollars</u>. Note that we should not calculate or simplify the value of 0.1/3 to 0.03 or 0.033, because in reality it is an unending decimal. In the solution, I multiplied the equation by 3 to get rid of this denominator.

10. Let m be the number of twenties Greg has. Then he has $3m$ tens, $m/2$ fifties, and two hundreds. The value of his bills is $m \cdot 20 + 3m \cdot 10 + m/2 \cdot 50 + 200$, which simplifies to $20m + 30m + 25m + 200 = 75m + 200$. Here is the equation and its solution:

$$
\begin{aligned}
75m + 200 &= 1{,}550 &&\mid -200 \\
75m &= 1{,}350 &&\mid \div 75 \\
m &= 18
\end{aligned}
$$

Greg has 18 twenties, 54 tens, 9 fifties, and two hundreds.

Equations with Fractions 1, pp. 167-169

Page 167

1. The solution processes may vary; check the student's solution. The final answer (the root) does not vary.

a. $\frac{3}{5}x + \frac{1}{2} = -3$ $\Big| \cdot 10$

$10(\frac{3}{5}x + \frac{1}{2}) = 10 \cdot (-3)$

$6x + 5 = -30$ $\Big| -5$

$6x = -35$ $\Big| \div 6$

$x = -35/6$

b. $\frac{2}{3}x - \frac{1}{6} = -\frac{1}{4}$ $\Big| \cdot 12$

$12 \cdot (\frac{2}{3}x - \frac{1}{6}) = 12 \cdot (-\frac{1}{4})$

$8x - 2 = -3$ $\Big| +2$

$8x = -1$ $\Big| \div 8$

$x = -1/8$

Page 168

2. a. $x - 20$ b. $x - 4$ c. $2(x - 3) = 2x - 6$ d. $2x - 15$

3.

a. $\frac{3}{4}x - 0.5 = \frac{1}{10}x + 1$ $\Big| \cdot 20$

$15x - 10 = 2x + 20$

$13x = 30$

$x = 30/13$

b. $2(y - 5) = \frac{3}{5}(y - 6)$ $\Big| \cdot 5$

$10(y - 5) = 3(y - 6)$

$10y - 50 = 3y - 18$ $\Big| -3y$

$7y - 50 = -18$ $\Big| +50$

$7y = 32$ $\Big| \div 7$

$y = 32/7$

Page 169

4. The solution processes may vary; check the student's solution. The final answer (the root) does not vary.

a. $4x - \frac{1}{10} = \frac{3}{5}x - 6$ $\Big| \cdot 10$

$10(4x - \frac{1}{10}) = 10(\frac{3}{5}x - 6)$

$40x - 1 = 6x - 60$ $\Big| -6x$

$34x - 1 = -60$ $\Big| +1$

$34x = -59$ $\Big| \div 34$

$x = -59/34$

b. $\frac{1}{6}x - 4 = \frac{2}{3}x - 1$ $\Big| \cdot 6$

$6(\frac{1}{6}x - 4) = 6(\frac{2}{3}x - 1)$

$x - 24 = 4x - 6$ $\Big| -x$

$-24 = 3x - 6$ $\Big| +6$

$-18 = 3x$ $\Big| \div 3$

$x = -6$

c. $\frac{1}{3}(x - 4) = -\frac{1}{8}$ $\Big| \cdot 24$

$8(x - 4) = -3$

$8x - 32 = -3$ $\Big| +32$

$8x = 29$ $\Big| \div 8$

$x = 29/8$

d. $-\frac{1}{2} = \frac{1}{10}(x + 5)$ $\Big| \cdot 10$

$-5 = x + 5$ $\Big| -5$

$x = -10$

e. $2(x - \frac{5}{8}) = x - \frac{1}{2}$

$2x - \frac{5}{4} = x - \frac{1}{2}$ $\Big| \cdot 4$

$8x - 5 = 4x - 2$ $\Big| -4x$

$4x - 5 = -2$ $\Big| +5$

$4x = 3$ $\Big| \div 4$

$x = \frac{3}{4}$

f. $\frac{1}{2}(x - \frac{3}{4}) = 10$ $\Big| \cdot 4$

$2(x - \frac{3}{4}) = 40$

$2x - \frac{3}{2} = 40$ $\Big| \cdot 2$

$4x - 3 = 80$ $\Big| +3$

$4x = 83$ $\Big| \div 4$

$x = 83/4$

Equations with Fractions 1, cont.

5. The solution process after the first step may vary; check the student's solution. The final answer (the root) does not vary.

$$\frac{1}{4}a + 5 = \frac{3}{8} \qquad \Big| -5$$

$$\frac{1}{4}a = \frac{3}{8} - 5$$

$$\frac{1}{4}a = -\frac{37}{8} \qquad \Big| \cdot 8$$

$$2a = -37 \qquad \Big| \div 2$$

$$x = -37/2$$

Equations with Fractions 2, p. 170

1. The solution processes may vary; check the student's solution. One way to start is to multiply both sides by a common multiple of the denominators. Another way is to treat these as proportions (which they are), and cross-multiply. Both ways lead to the same equation.

a.
$$\frac{3x-4}{2} = \frac{3x+1}{5} \qquad \Big| \cdot 10$$
$$\begin{aligned} 5(3x-4) &= 2(3x+1) \\ 15x - 20 &= 6x + 2 \qquad \Big| -6x \\ 9x - 20 &= 2 \qquad \Big| +20 \\ 9x &= 22 \qquad \Big| \div 9 \\ x &= 22/9 \end{aligned}$$

b.
$$\frac{15-2s}{8} = \frac{5s-1}{2} \qquad \Big| \text{cross-multiply}$$
$$\begin{aligned} 2(15-2s) &= 8(5s-1) \\ 30 - 4s &= 40s - 8 \qquad \Big| +4s \\ 30 &= 44s - 8 \qquad \Big| +8 \\ 38 &= 44s \qquad \Big| \div 44 \\ s &= 19/22 \end{aligned}$$

2. In (a), the error is that 5 does not get multiplied by 2. In (b), x and $2x$ do not get multiplied by 10.

a.
$$\frac{3x-4}{2} - 5 = 7 \qquad \Big| \cdot 2$$
$$3x - 4 - 10 = 14$$
$$3x - 14 = 14 \qquad \Big| +14$$
$$3x = 28 \qquad \Big| \div 3$$
$$x = 28/3$$

b.
$$3 - x = 2x + \frac{x-10}{2} \qquad \Big| \cdot 10$$
$$30 - 10x = 20x + 5x - 50$$
$$30 - 10x = 25x - 50 \qquad \Big| +10x$$
$$30 = 35x - 50 \qquad \Big| +50$$
$$80 = 35x \qquad \Big| \div 35$$
$$x = 80/35 = 16/7$$

Equations with Fractions 2, cont.

3. The only difference between these two equations is that in (b) there is a minus sign after $2x$. This, in turn, switches the minus sign in front of x to plus.

a.
$$2x + \frac{5-x}{6} = 4 \qquad | \cdot 6$$
$$6 \cdot 2x + \frac{6(5-x)}{6} = 6 \cdot 4$$
$$12x + 5 - x = 24$$
$$11x + 5 = 24 \qquad | -5$$
$$11x = 19 \qquad | \div 11$$
$$x = 19/11$$

b.
$$2x - \frac{5-x}{6} = 4 \qquad | \cdot 6$$
$$6 \cdot 2x - \frac{6(5-x)}{6} = 6 \cdot 4$$
$$12x - (5-x) = 24$$
$$12x - 5 + x = 24$$
$$13x - 5 = 24 \qquad | +5$$
$$13x = 29 \qquad | \div 13$$
$$x = 29/13$$

4.

a.
$$\frac{3x-8}{10} - 1 = x \qquad | \cdot 10$$
$$3x - 8 - 10 = 10x \qquad | -3x$$
$$-18 = 7x \qquad | \div 7$$
$$x = -18/7$$

b.
$$11 = 3y + \frac{5-5y}{3} \qquad | \cdot 3$$
$$33 = 9y + 5 - 5y$$
$$33 = 4y + 5 \qquad | -5$$
$$28 = 4y \qquad | \div 4$$
$$y = 7$$

c.
$$0 = \frac{3x-2}{4} + \frac{x+2}{5} \qquad | \cdot 20$$
$$0 = 5(3x - 2) + 4(x + 2)$$
$$0 = 15x - 10 + 4x + 8$$
$$0 = 19x - 2 \qquad | +2$$
$$2 = 19x$$
$$19x = 2 \qquad | \div 19$$
$$x = 2/19$$

d.
$$-x + \frac{1-3x}{2} = \frac{x}{3} + 2 \qquad | \cdot 6$$
$$-6x + 3(1 - 3x) = 2x + 12$$
$$-6x + 3 - 9x = 2x + 12$$
$$3 - 15x = 2x + 12 \qquad | -2x$$
$$3 - 17x = 12 \qquad | -3$$
$$-17x = 9 \qquad | \div (-17)$$
$$x = -9/17$$

5.

a.
$$\frac{3.2x - 1}{5} = 0.9x \qquad | \cdot 5$$
$$3.2x - 1 = 4.5x \qquad | -3.2x$$
$$-1 = 1.3x \qquad | \div 1.3$$
$$x \approx -0.77$$

b.
$$0.08x - \frac{0.1x}{4} = 0.2 \qquad | \cdot 4$$
$$0.32x - 0.1x = 0.8$$
$$0.22x = 0.8 \qquad | \div 0.22$$
$$x \approx 3.64$$

c.
$$\frac{20x - 4.3}{0.4} = \frac{3.89x}{2.5} \qquad \text{cross-multiply}$$
$$2.5(20x - 4.3) = 0.4 \cdot 3.89x$$
$$50x - 10.75 = 1.556x \qquad | -1.556x$$
$$48.444x - 10.75 = 0 \qquad | +10.75$$
$$48.444x = 10.75 \qquad | \div 48.444$$
$$x \approx 0.22$$

d.
$$5.4 - \frac{0.3 - x}{4} = \frac{x}{2} \qquad | \cdot 4$$
$$21.6 - (0.3 - x) = 2x$$
$$21.3 + x = 2x \qquad | -x$$
$$21.3 = x$$

Equations with Fractions 2, cont.

Page 172

6. $\frac{3}{5}\left(x + \frac{1}{2}\right) = -3$

$3\left(x + \frac{1}{2}\right) = -15$

$3x + \frac{3}{2} = -15$

$3x = -\frac{33}{2}$

$x = -\frac{11}{2}$

Puzzle corner. Guess and check. For example, $60 \cdot 60 = 3{,}600$, so a good starting point for guesses would be numbers that are slightly less than 60. Now, $56 \cdot 57 = 3{,}192$ which is low. And $57 \cdot 58 = 3{,}306$, so that fits. Since $58 \cdot 59 = 3{,}422$ is too high, 57 and 58 is the only solution.

Formulas, Part 1, pp. 173-174

Page 173

1. The height is $\dfrac{2A}{b} = \dfrac{2 \cdot 250 \text{ cm}^2}{50 \text{ cm}} = \underline{10 \text{ cm.}}$

2. a. $V = \pi r^2 h \quad \Big| \div \pi r^2$

$\dfrac{V}{\pi r^2} = h$

$h = \dfrac{V}{\pi r^2}$

b. An important thing to remember is that we need to either convert 8.00 m^3 into cubic centimeters, or convert 60.0 cm into meters, before doing the calculations. It will be easier to do the latter (the numbers will stay smaller). The final answer is given to three significant digits since the original quantities had three significant digits.

$h = \dfrac{8.00 \text{ m}^3}{\pi \, (60.0 \text{ cm})^2} = \dfrac{8 \text{ m}^3}{\pi \, (0.6 \text{ m})^2} = \dfrac{8 \text{ m}^3}{\pi \cdot 0.36 \text{ m}^2} \approx \underline{7.07 \text{ m}}$

3. a. $\qquad m = \dfrac{a_1 + a_2 + a_3}{3} \quad \Big| \cdot 3$

$3m = a_1 + a_2 + a_3 \quad \Big| -(a_2 + a_3)$

$3m - a_2 - a_3 = a_1$

$a_1 = 3m - a_2 - a_3$

b. Her third test score should be $3m - a_2 - a_3 = 3 \cdot 85 - 78 - 82 = \underline{95}$.

Page 174

4. See the solution of the equation on the right. After that, we then convert 3.769 hours into hours and minutes: 0.769 hours is $0.769 \text{ h} \cdot 60 \text{ min/h} \approx 46$ minutes.

It will take you 3 hours 46 minutes to travel 245 km with a speed of 65 km/h.

$d = vt$

$245 \text{ km} = 65 \text{ km/h} \cdot t \quad \Big| \div 65 \text{ km/h}$

$t = \dfrac{245 \text{ km}}{65 \text{ km/h}} \approx 3.769 \text{ h}$

5. The two halves need to be calculated separately. For the first half, it took her $t = d/v = 8 \text{ km}/(60 \text{ km/h}) = 8/60$ hours, which is 8 minutes. The second half took her $t = d/v = 8 \text{ km}/(40 \text{ km/h}) = 1/5$ hours, which is 12 minutes. It took her a total of $8 \text{ min} + 12 \text{ min} = \underline{20 \text{ minutes}}$.

6. $\qquad C = (1 - x/100)mp$

$\$1{,}200 = (1 - 8/100)m \cdot \$15$

$\$1{,}200 = 0.92m \cdot \$15$

$1{,}200 = 13.8m \qquad \Big| \div 13.8$

$m \approx 86.96$

You can get 86 items with $1,200.

Formulas, Part 2, pp. 175-176

Page 175

1. $A = \dfrac{a+b}{2}h$ $\Big|\ \cdot 2$

 $2A = (a+b)h$ $\Big|\ \div h$

 $\dfrac{2A}{h} = a+b$

 $a = \dfrac{2A}{h} - b$

2. a. $C = \dfrac{5}{9}(F-32)$ $\Big|\ \cdot 9$

 $9C = 5(F-32)$

 $9C = 5F - 160$ $\Big|\ +160$

 $9C + 160 = 5F$ $\Big|\ \div 5$

 $F = \dfrac{9C+160}{5}$

 or $F = \dfrac{9}{5}C + 32$

 b. $F = (9/5)(40) + 32 = 72 + 32 = \underline{104°F}$

3. $MPG = \dfrac{d}{g}$ $\Big|\ \cdot g$

 $g \cdot MPG = d$ $\Big|\ \div MPG$

 $g = \dfrac{d}{MPG}$

Page 176

4. a. $C = \dfrac{d}{MPG}\, p$ or $\dfrac{dp}{MPG}$

 b. We can plug in the given quantities to the formula $C = (dp)/MPG$, and then solve for MPG:

 $C = \dfrac{dp}{MPG}$

 $\$55 = \dfrac{267 \text{ mi} \cdot \$4.2/\text{gal}}{MPG}$ $\Big|\ \cdot MPG$

 $\$55 \cdot MPG = 267 \text{ mi} \cdot \$4.2/\text{gal}$ $\Big|\ \div \$55$

 $MPG = \dfrac{267 \text{ mi} \cdot \$4.2/\text{gal}}{\$55}$

 $MPG \approx 20.4 \text{ mi/gal}$

5. The interest accrued is calculated as $I = Prt$ = $\$5000 \cdot 0.06 \cdot 3 = \$900$. He withdrew the original principal plus the interest, or $\underline{\$5,900}$.

6. $I = Prt$

 $\$500 = \$2,500r \cdot 2$

 $500 = 5,000r$ $\Big|\ \div 5,000$

 $r = \dfrac{500}{5,000} = 0.1 = 10\%$

 The interest rate should be $\underline{10\%}$.

7. $I = Prt$

 $\$500 = \$12,000 \cdot 0.084 \cdot t$

 $500 = 1,008t$ $\Big|\ \div 1,008$

 $t = \dfrac{500}{1,008} \approx 0.496 \text{ years}$

 Lastly, we convert 0.496 years into days: 0.496 years = 0.496 years · 365 days/year ≈ 181 days.

 So, it will take about <u>half a year</u> for a $12,000 principal to earn $500 in interest, with 8.4% annual interest rate.

Puzzle corner. Let x be the amount Ann invests in the first opportunity (with the annual interest rate of 8%). Then, the amount she invests in the other opportunity is $10,000 - x$.

The first opportunity will earn her, in interest, $x \cdot 0.08 \cdot 8$, and the second $(10,000 - x) \cdot 0.06 \cdot 8$.

Simplifying those, we get $0.64x$ and $0.48(10,000 - x)$. She wants the sum of these to be $6,000, which gives us our equation:

 $0.64x + 0.48(10,000 - x) = 6000$

 $0.64x + 4,800 - 0.48x = 6000$

 $0.16x + 4,800 = 6000$ $\Big|\ -4,800$

 $0.16x = 1,200$ $\Big|\ \div 0.16$

 $x = 7,500$

She will invest $7,500 in the first opportunity, and $2,500 in the other.

69

More on Equations, pp. 177-179

1.

a. $2x + 2 = 2x - 7$ $\quad\mid -2x$ $\qquad 2 = -7$ No solutions.	b. $2x + 1 = 2(x - 3) + 7$ $\quad 2x + 1 = 2x - 6 + 7$ $\quad 2x + 1 = 2x + 1 \quad\mid -2x$ $\qquad 1 = 1$ An infinite number of solutions. This equation is an identity.	c. $5x + 1 = 2(x - 3)$ $\quad 5x + 1 = 2x - 6 \quad\mid -2x$ $\quad 3x + 1 = -6 \qquad\mid -1$ $\qquad 3x = -7 \qquad\mid \div 3$ $\qquad x = -7/3$ One solution.

2. He is not correct. This equation has one solution, namely $x = 0$. Subtracting 5 from both sides, the equation $9x + 5 = 5$ is transformed to $9x = 0$. Dividing both sides by 9, we get $x = 0$.

3.

a. $10 - 7x = x - 10 \quad\mid +7x$ $\quad 10 = 8x - 10 \quad\mid +10$ $\quad 20 = 8x \qquad\mid \div 8$ $\qquad x = 20/8 = 5/2$ One solution.	b. $4x + x = 5x - 3 + 7$ $\quad 5x = 5x + 4 \quad\mid -5x$ $\quad 0 = 4$ No solutions.	c. $2 + \dfrac{x-3}{4} = x - \dfrac{3x-5}{4} \quad\mid \cdot 4$ $\quad 8 + x - 3 = 4x - (3x - 5)$ $\qquad x + 5 = x + 5$ $\qquad 5 = 5$ An infinite number of solutions.

4. a. no solutions   b. one solution   c. one solution   d. an infinite number of solutions

5. a. It has zero solutions (no solutions).
   b. Answers will vary; check the student's answer. For example: $2y + 4 = 4 + 2y$ or $2y + 4 = 2(2 + y)$

6. a. It has one solution.
   b. Answers will vary; check the student's answer. For example: $3w - 4 = 3w$ or $3w - 4 = 5w - 2w$
   or $3w - 4 + 2w = 5w + 7$.

7.
$$2 = 5$$
$$3x + 2 = 5 + 3x$$
$$3x - 3 = 3x$$
$$3(x - 1) = 3x$$
$$3(x - 1) - 2x = x$$

8. Answers will vary; check the student's answer. For example:

   a. $20 - 5t = 10(3 - t/2)$ or $20 - 5t = 4 - 5t$
   b. $20 - 5t = 2t + 7$ or $20 - 5t = 4 - t$
   c. $20 - 5t = -5(t - 4)$ or $20 - 5t = 8 - 2t - 3t + 12$

9. a. Let n and $n + 1$ be the two consecutive numbers. The equation is $n + n + 1 = 5{,}437$. Solution: $n = 2{,}718$.
      The numbers are 2,718 and 2,719.

   b. We cannot say; there is an infinite number of solutions. The difference between *any* two consecutive numbers is 1.
      If we write an equation, we will get $(n + 1) - n = 1$, which simplifies to $1 = 1$ and is an identity.

More on Equations, cont.

Puzzle corner. For this equation to not have any solutions, the terms with x need to be identical on both sides of the equation (and the constant terms on both sides need to be different). Since the right side has $-10x$, then when $a = -10$, the equation has no solutions:

$$
\begin{aligned}
-10x + 9 &= 8 - 5(2x + 4) \\
-10x + 9 &= 8 - 10x - 20 \\
-10x + 9 &= -10x - 12 \qquad \Big| + 10x \\
9 &= -12
\end{aligned}
$$

Percent Word Problems, pp. 180-181

Page 180

1. (iii)

2. Let w be the workforce before the increase. Then $1.06w = 13{,}250$, from which $w = \underline{12{,}500}$.

3. The new price is $1.08 \cdot \$5.50 = \underline{\$5.94}$.

4. Percent of change is calculated as the quotient of (difference/original). The difference here is $0.5°C$, and the original temperature is $12.1°C$, so we get $0.5/12.1 \approx 0.041 = 4.1\%$.

Page 181

5.
$$
\begin{aligned}
0.85(150 + x) &= 150 \\
127.5 + 0.85x &= 150 \qquad \Big| - 127.5 \\
0.85x &= 22.5 \\
x &\approx 26.47
\end{aligned}
$$

Let's double check if this works. If the new price is $\$150 + \$26.47 = \$176.47$, then its 15% off sale price would be $0.85(\$176.47) = \$149.9995 \approx \$150$.

The store owner should <u>increase the price by $26.47</u> so that later on, the 15% off price would be $150. However, in reality the price might be set to some "nicer" number such as $176.50 or $177.

6. Let p be the original price. An increase of 12% means it gets multiplied by 1.12, and a decrease of 5% means it gets multiplied by 0.95. The equation is: $1.12 \cdot 0.95p = 154$, from which $p \approx \underline{\$144.74}$.

7. Let p be the original price. The equation is: $1.02 \cdot 1.05 \cdot 1.03p = 62.88$, from which $p \approx \underline{\$57.00}$.

8. Let x be the price increase. Then, we get the equation $0.8(400 + x) = 390$.

$$
\begin{aligned}
0.8(400 + x) &= 390 \\
320 + 0.8x &= 390 \qquad \Big| - 320 \\
0.8x &= 70 \\
x &= 87.5
\end{aligned}
$$

The price should be increased <u>by $87.50</u>.

9. Let r be the factor by which you multiply the price to get the price with the tax. For example, if sales tax is 7%, r would be 1.07. Then, $0.75 \cdot 92 \cdot r = 72.45$, from which $r = 72.45/(0.75 \cdot 92) = 1.05$. So, the sales tax is 5%.

Page 182

1. a. $6t$

   b. $3t + 20$

   c. $6t = 3t + 20$ from which $t = 20/3 = 6\ 2/3$ seconds. This tells us that 6 2/3 seconds after starting, Dad passes up Tommy (they are at the same position or distance).

2. a. Antonio's PipeFix will charge $3.5 \cdot \$45 + \$35 = \$192.50$. Paul the Plumber will charge $220. Antonio's PipeFix is the better deal.

   b. Let t be the number of hours for which both services cost the same amount. Then $45t + 35 = 220$, from which $t = 4.\overline{1}$ hours or 4 hours and about 7 minutes.

Page 183

3. a. At Express Print it costs $500 \cdot \$0.07 + \$2.50 = \$37.50$. At Sherry's Office Supplies, it costs $500 \cdot \$0.10 = \$50$. Express Print is the better deal.

   b. Let p be the page count at which both services cost the same. We get:

$$
\begin{array}{rcl|l}
0.07p + 2.50 & = & 0.1p & -0.07p \\
2.5 & = & 0.03p & \div 0.03 \\
p & = & 83.\overline{3} &
\end{array}
$$

   With 83 pages, both printing services cost about the same. (In fact, with less than 83 pages, Sherry's Office Supplies is cheaper, and more than 83 pages, Express Print is cheaper.)

4. a. $2.5m + 7.5$

   b. Seven hours.

   c.
$$
\begin{array}{rcl|l}
2.5m + 7.5 & = & 25 & -7.5 \\
2.5m & = & 17.5 & \div 2.5 \\
m & = & 7 &
\end{array}
$$

5. Let x be the new rate per hour. We can write the equation $3.5x + 9 = 20.90$, from which $x = 3.4$. The new rate is $\underline{\$3.40\ per\ hour}$.

Puzzle corner. In t hours since 2 PM, Henry has earned $30t$. At the same time, Sammy has earned $30(t - \frac{1}{2})$. We can write the equation $30t + 30(t - \frac{1}{2}) = 100$.

$$
\begin{array}{rcl|l}
30t + 30(t - \frac{1}{2}) & = & 100 & \\
30t + 30t - 15 & = & 100 & +15 \\
60t & = & 115 & \div 60 \\
t & = & 1.91\overline{6} &
\end{array}
$$

Lastly, we convert $0.91\overline{6}$ hours into minutes: $0.91\overline{6}$ hr $\cdot$ 60 min/hr = 55 minutes.
This means that at 3:55 PM their combined earnings reach $100.

Page 184

1. a. $8^6 = (2^3)^6 = 2^{18}$ b. $64^3 = (4^3)^3 = 4^9$ c. $100^5 = (10^2)^5 = 10^{10}$

2.

a.	b.	c.	d.
$5 \cdot 4^3 = 320$	$(-5)^3 = -125$	$-3^{-2} = -1/9$	$2 \cdot 5^{-2} = 2/25$
$(5 \cdot 4)^3 = 8{,}000$	$-5^3 = -125$	$(-3)^{-2} = 1/9$	$(2 \cdot 5)^{-2} = 1/100$

3. About $3 \cdot 10^5$ times the mass of the earth. (The calculation gives us $(1/3) \cdot 10^6$, and to give this in proper scientific notation, we write it as $3 \cdot 10^5$, with one significant digit, just like the numbers in the problem.)

4.

$\dfrac{a^2}{b^2}$	$9a^6$ —— $(3a^3)^2$	$-3a^2$
$3a^6$	$\dfrac{(3a)^2}{b}$	$\left(\dfrac{a}{b}\right)^2$	$\dfrac{3a^2}{b}$
$\dfrac{9a^2}{b}$	$\dfrac{a^2}{b}$	$3a^{3^2}$ —— $3a^9$

5. The third angle of the triangle is $180° - 39° - 74° = 67°$. The angle x is exterior to the $67°$ angle, so it measures $180° - 67° = \underline{113°}$.

6. The area of the bottom triangle is $10 \text{ cm} \cdot 15 \text{ cm} \div 2 = 75 \text{ cm}^2$. The volume is $36 \text{ cm} \cdot 75 \text{ cm}^2 = 2{,}700 \text{ cm}^3$.
(We do not need the fact that the longest side of the triangle is 18 cm.)

Page 185

7. In the image, angle 1 and angle x are supplementary, and therefore $\angle 1 = 180° - 76° = 104°$. Then, angle 1 and y are alternate interior angles, thus congruent. So, $y = 104°$.

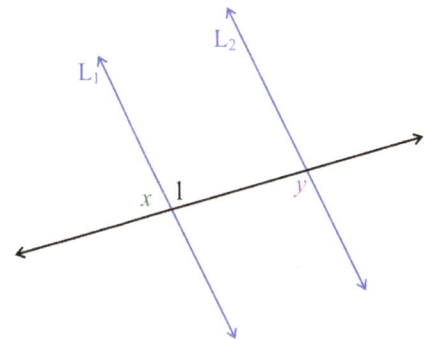

8. A"$(2, -1)$, B"$(2, -3)$, C"$(4, -4)$

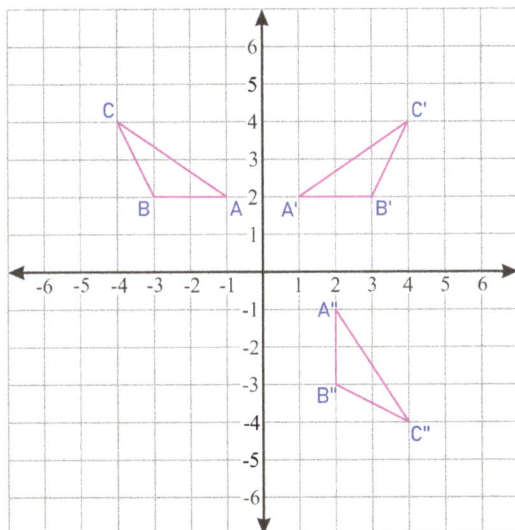

9.

a. scale factor 3	b. scale factor 1/2

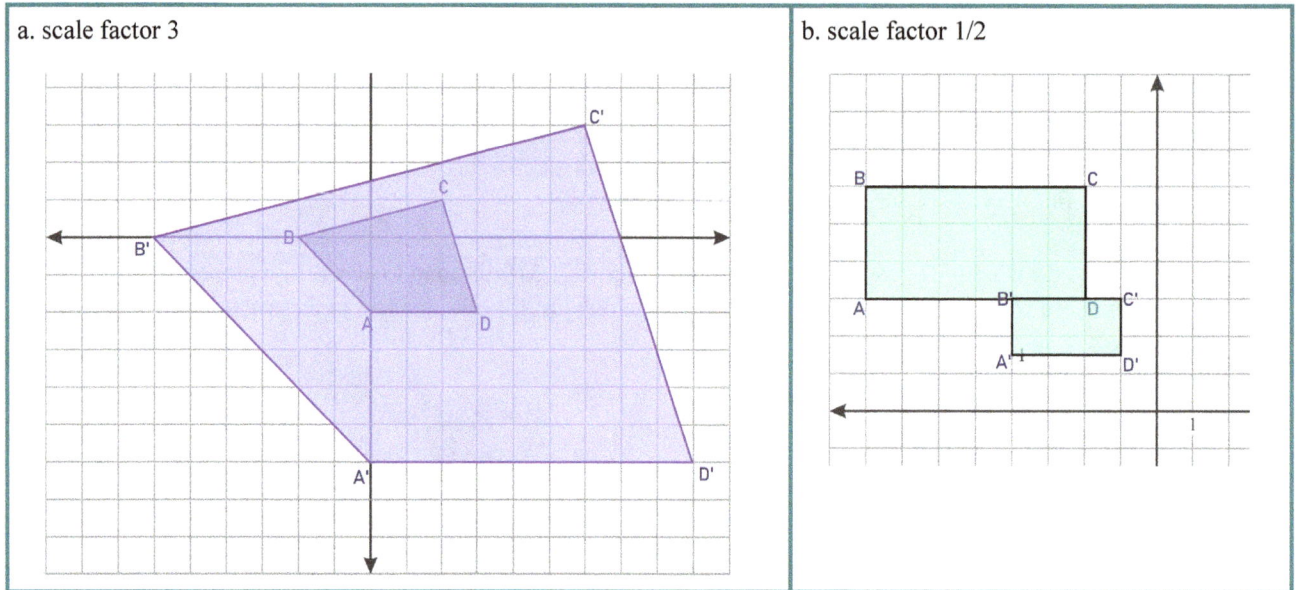

Chapter 3 Review, p. 186

Page 186

1.

a.
$$7x - 3(x - 5) - 2x = 10$$
$$7x - 3x + 15 - 2x = 10$$
$$2x + 15 = 10 \qquad | -15$$
$$2x = -5 \qquad | \div 2$$
$$x = -5/2$$

b. $x - \dfrac{1}{3} = \dfrac{3}{4}x - 2 \qquad | \cdot 12$

$$12x - 4 = 9x - 24 \qquad | -9x$$
$$3x - 4 = -24 \qquad | +4$$
$$3x = -20 \qquad | \div 3$$
$$x = -20/3$$

c.
$$20 - q = -q + 2(q - 5) - 6q$$
$$20 - q = -q + 2q - 10 - 6q$$
$$20 - q = -10 - 5q \qquad | +5q$$
$$20 + 4q = -10 \qquad | -20$$
$$4q = -30 \qquad | \div 4$$
$$q = -15/2$$

d.
$$-52 - 2(x + 14) = 80 - 11x + x$$
$$-52 - 2x - 28 = 80 - 10x$$
$$-80 - 2x = 80 - 10x \qquad | +10x$$
$$-80 + 8x = 80 \qquad | +80$$
$$8x = 160 \qquad | \div 8$$
$$x = 20$$

2. Let x be Heather's age now. The equation
is: $x - 5 = (1/3)(x + 45)$,
or it can also be written as $x - 5 = (x + 45)/3$.

See the solution on the right.

$$x - 5 = \dfrac{x + 45}{3} \qquad | \cdot 3$$
$$3x - 15 = x + 45 \qquad | -x$$
$$2x - 15 = 45 \qquad | +15$$
$$2x = 60 \qquad | \div 2$$
$$x = 30$$

Chapter 3 Review, cont.

Page 187

3.

a. $\dfrac{x-1}{2} = \dfrac{3x+2}{7}$ cross-multiply	b. $x - \dfrac{1-3x}{2} = 5$ $\cdot\,2$
$7x - 7 = 6x + 4$ $-6x$	$2x - (1 - 3x) = 10$
$x - 7 = 4$ $+7$	$2x - 1 + 3x = 10$
$x = 11$	$5x - 1 = 10$ $+1$
	$5x = 11$ $\div 5$
	$x = 11/5$

4. a.
$$\rho = \frac{m}{V} \quad \Big| \cdot V$$
$$V\rho = m \quad \Big| \div \rho$$
$$V = \frac{m}{\rho}$$

b. V = m/ρ = 4.6 kg/(850 kg/m³) ≈ 0.0054 m³ = 5,400 cm³.

The conversion factor between cubic meters and cubic centimeters:
1 m³ = 1 m × 1 m × 1 m = 100 cm × _100_ cm × _100_ cm = _1,000,000_ cm³

Page 187

5. Answers will vary; check the student's answers. For example:

a. $8x - 2 = 5 + 8x$ OR $8x - 2 = 4(2x + 6)$
b. $8x - 2 = 6x + 7$ OR $8x - 2 = 3(3x - 1)$
c. $8x - 2 = 3x + 5x + 8 - 10$ OR $8x - 2 = 2(4x - 1)$

Page 188

6.

a. $0.6 + 9.4x - 2 = x - 4.8x$	b. $4 - 0.3(2y + 2.8) = 1.5(y - 3)$
$9.4x - 1.4 = -3.8x$ $+3.8x$	$4 - 0.6y - 0.84 = 1.5y - 4.5$
$14.2x - 1.4 = 0$ $+1.4$	$3.16 - 0.6y = 1.5y - 4.5$ $+0.6y$
$14.2x = 1.4$ $\div 14.2$	$3.16 = 2.1y - 4.5$ $+4.5$
$x \approx 0.10$	$7.66 = 2.1y$ $\div 2.1$
	$y \approx 3.65$

7. Let n be the number of cheaper cartons of almond milk. Then, $16 - n$ is the number of the more expensive ones.

$$4.5n + 5.2(16 - n) = 76.2$$
$$4.5n + 83.2 - 5.2n = 76.2$$
$$83.2 - 0.7n = 76.2 \quad \Big| -83.2$$
$$-0.7n = -7 \quad \Big| \div(-0.7)$$
$$n = 10$$

8. Let x be the increase in price.

$$
\begin{aligned}
0.75(8.5 + x) &= 7.5 \\
6.375 + 0.75x &= 7.5 \qquad \big| -6.375 \\
0.75x &= 1.125 \qquad \big| \div 0.75 \\
x &= 1.5
\end{aligned}
$$

You should increase the price by $1.50.

9. Let p be the original price.

$$
\begin{aligned}
0.65 \cdot 1.07p &= 53.90 \\
0.6955p &= 53.9 \qquad \big| \div 0.6955 \\
x &\approx 77.4982
\end{aligned}
$$

The original price was $\underline{\$77.50}$.

Chapter 4: Introduction to Functions

Functions, pp. 191-192

Page 191

1. It is not a function because the input "dog" is mapped to two outputs ("meat" and "milk"), and similarly, the input "cat" is also mapped to two outputs.

2. a. −27   b. 4

3. See the tables on the right. Yes, each table represents a function. For each input, there is exactly one output.

#1	
(Input) **Weight**	(Output) **Cost**
1 kg	$3
2 kg	$6
3 kg	$9
5 kg	$15
12 kg	$36

#2	
(Input) **Cost**	(Output) **Weight**
$12	4 kg
$30	10 kg
$48	16 kg
$72	24 kg
$90	30 kg

Page 192

4. No, this is not a function. Bob has no favorite color.
   ("Bob" is an input that is not mapped to any output.)
   Also, Allie and Pete have two favorite colors (mapped to two outputs).

We can change the function to fix those issues, for example like this:

Input	Allie	Julie	Danny	Juan	Pete	Bob	Samantha
Output	pink	blue	gray	yellow	blue	black	purple

5. a.   INPUT   OUTPUT

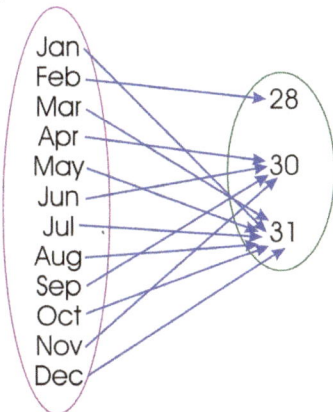

b. No, it isn't. For example, for the input 30, there would be several outputs (April, June, September, November).

6. a.

Input (x)	−4	−3	−2	−1	0	1	2	3	4
Output (y)	15	8	3	0	−1	0	3	8	15

b. See the plot on the right.

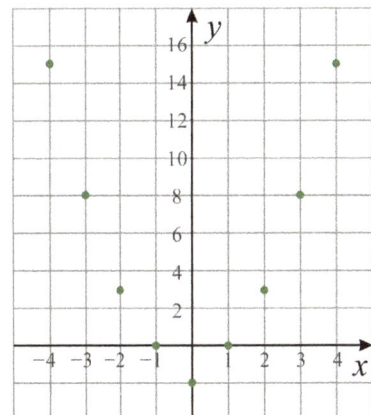

c. No, it wouldn't, because for most inputs, there would be two outputs. For example, 3 would be mapped to both 2 and −2.

Functions, cont.

Example 4. No. If we consider the time as a function of distance, it is no longer true that for each input, there is exactly one output. For example, for the distance 1.9 km, there would be three outputs: 12 minutes, 13 minutes. and 15 minutes.

7. a. Yes. Each name is mapped into an age.

  b. No. The age "14" is mapped to two different names.

8. (1) is a function if the inputs are the days of the month, and outputs are the amounts of rainfall, and otherwise not. In other words, rainfall is a function of the day of the month, but not vice versa.
  (2) is a function.
  (3) is not a function. For any particular zip code, there will be several people living there.
  (4) is not a function. Some first names, such as "Jack", would be mapped to several bank account numbers, because there are several people with that first name that have bank accounts. An additional problem is that some people, even if their first name was unique, have several bank accounts.

9.

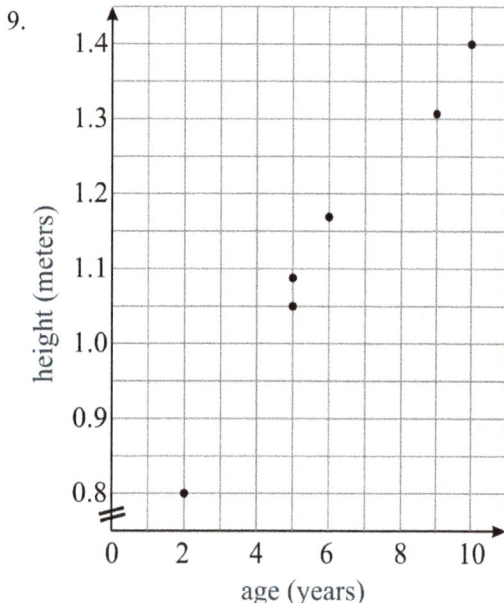

  a. No. For the input 5, there are two outputs.

  b. If it was a function, the age would be the independent variable, and the height the dependent variable.

Input	Output
Name	**Grade level**
Jenny	8
Pedro	7
Ann	8
Marsha	6
Rob	9

10. a. Answers will vary. Check the student's answer. Two things are necessary to change to make this a function:

  (1) One needs to add a grade level for Marsha.
  (2) Ann cannot be in two grade levels. One of the rows with "Ann" needs removed, for example the last row.

  b. The domain is {Jenny, Pedro, Ann, Marsha, Rob}

  c. Answers will vary. Check the student's answer. For example: the range is {6, 7, 8, 9}. This set needs to include Marsha's grade level as added in part (a).

11. The range is {1, 3, 5, 7, 9, 11}.

12. a. Domain: {−3, −2, 0, 1, 4}. Range: {−1, 1, 3}
  b. Domain: {−4, −3, −2, 1, 3}. Range: {−2, −1, 1}

13. {1, 2, 3, 4, 5, 6, 7, 8}

14. {5, 10, 15, 20, 25}

Linear Functions and the Rate of Change 1, pp. 195-198

Page 195

1. a. Rate of change = ($25 − $12.50)/(10 kg − 5 kg) = $12.50/5 kg = $2.50/kg. Yes, I get the same rate of change as calculated in the example.

   b. Rate of change = ($37.50 − $25)/(15 kg − 10 kg) = $12.50/5 kg = $2.50/kg.

Page 196

2. a. $15/hour   b. 2.8 kg/(2 L) = 1.4 kg/L

3. 3/5

4. a.

t (hours)	0 hrs	1 hr	2 hrs	3 hrs	4 hrs	5 hrs	6 hrs
d (km)	0 km	120 km	240 km	360 km	480 km	600 km	720 km

   b. 120 km/hr

5. a. $400   b. 9 hours   c. $20 per hour

6. a.

Input (t)	0 min	10 min	20 min	30 min	40 min	50 min	60 min
Output (d)	0.5 m	4 m	7.5 m	11 m	14.5 m	18 m	21.5 m

   b. Answers vary; check the student's answer. For example, a snail was at 0.5 meters from a house, and started crawling away from the house at the speed of 7 meters every 20 minutes.

7. The two are the same ($2/kg).

Page 197

8. a. 1/2   b. −1/3   c. 2/3   d. −2   e. −3/4   f. 1/4

Page 198

9. a. (15 L)/ (5 min) = 3 L/min   b. $30/(10 mi) = $3.00/mi
   c. −$1,400/(7 mo) = −$200/mo   d. (−200 km)/(4 hr) = −50 kph

10. a. $170   b. $250   c. $40/hr

Linear Functions and the Rate of Change 2, pp. 199-200

Page 199

1. Yes, it is linear. The rate of change (the speed) remains constant during that period, and the graph of the function is a line.

2. (1.5 mi)/20 minutes = 0.075 miles/min. Using hours, it would be (1.5 mi)/(1/3 hour) = 4.5 mph.

Page 200

3.

Point	A	B	C	D	E	F	G	H
Time (sec)	0	1	2	3	4	5	6	7
Distance (m)	0	20	33	38	39	40	40	40

   b. 20 m/s   c. 13 m/s   d. 0 m/s

   e. No, it is not a linear function, because the rate of change (speed) changes.

   f. For example, a car slowing down to a complete stop. The initial speed of 20 m/s corresponds to 72 km/h, which is a typical speed for a car on a highway.

Linear Functions and the Rate of Change 2, cont.

Page 200

4. a. No, it isn't. The rate of change is not constant. We can see that by checking how the y-values change in comparison to how the *x*-values change. As the x-values increase by ones, initially the y-values decrease by twos, but after reaching zero, they start increasing.

   b. Yes. From $x = -5$ to $x = -1$, the function is linear, and also from $x = -1$ to $x = 5$ (or on any subintervals of these).

Page 201

5. a.

Input (*t*)	0	0.5	1	1.5	2	2.5	3	3.5	4
Output (T)	20	20.25	20.5	20.75	21	21.25	21.5	21.75	22

   b. Student graphs will vary because the scaling of the *t*-axis was not given in the problem; check the student's graph. For example:

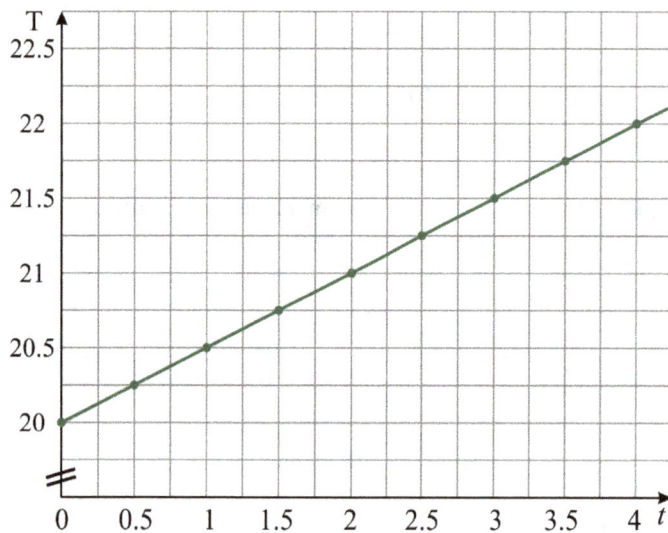

   c. Yes. The points fall on a line in the graph. Also, as the time increases by 1/2 hour, each time, the temperature increases by 0.25 degrees, so the rate of change is constant.

   d. (1.5 degrees)/3 hr = 0.5 degrees/hr

8. Function in (a) is linear, because each time the x-values increase by one, the *y*-values decrease by three. In contrast, in function in (b), while the *x*-values always increase by one, the *y*-values don't change by the same amounts.

Linear Functions as Equations, p. 202

Page 202

1. a. C = 2.5*d*
   b. Zero dollars. It means that when you hop into his taxi, in the beginning of the ride, you owe nothing.
   c. 20 km

2. a. C = 2*d* + 2.5
   b. $2.50
   c. 23.75 km

3. a. $5
   b. $3/(2 km) = $1.50/km
   c. The cost is $5 plus $1.50 for each kilometer driven. As an equation, C = $1.5*d* + $5.

4. Joe's service has the largest rate of change, at $2.50/km. For a 10-km ride, Andrew provides the best deal with $20.

Linear Functions as Equations, cont.

Page 203

5. a. Yes, it is. Its equation is in the format $y = mx + b$.
   b. The rate of change is $50/week.
   c. The initial value is $32.
   d. After 12 weeks of saving (March 26 or later).

6. a. Initial value: 60 km. Rate of change: 20 km/hr.
   b. $d = 20t + 60$ (or $d = 60 + 20t$)

Page 204

7. a. 2.5 gal/min. it means 2 ½ gallons of water are being added to the tank per minute.
   b. $V = 2.5t + 20$ (or $V = 20 + 2.5t$)
   c. Student graphs will vary because the scaling is not given in the problem; check the student's graph. For an example, see the graph on the right.
   d. 70 gallons
   e. After 32 minutes

Puzzle corner. The rate of change for F is
$((-1) - (-7))/((-4) - (-9)) = 6/5$.

The rate of change for G is
$((-3) - (-9))/((-4) - (-8)) = 6/4$.

G has the larger rate of change.

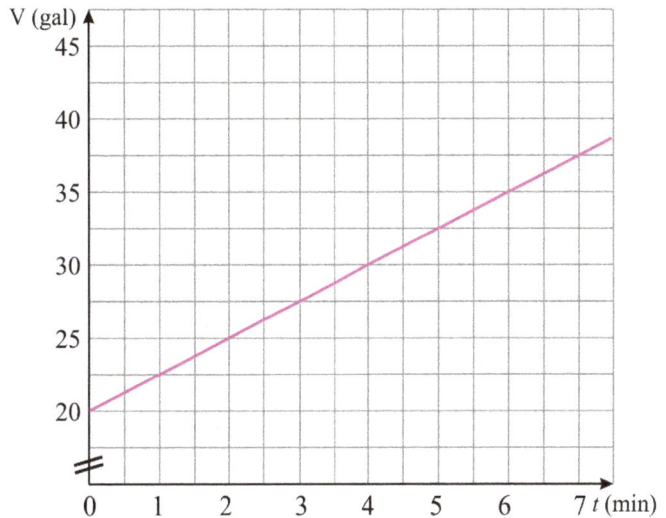

Linear Versus Nonlinear Functions, pp. 205-206

Page 205

1. a. Yes. For each time stamp, there is a corresponding temperature (and only one).

   b. It is not a linear function. Its graph is not a line.

   c. About $-0.8°/(4 \text{ hr}) = -0.2°/\text{hr}$

   d. $10°/(3 \text{ hr}) \approx 3.3°/\text{hr}$

   e. From 4:00 to 6:00 and from 12:00 to 15:00

   f. From 0:00 till 4:00 and from 14:00 till 18:00

   g. from 7:00 till 9:00

Page 206

2. Answers will vary; check the student's answers. For example:

Function 1:

time (hours)	0	1	2	3	4	5	6	7	8	9	10
Cost ($)	5	6	7	8	9	10	11	12	13	14	15

Function 2:

time (hours)	0	1	2	3	4	5	6	7	8	9	10
Cost ($)	10	10	10	10	12	12	15	15	15	20	20

Linear Versus Nonlinear Functions, cont.

3. Answers will vary; check the student's answers. For example:

The equation $A = s^2$ is not of the form $y = mx + b$; it contains a squaring of the variable. Therefore the function is not linear.

Or: The rate of change is not constant. From s = 0 to s = 1, the rate of change is 1, but from s = 3 to s = 4, the rate of change is $(16 - 9)/1 = 7$.

Or: When you graph it, you will not get a line. (See the graph on the right.)

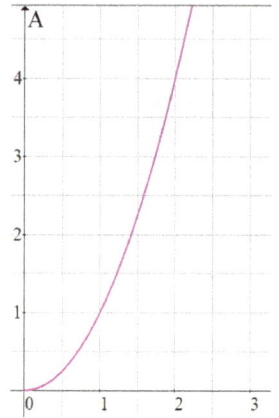

4. Yes, this is a linear function. Equation: $C = 2.5w$, where C is the cost and w is the weight.

5. a. If there are 10 workers, it takes 8 hours. If there are 5 workers, it takes 16 hours. It takes half the time if there are 10 workers versus if there are five.

   b. No. The equation is not of the form $y = mx + b$. Also, the rate of change is not constant. For example, from N = 1 to N = 2, the rate of change is 40 hours per worker, but from N = 4 to N = 5, it is four hours per worker.

6. Answers will vary; check the student's answer. For example: $y = 5/x$, $y = x^2 + 5$ or $y = 2x^3$. The independent variable could be raised to any exponent except one. The independent variable could be in the denominator. There are many other functional relationships, too, such as $y = \sin(x)$, but most 8th graders probably don't know about them.

7. Answers will vary; check the student's answers. For example:

time (minutes)	0	5	10	15	20	25	30	35	40	45	50	55	60
Distance (km)	0	0.5	1	1.5	1.5	2.2	2.9	3.6	4.1	4.6	5.2	5.8	6.4

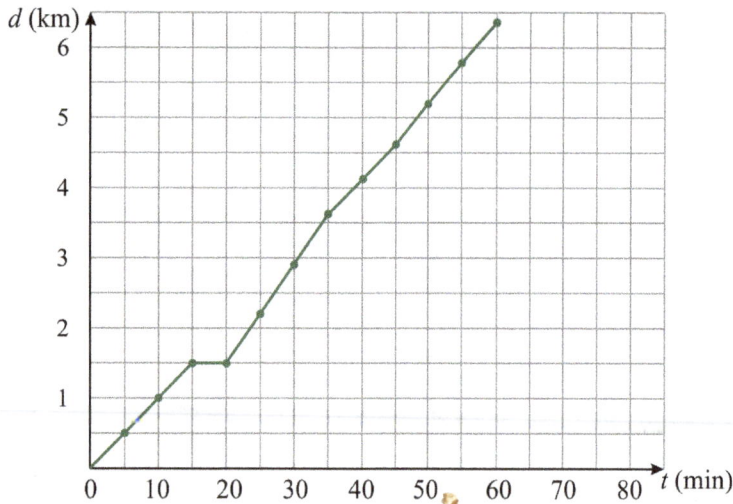

Puzzle corner. (i) - (b), (ii) - (a), (iii) - (c), (iv) - (d), (v) - (f). Graph (e) does not get matched, and is not even a plot of any function (since for most input values, there are two corresponding output values).

Modeling Linear Relationships, p. 208

1. a. the rate of change is $4/(20 miles) = $1 per 5 miles = $0.20/mile. Each mile you drive over the 200-mile allowance will add $0.20 to the base cost.
   b. The initial value is $90. This is the base cost to rent the car.
   c. $C = 0.20m + 90$ where m is the number of miles over 200 miles
   d. $113.70
   e. 235 miles

Modeling Linear Relationships, cont.

Page 208

2. a. savings = 25t + $140 where t is time in weeks
   b. 14 weeks after he started putting $25 into his savings every week

Page 209

3. a. (−100 gal)/(5 min) = −20 gal/min. The tank is emptying at a rate of 20 gallons per minute.
   b. 400 gallons. The tank had 400 gallons of water in it before it started emptying.
   c. $V = 400 − 20t$

4. a. $D = 6,500 − 350t$

   b. Student graphs will vary because the scaling on the axes may vary; check the student's graph. See the graph below for an example. The point (10, 3000) is marked because it is a point on the gridlines, and is useful for drawing the line. The point corresponding to *time* = 12 months is also marked.

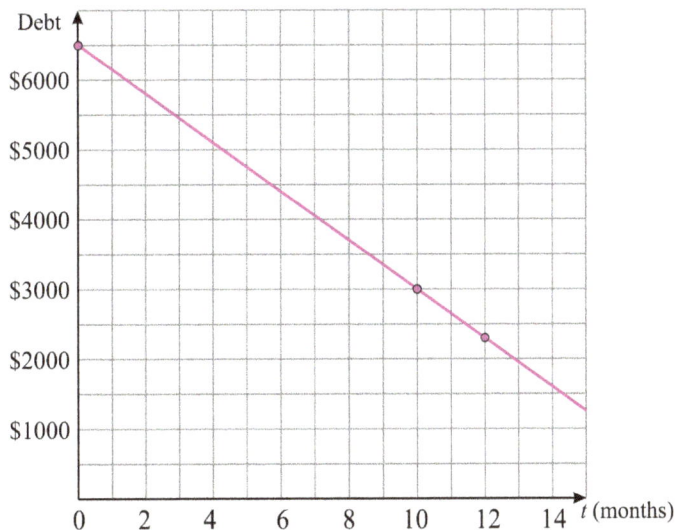

Page 210

5. a. The rate of change is 25 kg/month. It shows how much weight the heifer gains each month.

   b.

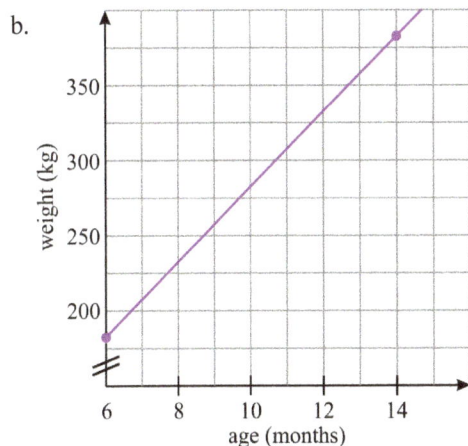

   c. To write the equation, you need to find the weight at zero months of age, if the heifer had been following the same pattern of weight gain prior to 6 months of age. (In reality, the weight gain of heifers is not linear in the first few months, but we can use this technique to find the initial value for this equation.) Gaining 25 kg/month and weighing 180 kg at 6 months means that at zero months, the weight would have been 30 kg. So, 30 kg is the initial value, and the equation is $W = 25A + 30$.

   Once again, this equation is only valid, or only models the weight accurately, from 6 to 14 months of age.

6. a. 32°. It means that when the temperature in Celsius degrees is zero, the temperature in Fahrenheit is 32°.

   b. The rate of change is 9/5. Each 9-degree change in Fahrenheit degrees corresponds to a 5-degree change in Celsius degrees.

   c. F = (9/5)C + 32

   d. F = (9/5)18° + 32° = 64.4°

   e. We solve the equation 100 = (9/5)C + 32 for C:

   $$\begin{aligned} 100 &= (9/5)C + 32 \\ 68 &= (9/5)C \\ 340 &= 9C \\ C &\approx 37.8° \end{aligned}$$

7. a. The initial value is 0 cm^3. (When the height is 0 cm, the volume is zero.)

   b. From the given points, we can calculate the rate of change as (385 − 231)/(10 − 6) = 154/4 = 38.5.
   The equation is then V = 38.5h.

   c. We solve the equation 500 = 38.5h, and get h = 500/38.5 ≈ 13.0 cm.

   d. V = 38.5 · 7.4 ≈ 285 cm^3.

   e. V = π · (3.5 cm)^2 · h ≈ 38.4845h

Puzzle corner. a. Line 1 is Car 1, and Line 2 is Car 2.
b. We will first figure out the initial value using the given information. At 10 months, she has $20,000 left to pay, and she has paid 10 payments of $1,800. This means the original price was $20,000 + 10 · $1,800 = $38,000.
Let D signify the amount of debt, and t time in months. The equation is then D = 38,000 − 1,800t.

Describing Functions 1, pp. 212-213

1. It is decreasing and nonlinear in the interval [−15, −5]. It is increasing and linear in the interval [−5, 10].
   It is decreasing and nonlinear in the interval [10, 16]. It is constant in the interval [16, 25].

2. a. from 2015 to 2019
   b. from 2019 to 2020
   c. It has been decreasing in a nonlinear manner.

3. a. from 2.25 till 3 years
   b. From 4 to 4.75 years
   c. −0.5%
   d. From 0 to 1 year: −1%/year; from 2 to 3 years: 1.5%/year

Describing Functions 1, cont.

4. Answers will vary; check the student's graph. Two possibilities are shown here:

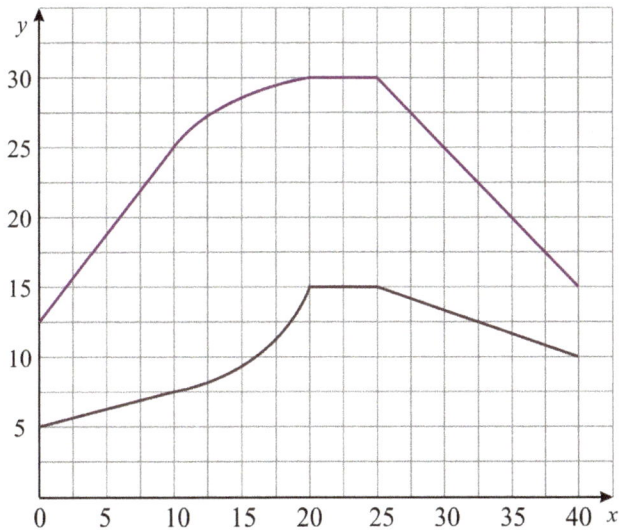

5. Answers will vary; check the student's story. For example:
   In the spring, the level of water was steady until June, because there was enough rain. During the summer and through September, farmers needed water from it, but there was no rain, so the level was declining. The water level stabilized till December, and then the winter rains filled the reservoir up to 90 meters.

Describing Functions 2, pp. 215-216

1. a. 3.5 km
   b. Zero. (She was not moving.)
   c. From 20 to 35 minutes, and from 45 to 60 minutes
   d. She started out with a brisk speed (about 1 km/5 minutes or 12 km/hr) but then slowed down, and eventually stopped at 15 minutes.

2. Answers will vary; check the student's answer. For example: John throws a stick, and Max dashes out 15 meters from John, and quickly comes back, taking a total of 11 seconds for the trip. Then he stays with John for nine seconds, until John throws the stick again. Max runs out for 6 seconds, to 20 meters away, and stops there for a bit (7 seconds), then starts back towards John. But after just a few seconds, he stops again to sniff the ground for five seconds. Then he slowly comes back to John.

3. Answers will vary somewhat; check the student's graph. For example:

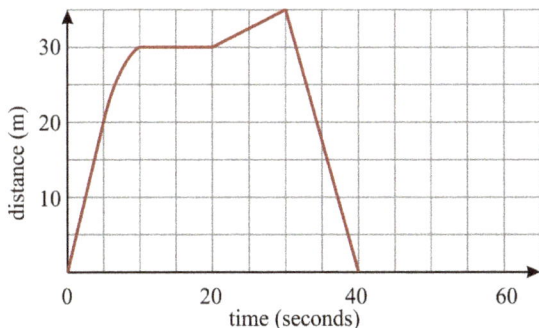

85

Describing Functions 2, cont.

Page 216

4. The graph should look approximately like the graph on the right:

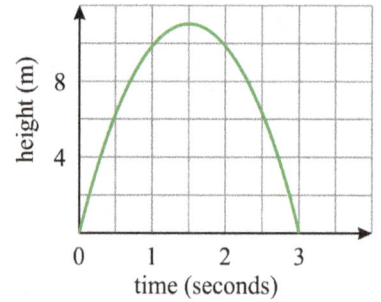

Page 217

5. Graph 1 is the rabbit, and graph 2 is the turtle. Student stories will vary; check the student's story. For example:

Rabbit starts out very fast, and quickly outruns the turtle, which is going at a steady pace. After a bit, the rabbit feels he's quite a bit ahead and stops to sleep. When he wakes up, he notices the turtle has just passed him, so he starts running really fast again for a short while, but tires, and stops again. Once again, he takes a nap for a bit. When he wakes up, the turtle is not in sight... so he quickly finishes the race, but the turtle had reached the finish line moments earlier.

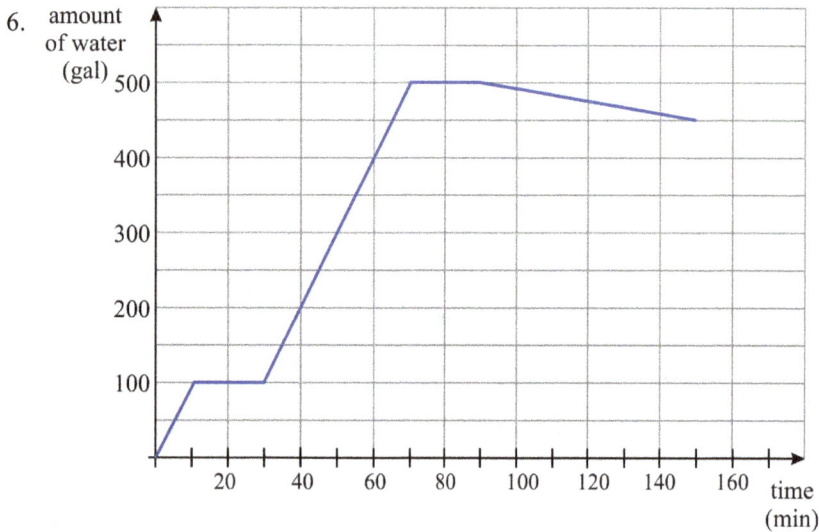

6.

Describing Functions 3, p. 218

Page 218

1. Student graphs will vary because the scaling for the vertical axis was not given in the problem. Check the student's graph.

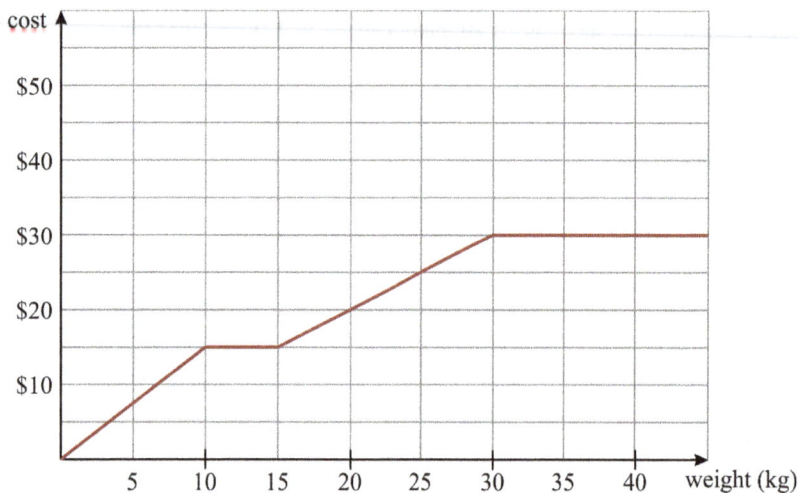

Describing Functions 3, cont.

2. a. He changes speed and starts going faster.

   b. Mason starts up the river to the north. After a bit, he starts going faster. Then he stops for a bit (maybe at a dock). Then he comes back to his starting point at a constant speed, fairly fast. Then he stops for a while (maybe to eat lunch?). After that, he goes southward, with constant speed, until he stops somewhere for a bit, and then comes back.

3. Student graphs will vary because the scaling for the vertical axis was not given in the problem. Check the student's graph.
   See the graph on the right for an example.

4. a. She is chewing.

   b. She takes a cracker or crackers out of the package.

   c. No. Shortly before 2 1/2 minutes, the line becomes a horizontal line at 50 grams. So, she left 50 grams of crackers in the package.

   d. Looking at the height of the vertical lines, they are 2.5 g, 5 g, or 7.5 g. So, most likely one cracker weighs 2.5 g.

5. a. (2). Axes: time and price
   b. (6). Axes: time and distance
   c. (4). Axes: weight and cost
   d. (5). Axes: time and volume
   e. (8). Axes: time and speed
   f. (7). Axes: time and volume

Puzzle corner.

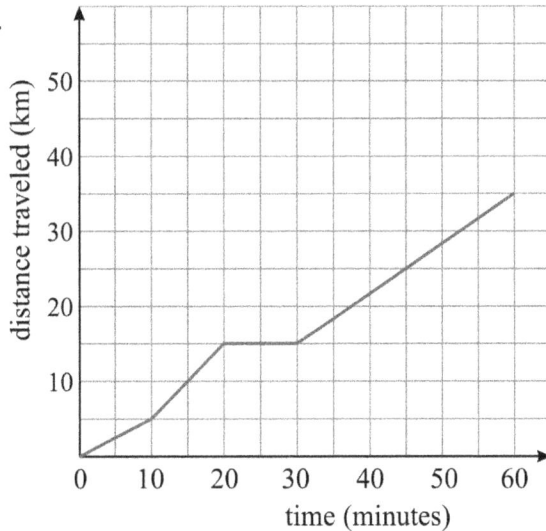

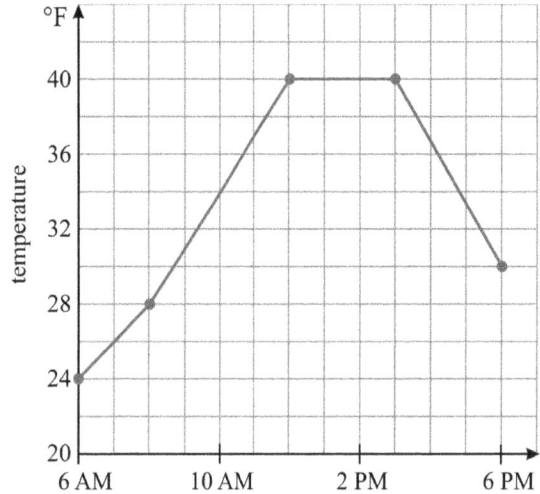

87

Comparing Functions 1, pp. 222-224

Page 222

1. a. City Round ($4.50/hr vs. $4/hr)
   b. Mike's Bikes ($4 versus $3)
   c. Mike's Bikes: $C = 4t + 4$. City Round: $C = 4.5t + 3$, where C is the cost in dollars, and t is time in hours.
   d. The cost for renting from Mike's Bikes is $14, and from City Round $14.25, so Mike's Bikes is the better deal.

2. a. No, she is not correct, because the table of values for Function 2 does not list a value for $x = 0$.
      We don't know the initial value for Function 2.

   b. No, he is not correct. The rate of change for Function 2 is indeed 3, but for Function 1 it is *negative* four.
      So, the rate of change for Function 2 is greater.

Page 223

3. a. Debt 1 (at $350 per month, vs. $250 per month for Debt 2.).
   b. Debt 1 was $4,000 originally.
   c. In eight more months. He will make seven more $350 payments, and the last payment will be only $150.
      Debt 2 will take eight additional months to pay off, as well: seven payments of $250, and a final payment of $150.

4. a. For the first hour, Airplane 2 travels at a speed of 225 km/h, so Airplane 1 travels faster (230 km/h).
   b. The speed of Airplane 2 in this time frame is (525 km − 300 km)/0.75 hr = 300 km/hr, so it travels faster.
   c. In three hours, Airplane 1 covers 690 km, and Airplane 2 covers 675 km, so <u>Airplane 1 covers a greater distance</u>.

Page 224

5. a. Service 1. The rate of change for Service 2 is $16 for 1/2 kg.
   b. Service 1, at $105.
   c. The cost with Service 1 is $231, and with Service 2, $233. Service 1 is the better deal.

   d. Let m be the weight of the package, and C be the cost. The equation for Service 1 is $C = 36m + 15$ and for
      Service 2, $C = 32m + 41$. (Note that the question specifies the cost for Service 1 for *half* a kilogram.)

   We set these to be equal:

$$36m + 15 = 32m + 41$$
$$4m = 26$$
$$m = 6.5$$

For a package weighing 6.5 kg, the cost of both services is equal ($249).

6. a. (4) and (6)
   b. (1) and (6) or (1) and (4)
   c. (2) and (3)

Comparing Functions 2, p. 225

Page 225

1. a. Function 3: its rate of change is 3, whereas Function 1 and Function 2 have a rate of change of 1.
   b. Function 1: its rate of change is 3, whereas Function 2 has a rate of change of 1, and Function 3 has 2.

2. a. Function 2 (its initial value is 20).
   b. Functions 2 and 3
   c. The rates of change are: Function 1: −1, Function 2: 1.5, and Function 3: −2.5.
      Function 3 has the smallest rate of change.
   d. Function 1: decreasing. Function 2: increasing: Function 3: decreasing.

Page 226

3. (1) and (5)
   (2) and (8)
   (3) and (7)
   (4) and (6)

Puzzle corner. The rates of change are:
(1) $4/3 - 1 = 1/3$
(2) $3\ 1/2 - 3 = 1/2$
(3) $1/3 - 0 = 1/3$

Function 2 has the largest rate of change from $x = 2$ to $x = 3$.

Mixed Review Chapter 4, pp. 227-228

1. d, e, f.

2.

a. $\dfrac{1}{b^8}$	b. $16y^4$	c. $\dfrac{1}{49a^2}$	d. $-6x^9y^6$
e. $\dfrac{2x^6}{3}$	f. $5s^5$	g. $\dfrac{4x^2}{9}$	h. $\dfrac{a^4}{16b^8}$

3.

a. $0.0005 + 0.002 = 0.0025 = 2.5 \cdot 10^{-3}$	b. $9{,}000{,}000 + 20{,}000{,}000 = 29{,}000{,}000 = 2.9 \cdot 10^{7}$
c. $0.05 - 0.008 = 0.042 = 4.2 \cdot 10^{-2}$	d. $8{,}000{,}000 - 700{,}000 = 7{,}300{,}000 = 7.3 \cdot 10^{6}$

4. a. $1.6 \cdot 10^{17}$ b. $4 \cdot 10^{2}$

5. a. $h = \dfrac{3V}{A_b}$

   b. 31.9 cm

6.

a.

b.

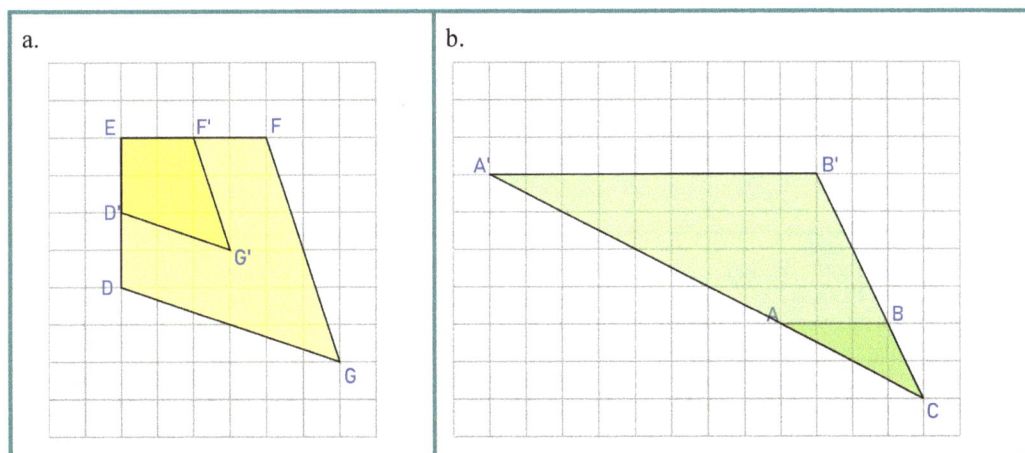

7. Let n be the number by which you multiply the current price to get the new price. Then: $n \cdot 2.76 = 2.98$, from which $n = 2.98/2.76 \approx 1.0797$, and the percent of increase is 7.97%.

   Another way to solve this is to use the basic formula for the percent of increase: (difference/reference). In this case, we get: (difference/reference) = $(2.98 - 2.76)/2.76 \approx 0.0797$ or 7.97%.

Page 228

8. The discounted price is $0.7 \cdot \$72.95$. Now, let t be the number by which you multiply the price to get the price with tax. Then: $t \cdot (0.7 \cdot \$72.95) = \$53.92$, from which $t = \$53.92/(0.7 \cdot \$72.95) \approx 1.056$. So, the sales tax is <u>5.6%</u>.

9.

a.
$$3x + \frac{x+3}{5} = 1 \qquad \Big| \cdot 5$$
$$15x + x + 3 = 5$$
$$16x + 3 = 5 \qquad \Big| -3$$
$$16x = 2 \qquad \Big| \div 16$$
$$x = 1/8$$

b.
$$3x - \frac{x+3}{5} = 1 \qquad \Big| \cdot 5$$
$$15x - (x+3) = 5$$
$$15x - x - 3 = 5$$
$$14x - 3 = 5 \qquad \Big| +3$$
$$14x = 8 \qquad \Big| \div 14$$
$$x = 4/7$$

Page 229

10. a. One solution.
   b. No solutions.
   c. An infinite number of solutions.
   d. One solution.

11.

a.
$$10 + 3(a+5) = 2(a-6) - 4a$$
$$10 + 3a + 15 = 2a - 12 - 4a$$
$$3a + 25 = -12 - 2a \qquad \Big| +2a$$
$$5a + 25 = -12 \qquad \Big| -25$$
$$5a = -37 \qquad \Big| \div 5$$
$$a = -37/5$$

b.
$$20x - 2(x+1) = 10 - (x-5)$$
$$20x - 2x - 2 = 10 - x + 5$$
$$18x - 2 = 15 - x \qquad \Big| +x$$
$$19x - 2 = 15 \qquad \Big| +2$$
$$19x = 17 \qquad \Big| \div 19$$
$$x = 17/19$$

c.
$$\frac{1}{6}x - 1 = 1 + \frac{4}{5}x \qquad \Big| \cdot 30$$
$$30(\frac{1}{6}x - 1) = 30(1 + \frac{4}{5}x)$$
$$5x - 30 = 30 + 24x \qquad \Big| -5x$$
$$-30 = 30 + 19x \qquad \Big| -30$$
$$-60 = 19x \qquad \Big| \div 19$$
$$x = -19/60$$

d.
$$2z + \frac{2}{5} = \frac{1}{4}z - 1 \qquad \Big| \cdot 20$$
$$40z + 8 = 5z - 20 \qquad \Big| -5z$$
$$35z + 8 = -20 \qquad \Big| -8$$
$$35z = -28 \qquad \Big| \div 35$$
$$z = -4/5$$

Chapter 4 Review, p. 230

Page 230

1. Answers will vary; check the student's answer. The problems are: One needs to add an age for Max, and Luna should not have two ages (one needs removed). For example:

2. Because for the input 2, there are two outputs (10 and 20).

3. In the x-interval $[-10, -5]$, the function is increasing in a nonlinear manner.
   In the interval $[-5, 1]$, the function is decreasing in a nonlinear manner.
   In the interval $[1, 4]$, the function is increasing in a linear manner.
   In the interval $[4, 9]$, the function is constant.

Input	Output
Name	**Age**
Fifi	2
Bella	5
Max	3
Luna	2
Charlie	6

Chapter 4 Review, pp. 230-233

Page 230

4. Answers will vary; check the student's answer. For example:

Time (hours)	0	1	2	3	4	5	6	7	8	9	10	11	12
Cost ($)	0	5	6	7	10	12	14	17	18	18	20	20	20

Page 231

5. a. No, it is not linear, because the graph is not a line.
   b. $5,000/yr
   c. $2,000/yr

6. a. The rates of change are as follows. Function 1: 0.22 miles/minute.
      Function 2: (0.45 mi)/(2 min) = 0.225 miles/minute. Function 2 has the greater rate of change.
      This means Jayden's sister was bicycling with a faster speed during that time.
   b. Function 1 is linear, and function 2 is nonlinear.
   c. Jayden takes (1.5 mi)/(0.22 mi/min) ≈ 6.82 minutes to reach home. His sister is bicycling at a rate of
      0.12 miles/minute between 5 and 6 minutes, which means that at 7 minutes, she will have covered 1.36 miles, and
      at 8 minutes, 1.48 miles. But Jayden has already reached home by 7 minutes of time, so he reached home first.

Page 232

7. a. 3 inches/hour. It means the snow is falling at that rate.
   b. 17 inches. Before the blizzard started, there was already 17 inches of snow on the ground.
   c. S = 17 + 3*t* where S is the amount of snow, and *t* is time in hours
   d. From the equation 26.5 = 17 + 3t, we can solve that *t* = 3 1/6 hours, or 3 hours 10 minutes.
      At 5:40, the snow was 26.5 inches deep.
   e. 9 PM is 6.5 hours after the blizzard started, so the depth of snow is S = 17 + 3(6.5) = 36.5 inches deep.
   f. Answers will vary; check the student's graph. For example:

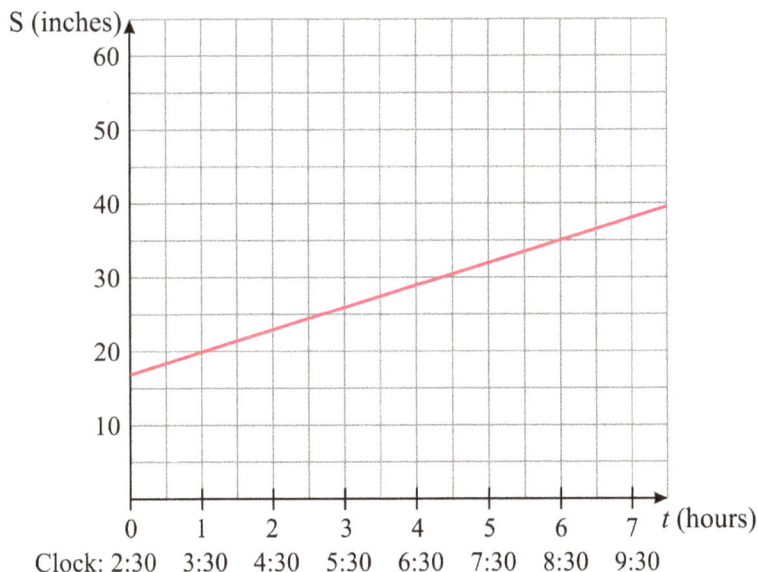

S (inches)
Clock: 2:30 3:30 4:30 5:30 6:30 7:30 8:30 9:30

Page 233

8. Answers will vary; check the student's story. For example: Maria sits at 12 feet from the blackboard. Two minutes
   into the class, the teacher calls her to the board to solve a math problem, so Maria quickly walks there. She solves
   the problem, and after two minutes, goes back to her desk. Then, 6 minutes into the class, she goes to the back of
   the room to sharpen her pencil, which takes less than a minute. She then returns to her desk and stays seated.

9. a. Greg.
   b. They meet from 20 to 25 minutes, and also at about 32 minutes.
      Between 20 and 25 minutes, they have traveled 1.5 miles. At 32 minutes, they have traveled about 1.9 miles.
   c. Trevor finishes first, about 2 minutes quicker.

Math Mammoth
Grade 8-B
Answer Key

By Maria Miller

Contents

Chapter 5: Graphing Linear Equations

Graphing Proportional Relationships 1, pp. 13-15

Page 13

Example 2. The unit rate from $6/(5 kg) is $1.20/kg, and from $22/(20 kg), it is $1.10/kg. So, they are not equal.

1. a. No. If we look at the rates as distance over time, and calculate the unit rates, using 1 hour and 50 km we get 50 km/h, but using 2 hours and 90 km we get 45 km/h.
   b. Yes. Each unit rate is 45 km/hr.
   c. No. For example, when the age doubles from 2 days to 4 days, the height does not double. The same happens when the age doubles from 3 days to 6 days: the height goes from 1 in to 4 in (not doubling).
   d. Yes. Here the unit rate ends up being $6 per meter, and it is consistent for all the rates in the table.

2. Concerning 1b, the rate of change is 45 km/hour. This is the same as the unit rate.
   Concerning 1d, the rate of change is $6/m. This is the same as the unit rate.

Page 14

3. In the equation $y = 3x$, the variables are proportional.

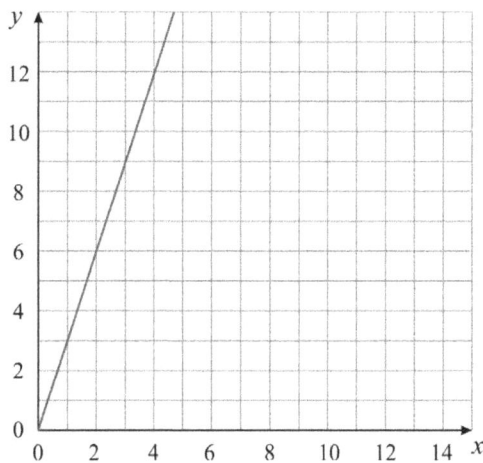

4. b, f, and g.

5. f and g.

Page 15

6. a. $y = (5/7)x$
   b. See the graph below. The easiest way is to plot the point (7, 5) and draw a line through that and the origin.

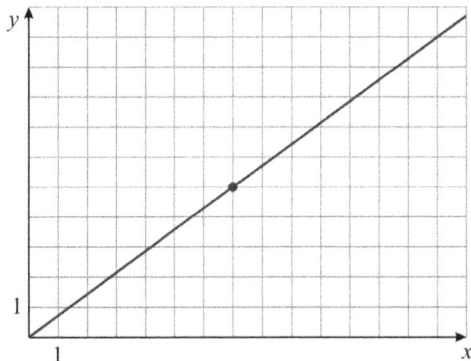

   c. $40 = (5/7)x$, from which $x = 40 \cdot 7 / 5 = \underline{56}$.

Graphing Proportional Relationships 1, cont.

7. a. $C = (5/2)w$

b.

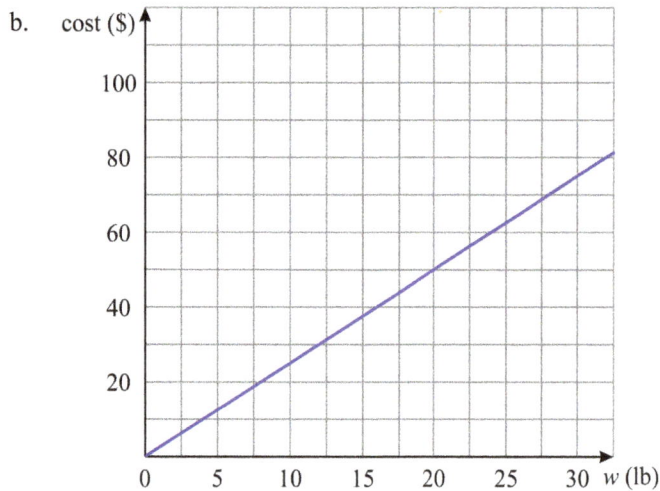

c. $C = (5/2) \cdot 36 = \$90$

8. y is $120/5 = \underline{24}$.

Graphing Proportional Relationships 2, pp. 16-18

1. a. Yes. The graph is a line through the origin.
   b. 1.5 m/(2 minutes) = 0.75 m/min. The unit rate is 0.75 m/min. The slope is 0.75 (or 3/4).
   c. $d = 0.75t$
   d. $9.5 = 0.75t$, from which $t = 9.5/0.75 \approx 12.667$ minutes. It takes him 12 minutes 40 seconds to travel 9.5 meters.

2. The second caterpillar is faster. The line depicting the distance it has gone is steeper.

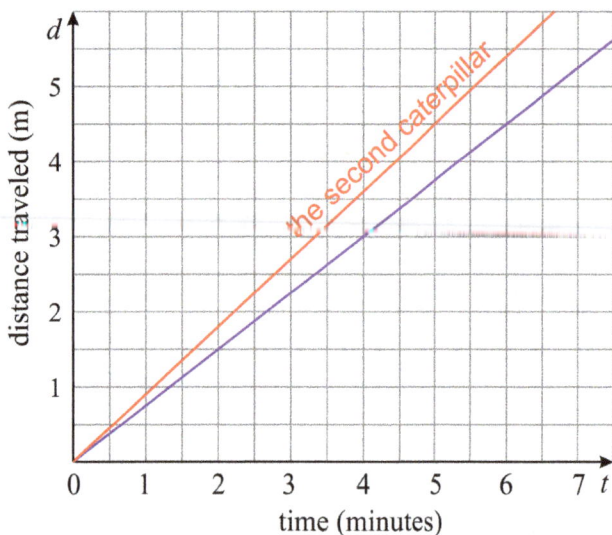

3. a. C = 5.5*l*, where C is the cost and *l* is the length of the fabric. The student may use other variables than C and *l*.

b. Graphs will vary; check that the student's graph is scaled in such a manner that it includes the point for 5 meters of fabric (5, 27.5). For example:

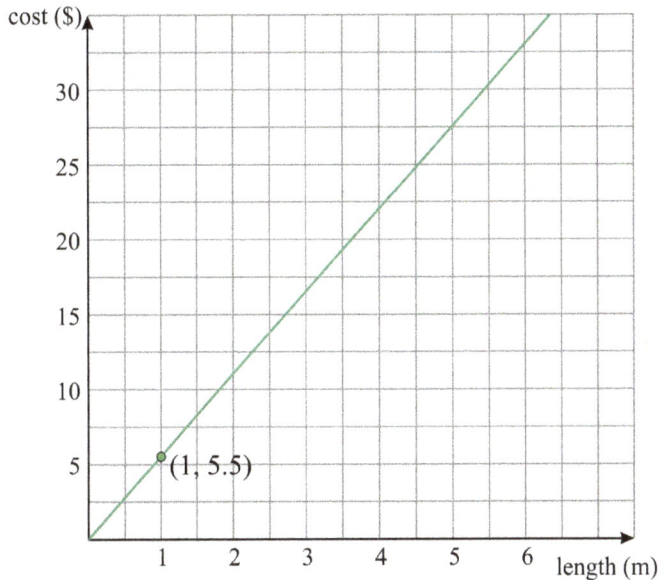

c. The slope is 5.5. The rate of change is $5.50 per meter.
d. See the image.

4. a. Yes. He rode with a constant speed, which means the distance and time are in proportion.
b. (6 miles)/(12 min) = 0.5 miles/minute. Alternatively, in one hour he would cover 30 miles, so his speed is 30 mph.
c. The unit rate is the same as the speed: 0.5 miles/minute. The slope is 0.5.
d. $d = 0.5t$
e. Graphs will vary; check that the student's graph is scaled in such a manner that it includes the point for 30 minutes of time (30, 15). Two possibilities are shown below:

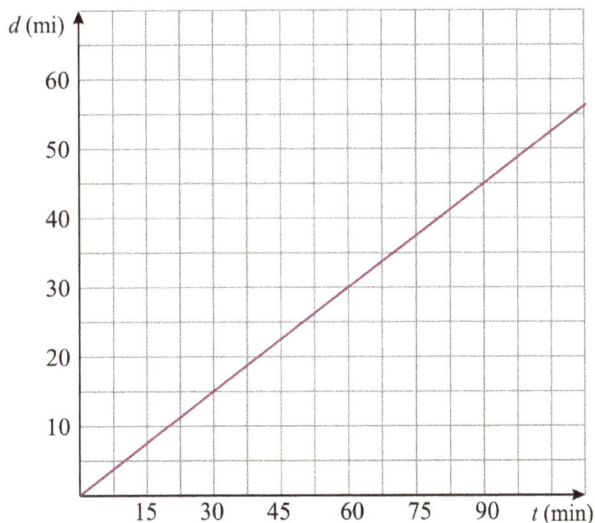

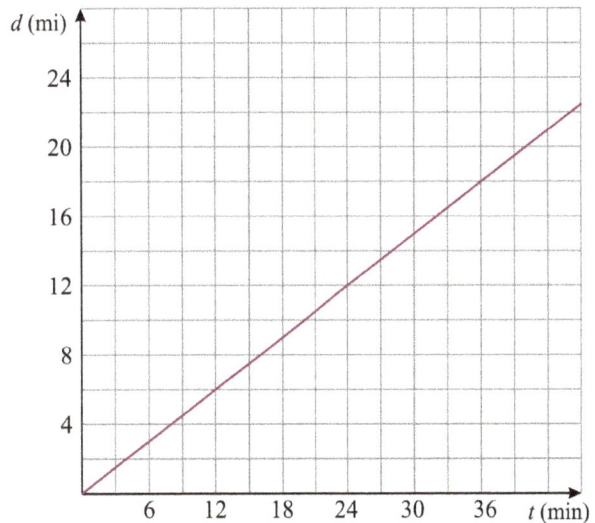

f. 30 minutes

5. a. 0.25 L/month
   b. Using t for time in months, and S for the amount of shampoo in milliliters, $S = 0.25t$.

   c.

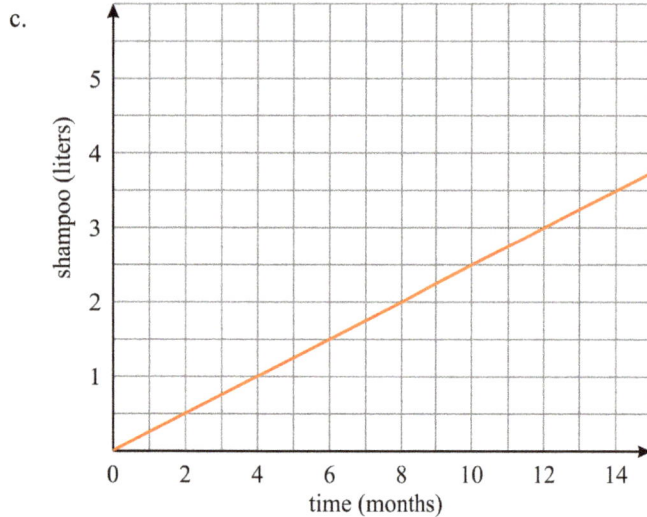

   d. The slope is 0.25.
   e. They use one liter of shampoo in four months, so it will take them 20 months to use five liters of shampoo.

6. a.

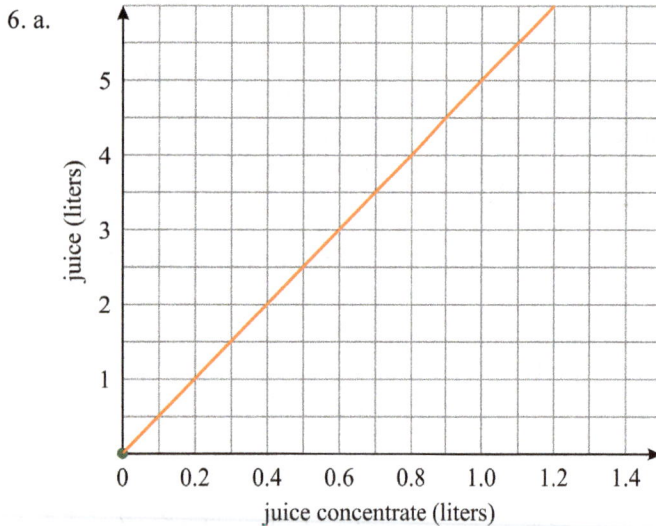

   b. $J = 5C$
   c. 5 L/L, or five liters per liter. This means you get 5 liters of juice per 1 liter of concentrate. It looks a bit odd when the unit for both quantities is the same, but this is how it works.

   d. We can use the equation from (b): $6 = 5C$, from which $C = 6/5$ L or 1.2 liters.
      She will need 1.2 liters of concentrate and 4.8 liters of water to get 6 liters of juice.

Comparing Proportional Relationships, pp. 19-22

1. a. Bananas cost $2.20 more for 10 lb than watermelon. Bananas cost $7.70 for 10 lb, whereas watermelon costs
      $5.50 for 10 lb.
   b. The unit price for bananas is $0.77/lb, and for watermelon, $0.55/lb.

Page 19

2. a. Paint 2.

   b. For Paint 1, the unit rate is 350 ft^2 per gallon. Therefore, using Paint 1, you would need 320 ft^2/(350 ft^2/gal) = <u>0.914 gallons of paint </u>to paint 320 square feet.

   For paint 2, the point B on the graph allows us to calculate the unit rate exactly:
   2000 ft^2 / (5 gallons) = 400 ft^2 per gallon. With Paint 2, you would need 320 ft^2/(400 ft^2/gal) = <u>0.8 gallons of paint</u>.

   c.

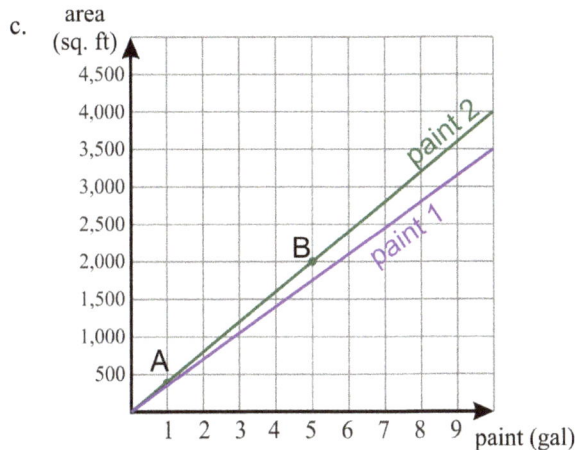

   d. Paint 2 covers about 3,200 sq. ft. and Paint 1 about 2,800 sq. ft. So, Paint 2 covers about 400 sq. ft. more.
   [Calculating this exactly, we get 8(400 ft^2 − 350 ft^2) = 400 ft^2.]

Page 20

3. a. Reading the graph, Jerry travels 30 km in an hour. So, the moped travels faster, 5 km/h faster than the tractor.
   b.

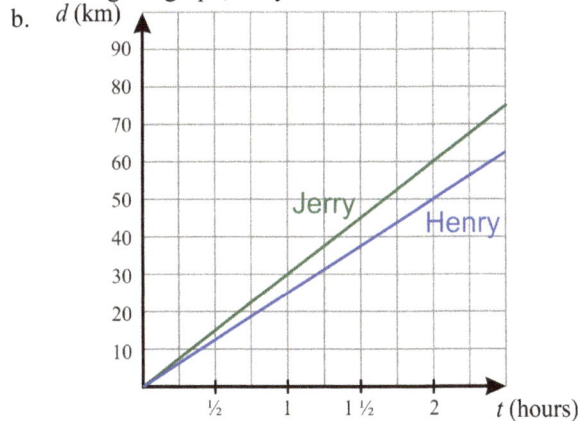

   c. Jerry: 30. Henry: 25.
   d. The moped travels 10 km <u>four minutes faster</u> than the tractor. This can be solved in various ways. For example:

   For the moped, from the equation $d = 30t$ we can solve that $t = d/30$. The moped takes 10/30 h = 1/3 h = 20 minutes to travel 10 km.

   Similarly, the tractor takes 10/25 h = 0.4 h = 24 minutes to travel 10 km.

Comparing Proportional Relationships, cont.

Page 20

4. a. William.

   b. See the graph on the right.

   c. The units in the graph are kilometers and minutes, and to find the slope, we need to use those units. The slope for William is therefore 18/30 = 3/5. The slope for Jeanine has to be figured out using the graph, and therefore the student answers may vary. The line goes through one grid point: (45, 15), which gives us the slope of 15/45 = 1/3.

   d. In kilometers per hour, William's speed is 36 km/h. He therefore takes 14/36 hr = 7/18 hr ≈ 0.3888 hr ≈ 23 minutes to reach Grandma's place. Jeanine's speed is 1/3 kilometers per minute, or 20 km/h. She will take 14/20 = 0.7 hours = 42 minutes to reach Grandma's place. So, William will get there in <u>19 minutes</u> less time.

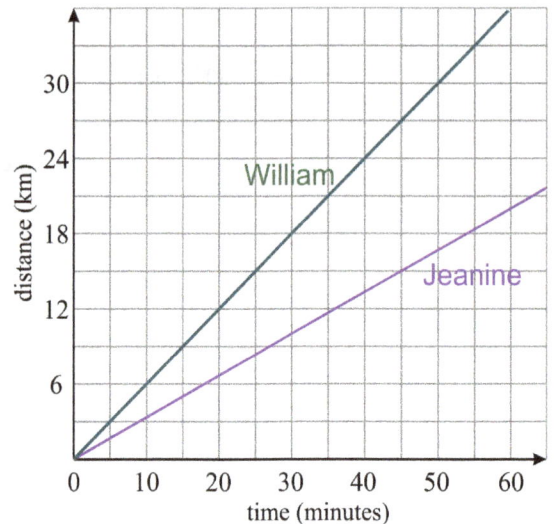

Page 21

5. a. Answers will vary since the scaling for the vertical axis will vary. For example:
   b. Strawberries: C = 1.8w. Blueberries: C = 3w.
   c. The slopes are 1.8 and 3.
   d. The line for the blueberries is steeper.
   e. The difference is a little bit less than $10, maybe $9.50.

6. a. The slope has to be slightly less, so (i) is the only reasonable answer.
   b. Check the student's drawing. The line should start at the origin and be somewhat less steep than line 2.

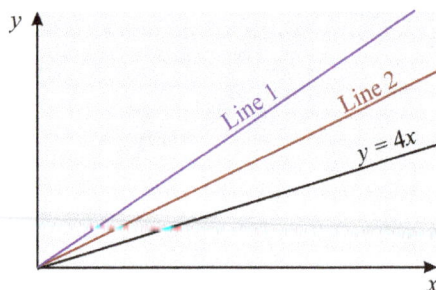

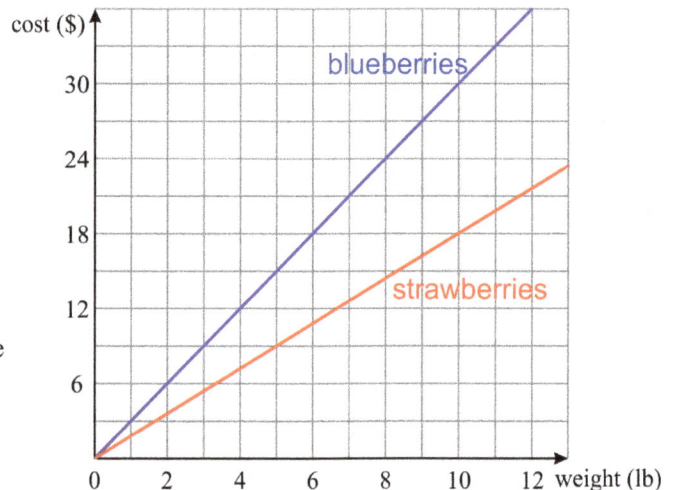

Page 22

7. a. Looking at the graph, the line goes through the grid point (50, 2,250), from which we can calculate the gas mileage for Car 2 as 2,250/50 mpg = 45 mpg. <u>Car 2 has the better gas mileage</u>.

   b. Car 1 uses 1,000 mi/(36 mpg) ≈ 27.778 gallons for 1,000 miles, and Car 2 uses 1,000 mi/(45 mpg) ≈ 22.222 gallons. The difference is about <u>5.6 gallons</u>.

   c. For 14,000 miles, the amount of gasoline saved (in gallons) is 14,000/36 − 14,000/45. We multiply this by $3.00 to get the savings: $3(14,000/36 − 14,000/45) ≈ <u>$233.30</u>.

Puzzle corner. From the graph, the cheetah's speed is 500 m/(20 sec) = 25 m/s. We now convert the lion's speed of 72 km/h to meters per second: 72 km/h = 72,000 m/(3,600 s) = 20 m/s. <u>The cheetah is faster</u>.

Slope, Part 1, pp. 23-26

Page 23

1. a. Roof 1.
   b. We can take the ratio of the height to the half of the width, like given in the illustrations. For Roof 1, this ratio is 8/10 = 4/5, for roof 2, 8/12 = 2/3, and for roof 3, 6/10 = 3/5.

2. We can draw a triangle for each line, and write the ratio of the height of the triangle to its base. This ratio measures the steepness of the line.

   Line b is the steepest, and line f is the least steep.

   These ratios are:

   Line a: 2/3. Line b: 3/3 = 1. Line c: 4/7.
   Line d: 3/5. Line e: 3/4. Line f: 2/4 = 1/2.

   Lines in order of steepness: f, c, d, a, e, b.

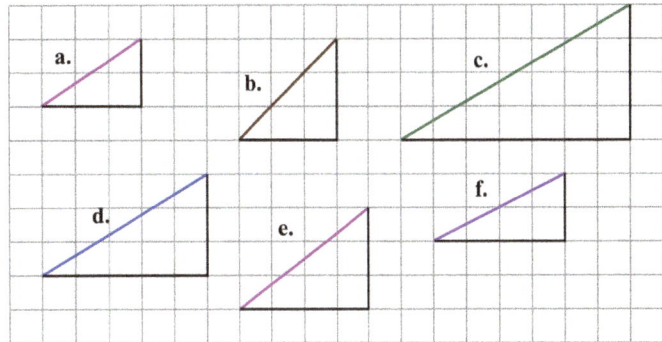

Page 24

3. a. 1   b. 1/4   c. 3   d. 2/5

Page 25

4. a. 2/4 = 1/2   b. 1/2   c. 4/8 = 1/2

5. a. −2   b. −2/3   c. −1/4   d. −5/2

Page 26

6. a. 1/3   b. 4   c. −2/3

7. a. −2/5   b. 5/7   c. −5/9

8. Calculate the difference in the *y*-coordinates, and the difference in the *x*-coordinates, and write the ratio of the two.
   a. 3/4   b. −7/5

Slope, Part 2, pp. 27-29

Page 27

Teaching box:

> Consider the angles BAC and EDF (marked with a single arc). The line segments AC and DF are _parallel_, and therefore, the angles BAC and EDF are _corresponding_ angles, thus they are congruent.
>
> Since ∠BAC = ∠EDF, and the angles BCA and EFD are equal (being right angles), this means the third angles of the triangles ABC and DEF are equal too, and the triangles are _similar_.
>
> In _similar_ triangles, corresponding side lengths are in the same ratio. More than that, the ratio of any one side length to another in *one* triangle equals the ratio of the corresponding sides in the other triangle.
>
> Slope is the ratio rise/run. If calculated using points A and B, it is the ratio $\frac{BC}{AC}$.
>
> If calculated using points D and E, it is the ratio $\frac{EF}{DF}$.
>
> What can we therefore conclude?
>
> The slope is the same, no matter what points we use to calculate it.

Page 28

Teaching box:

> Now, the slope of this line is *m*, and it is also rise/run,
>
> or using the symbols in our illustration, it is $\frac{y}{x}$.
>
> So, we can write the equation $m = \frac{y}{x}$.
>
> Solving this equation for *y*, we get *y* = __*mx*__.
>
> Since we used a generic point on a line through origin, **this equation holds true for any point on the line in the first quadrant.**
>
> In other words, in this case, $\frac{|y|}{|x|} = \frac{y}{x} = $ __*m*__
>
> And therefore, the equation *y* = *mx* still holds.

1. a. The slope is 28/20 = 7/5, so the equation is *y* = (7/5)*x*.

   b. The slope is −14/18 = −7/9, so the equation is *y* = (−7/9)*x*.

Page 29

2. The two triangles in the image are similar, so we can write the proportion 8/6 = *x*/4. From that, *x* = 4 · 8/6 = 32/6 = 16/3 = 5 1/3.

   Then, to find the coordinates of the point B, we add 4 and 16/3 to the coordinates of point A, to get (10, 15 1/3).

3. Since the line goes through the origin, we can calculate the slope just using the point (12, 10). The slope is 10/12 = 5/6.

   The point (30, *b*) must produce the same value for the slope, so we get the equation *b*/30 = 5/6, from which *b* = (5/6) · 30 = __25__.

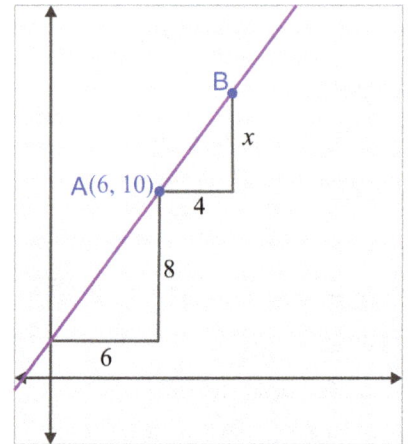

4. a. c. d. See the image on the right.

   b. The slope is 4/10 = 2/5.

   e. From the two similar triangles in the illustration, we can write the proportion 2/5 = 1/*s*. Cross-multiplying, we get 2*s* = 5, so *s* = 2.5. The *x*-coordinate of point B is therefore 5 + 2.5 = __7.5__

   Another way to solve this is to note that this line goes through the origin. Let the coordinates of B be (*x*, 3). Then, the slope calculated using B is 3/*x* which also 2/5. From this proportion, *x* = 7.5.

   f. The two triangles formed are __similar__ triangles.

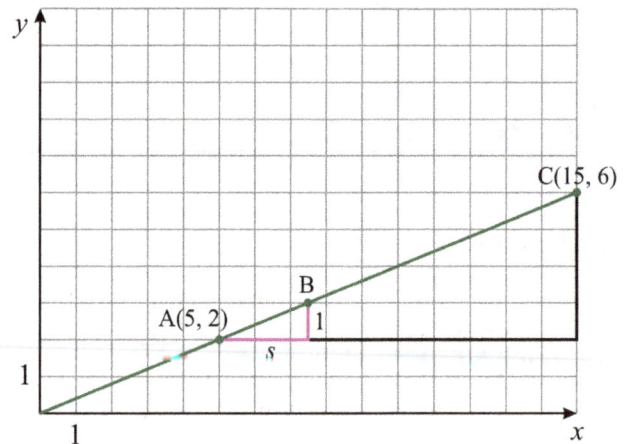

5. We can calculate the slope of the line using any two of the points, and check whether we get the same slope. If yes, the points fall on the same line, and if not, they don't.

   Using (4, 3) and (7, 5), we get slope = 2/3. Using (7, 5) and (10, 7), we get slope = 2/3. And lastly, using (4, 3) and (10, 7), we get slope = 4/6 = 2/3. So, yes, these points fall on the same line.

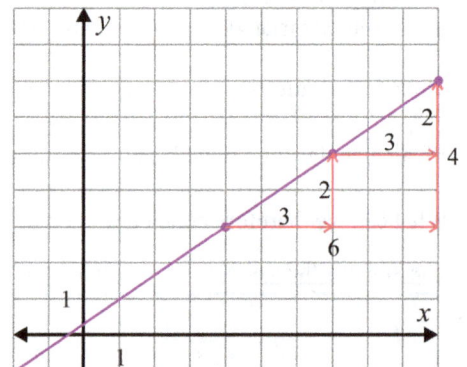

Slope, Part 2, cont.

6. a. Since the points are located symmetrically around the origin, the line goes through the origin. The slope is 4/7, and the equation is $y = (4/7)x$.

   b. Since the line goes through the origin, the point $(s, -5)$ can be used to calculate the slope: $-5/s = 4/7$, from which $4s = -35$, and <u>$s = -35/4$ or -8.75</u>.

   Alternatively, you can reason this way. The point $(s, -5)$ satisfies the equation of the line, $y = (4/7)x$. Substituting the coordinates s and -5 to the equation of the line, we get $-5 = (4/7)s$, from which, again, $s = -35/4$.

Slope, Part 3, pp. 30-34

Page 30

1. a. 1.5   b. -2 c. $-1/2$

2. The y-values change by 5 units, while the x-values change by 2 units. The slope is 5/2.

3. He changed the order of the x-values in his calculation. In reality, the change in the x-values is $-2 - (-4) = 2$ units, not -2. The slope is $-5/2$.

Page 31

4. a. It doesn't matter which points are chosen. The slope ends up being 0. Using the points in the image below for an example, the slope is $0/4 = 0$.

   b. It doesn't matter which points are chosen. The x-values do not change, and we get zero in the denominator, which means we cannot divide. There is no slope. Using the points in the image, we get 4/0.

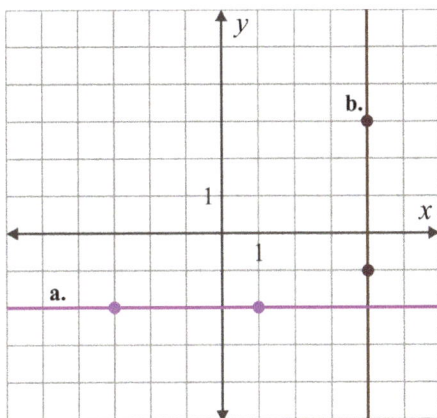

5. a. Slope: 0.
   b. Slope: -2.
   c. No slope.

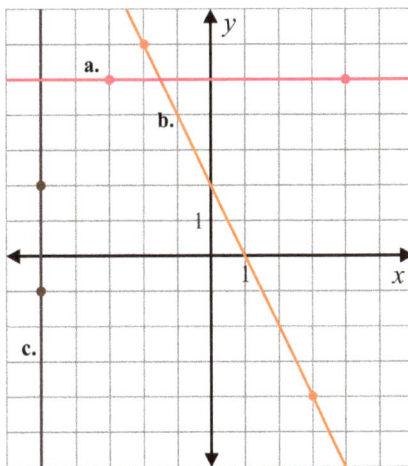

6. a. $10/15 = 2/3$ b. $-20/5 = -4$
   c. 10   d. $-10/2 = -5$

Page 32

7.

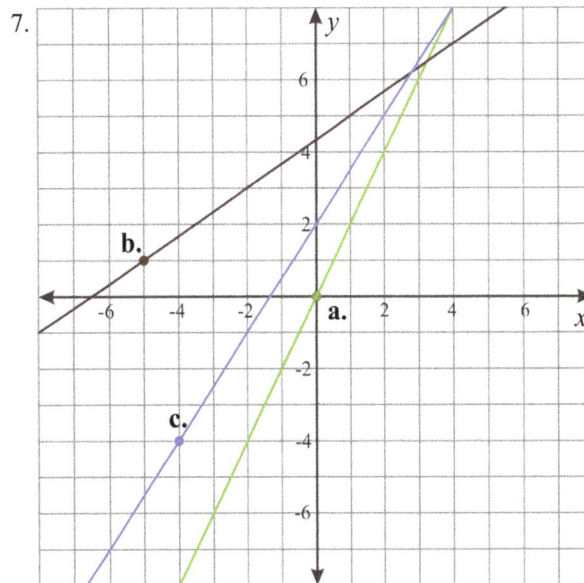

105

Page 32

8. and 9. Answers will vary. Some examples:

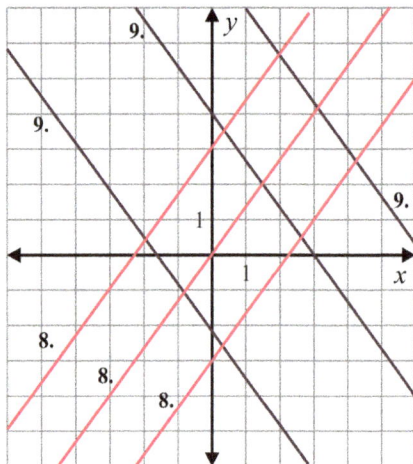

Page 33

10.

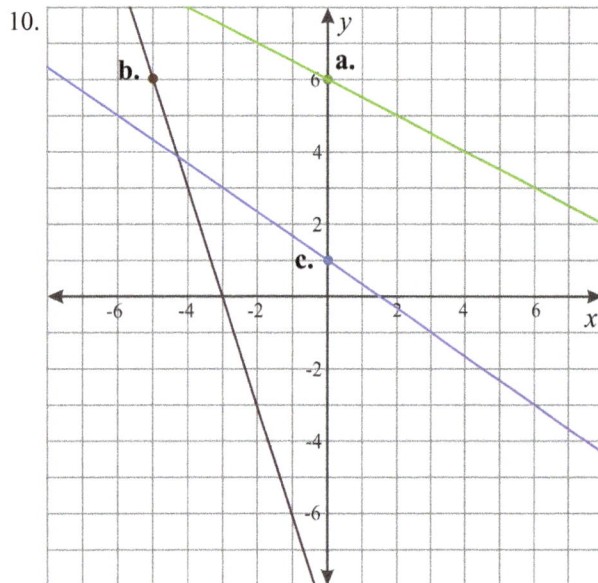

11. and 12: see the image on the right. For 11, answers will vary but the student's line should have the same steepness as in the line in this image.

13. Alice needs to look at two points to find the slope, not just one. Her method works IF the line goes through the origin, because in that case, the other point we are using is actually (0, 0). To use two points, we can use (10, 20) and for example (30, 50). The slope is 30/20 = 3/2.

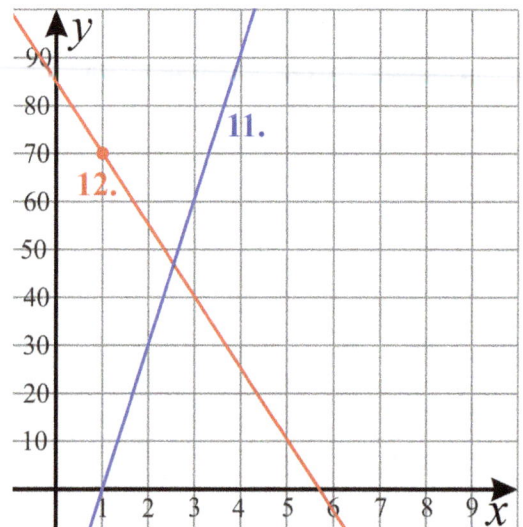

Slope, Part 3, cont.

Page 34

14. a. b. See the image below. For (c), the answers will vary. Two examples are given in the image.

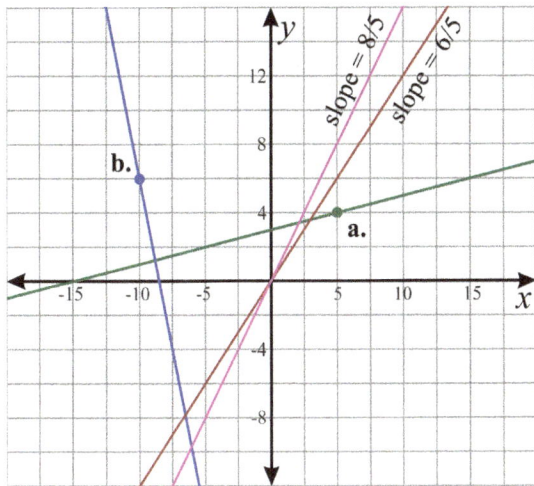

15. a. See the image on the right.

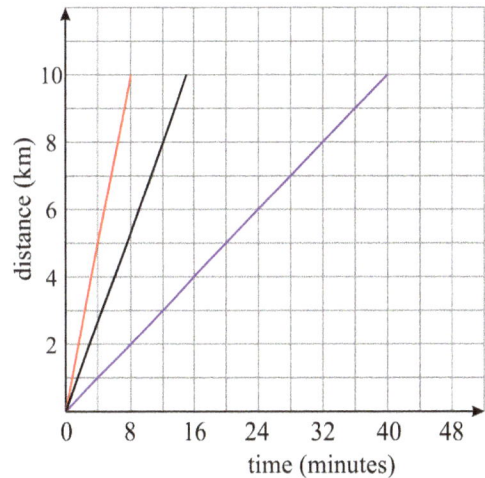

   b. The car takes (10 km)/(75 km/h) = $0.1\overline{3}$ hr = 8 minutes.
   The moped takes (10 km)/(40 km/h) = 1/4 hr = 15 minutes.
   The car is <u>7 minutes faster</u>.

   c. You can check the length of the horizontal line segment marked in the image below as the "time difference". It is about 7 minutes.

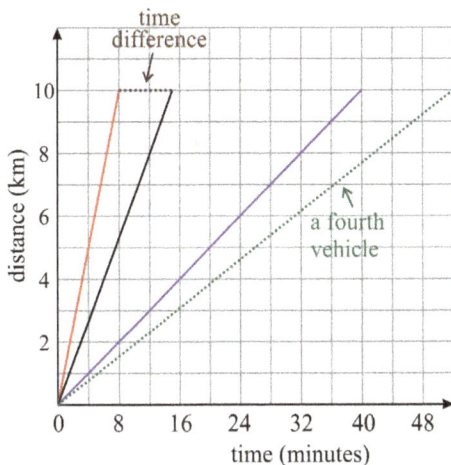

   d. Answers will vary. See the image above for an example. The slope of this fourth line should be less than of all the other three.

16. If they fall on one line, then it won't matter which two points we use to calculate the slope of that line; we will get the same result.

   Using (1, 3) and (2, 7), the slope would be 4/1 = 4.

   Using (2, 7) and (4, 18), the slope would be 11/2 = 5.5.

   Using (1, 3) and (4, 18), the slope would be 15/3 = 5.

   The points do not fall on the same line.

107

17. The slope is the ratio (difference in y-values)/(difference in x-values), and in this case, that ratio equals -3. Using these two points, we can write an equation. See its solution on the right:

So, s equals -10.

You could also use mental math and logical reasoning. The y-values have a difference of 15, and thus, for the slope to be -3, the x-values have to have a difference of 5. Thus, the point $(-10, -5)$ fits the bill.

$$\frac{-5 - 10}{s - (-15)} = -3$$

$$\frac{-15}{s + 15} = -3 \qquad \Big| \cdot (s + 15)$$

$$-15 = -3(s + 15)$$

$$-15 = -3s - 45 \qquad \Big| + 45$$

$$30 = -3s \qquad \Big| \div (-3)$$

$$-10 = s$$

Slope-Intercept Equation 1, pp. 35-38

1. a. The slope of both lines is -2.
   b. See the image on the right.
   c. At $(0, 3)$.
   d. The y-values for Line 2 are three more than the y-values for Line 1 (for the same x-values). We see that in the image in the fact that Line 2 is three units higher than Line 1. You can use a translation of 3 units up to transform Line 1 to Line 2.

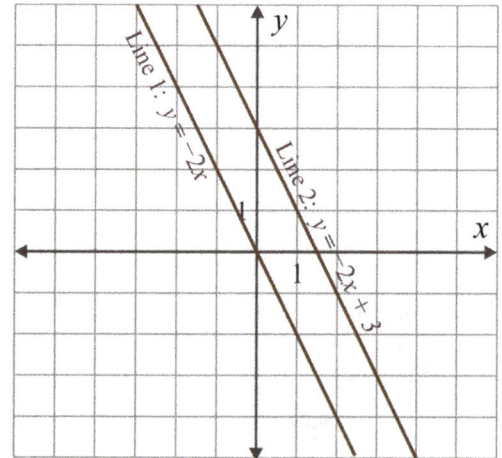

   Note that the student may say a translation 1.5 units to the right. This is geometrically correct; however, it is better to view this as a translation towards UP, because the equation of Line 2 has "+ 3" added to it as compared to Line 1, and that addition of 3 affects the y-values.
   e. $y = -2x + 3$.
   f. The y-values are two less than the corresponding y-values of Line 1. A translation of 2 units down would transform Line 1 to Line 3.

2. a. A translation of b units up, if b is positive, and a translation of $|b|$ units down if b is negative.
   b. The equation becomes $y = mx + b$.

3.

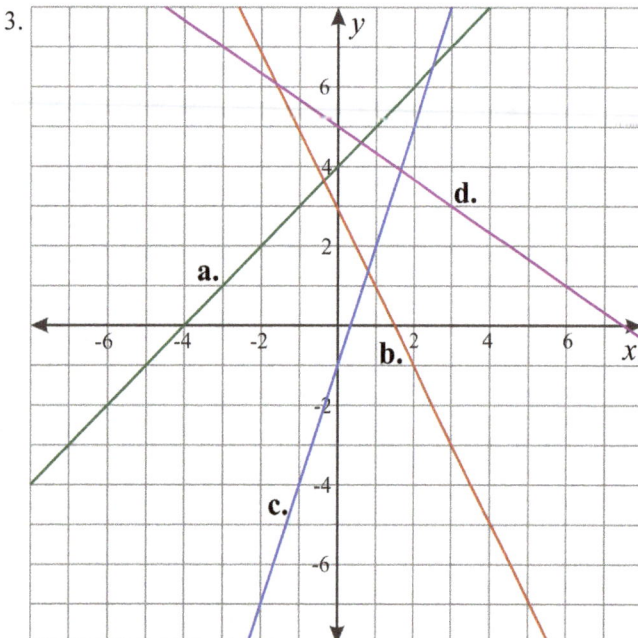

Slope-Intercept Equation 1, cont.

4. a.

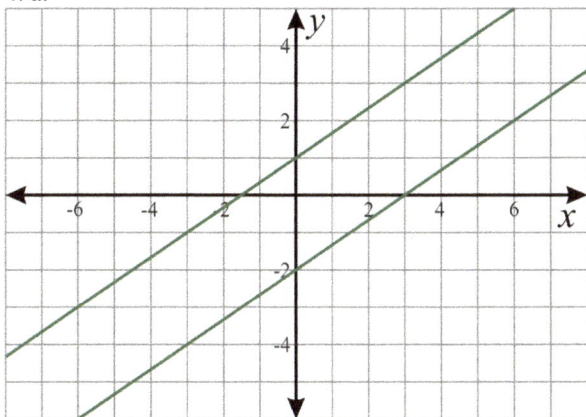

b. They differ by three units. When $x = 58$, the y-values of the two functions differ by three.

5.

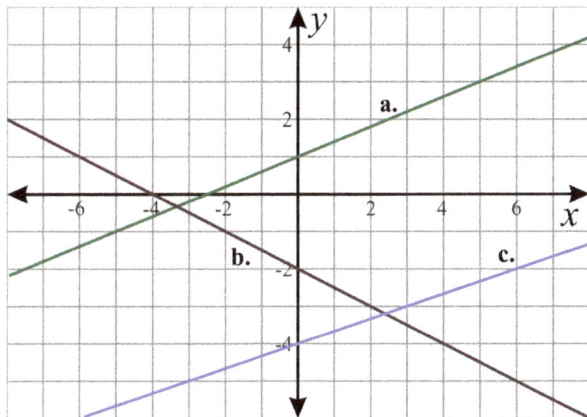

6.

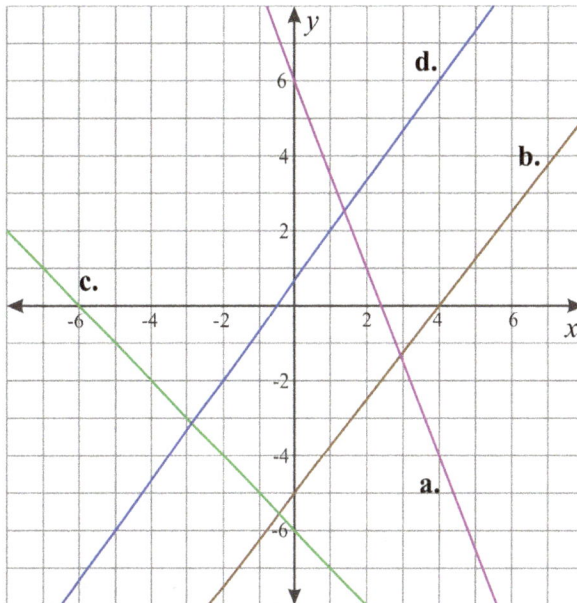

7. a. $y = (-2/5)x + 16$
   b. $y = (4/5)x + 3$
   c. $y = (1/4)x - 6$

Puzzle corner.
a. The equation $y = ax + 3$ tells us that the line goes through the point $(0, 3)$. It also goes through the point $(2, -2)$. So, we can calculate the slope as $-5/2$. The equation is then $y = (-5/2)x + 3$. This can also be solved by graphing, and determining the slope from the graph.

b. Since the slope is 2, we can calculate other points the line goes through by adding 2 to the y-values and 1 to the x-values, starting from the given point $(-6, -3)$:

x	−6	−5	−4	−3	−2	−1	0
y	−3	−1	1	3	5	7	9

We can see that the line would cross the y-axis at the point $(0, 9)$, so the y-intercept is 9. The equation is $y = 2x + 9$. One can also solve this algebraically, and in fact, this method is explained in an upcoming lesson.

Slope-Intercept Equation 2, pp. 39-41

Page 39

1. $y = (1/6)x + 1$

2. a. $y = (1/3)x + 2$ d. $y = (-2/3)x + 5$
   b. $y = (2/3)x - 3$ e. $y = -x + 2$
   c. $y = 3x$ f. $y = -2x - 1$

Page 40

3. a. slope $= (-3 - 5)/(6 - (-4)) = -8/10 = -4/5$
   b. slope $= (13 - 42)/(85 - 20) = -29/65$
   c. slope $= (2 - (-7))/(-6 - (-13)) = 9/7$
   d. slope $= (-5 - 13)/(-5 - (-20)) = -18/15 = -6/5$

4. See the graphs of the lines on the right.
   a. $y = (7/4)x + 1$ b. $y = -x + 5$

5. a. Let V be the value, and t be the time in years. This line has
   the slope of $(16,300 - 28,000)/3 = -11,700/3 = -3,900$.
   The equation of the line is therefore $V = -3900t + 28000$.

   b. At 5 years, the value of the car would be $8,500

   c. At 10 years, the value of the car would be $-$11,000,
   which is not possible. In reality, the value of a car does
   not diminish in a linear fashion, but it is a curve that tapers
   down and gets less steep as it goes.

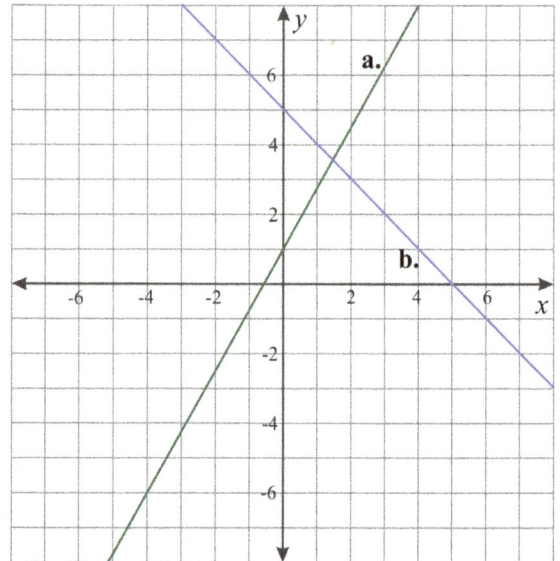

Page 41

6. a. Both are lines, but a line depicting a proportional relationship always
   goes through the origin.
   b. Yes. Its graph is a line through origin, which is a line, or linear.

7. a. A linear relationship.
   b. A proportional and linear relationship. The unit rate is 3 L/min.
   See the point for the unit rate plotted on the graph on the right.

8. a. Joe: Cost $= \$5 + 4.5m$, where m is the number of miles driven.
   Eric: cost $= 5m$.
   b. Yes. The cost for Eric's taxi service
   is a proportional relationship.

   c. See the image on the right.

   d. The lines meet at the point (10, 50).
   It means that for 10 miles, the cost
   of both services is the same: $50.

   e. For distances from 0 to _10_ miles,
   Eric's service is the better buy.

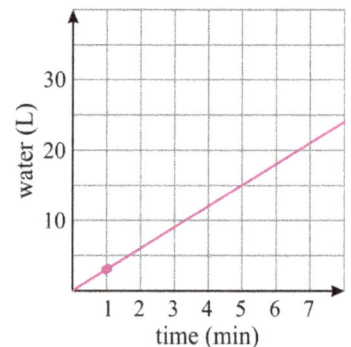

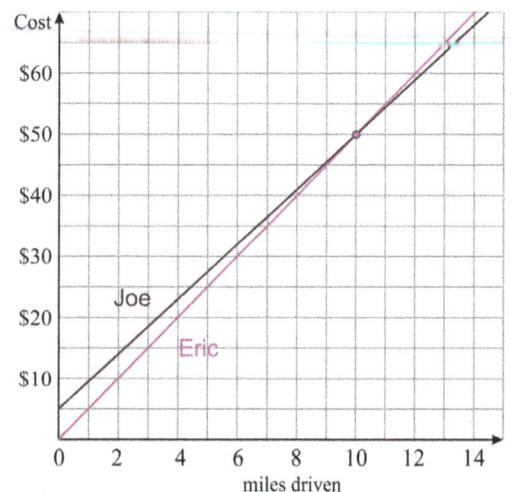

Write the Slope-Intercept Equation, pp. 42-44

Page 42

1. a. We check whether $5 = -4(1) + 1$ is a true equation. It simplifies to $5 = -3$, so, no, it isn't. So, the point does *not* fall on the line.

   b. We check whether $-8 = (-2/3)6 - 4$ is a true equation. It simplifies to $-8 = -4 - 4$, so, yes, it is. The point does fall on the line.

2. a. We substitute $x = 20$ and $y = 7$ in the equation $y = (1/2)x + b$, to get $7 = 10 + b$, from which $b = -3$. The equation of the line is $\underline{y = (1/2)x - 3}$.

   b. We substitute $x = -6$ and $y = 5$ in the equation $y = -2x + b$, to get $5 = 12 + b$, from which $b = -7$. The equation of the line is $\underline{y = -2x - 7}$.

   c. We substitute $x = 33$ and $y = 40$ in the equation $y = (1/3)x + b$, to get $40 = 11 + b$, from which $b = 29$. The equation of the line is $\underline{y = (1/3)x + 29}$.

   d. We substitute $x = -20$ and $y = -30$ in the equation $y = (-2/5)x + b$, to get $-30 = 8 + b$, from which $b = -38$. The equation of the line is $\underline{y = (-2/5)x - 38}$.

Page 43

3. a. From the graph, the slope is -1. Using the same technique as in #2, the equation of the line is $\underline{y = -x + 22.}$

   b. From the graph, the slope is $2/3$. Using the same technique as in #2, the equation of the line is $\underline{y = (2/3)x - 61/3.}$

   c. From the graph, the slope is $2/5$. The equation of the line is $\underline{y = (2/5)x + 7/5}$.

   d. From the graph, the slope is $-1/3$. The equation of the line is $\underline{y = (-1/3)x - 10}$.

4. a - (iv). b - (ii). c - (i) d. - (iii)

Page 44

5. a. This situation produces a linear function, so, the equation for the cost (C) is of the form $C = mw + b$.

   From the two points (1.6, 83.40) and (2, 97), we can calculate the rate of change (slope) to be $(97 - 83.4)/(2 - 1.6) = 34$. So, the cost equation is $C = 34w + b$ and each kilogram costs \$34.

   Then, to solve for the initial value b, we substitute (2, 97) to the equation $C = 34w + b$:

   $97 = 34 \cdot 2 + b$
   $b = 29$

   So, the fixed fee is \$29.

   This can be solved with logical reasoning also, without an equation. Once you know that each kilogram costs \$34, and that a 2-kg package cost \$97, one can subtract two times \$34 from \$97 to get the fixed fee.

   The final equation is $C = 34w + 29$ (or $C = \$29 + \$34w$).

   b. To send a 1-kg package costs $34 + \$29 = \$63$.

6. a. First, we calculate the fixed delivery fee, using the given facts. At \$6.25 per jar, the cost of three jars would be \$18.75. Louise paid \$23.50, so the delivery fee is $\$23.50 - \$18.75 = \$4.75$.

   To purchase seven jars would cost $7 \cdot \$6.25 + \$4.75 = \$48.50$.

   b. $C = 6.25t + 4.75$

7. a. $\$2710 - 13 \cdot \$140 = \$890$

   b. Total $= 140m + 890$ where m is the number of months after she started saving at the steady rate.

   c. $\$140 \cdot 36 + \$890 = \underline{\$5,930}$

Puzzle corner. Calculating the speed between the times we are given:

From 4 to 6 hours: speed = 32 km/h
From 6 to 13 hours: speed = 29.714 km/h
From 13 to 20 hours: Speed = 28 km/h

So, the change in speed must have happened between 6 and 13 hours, and the change was from 32 km/h to 28 km/h.

The question is now, how many hours did the boat travel with 32 km/h? This can be solved with guess & check.

For example, let's guess that the switch happened at 8 hours, and make a table of the distances hour-by-hour, and see if our data will match what is given, that at 13 hours the distance covered will be 400 km. Up through 8 hours, we add 32 km to the distance per hour, and after that, 28 km per hour.

This didn't work, because at 13 hours we have 396 km, not 400 km. By adjusting the guess, you will find that making the change of speed <u>at 9 hours</u> works.

Another way is to think of how many whole-number increments of 32 km and whole-number increments of 28 km will span the difference from 192 km to 400 km (208 km). Again, guess and check will help you find that $3 \cdot 32 + 4 \cdot 28$ will work.

Time (hours)	Distance (km)
6	192
7	224
8	256
9	284
10	312
11	340
12	368
13	396

Horizontal and Vertical Lines, pp. 45-47

1.

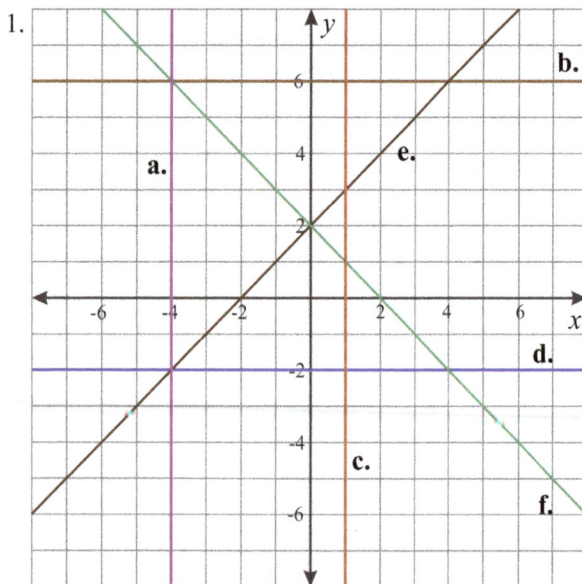

2. a. $x = -14$ b. $y = 30$

3. a. $x = -34$ b. $y = (-2/3)x + 2$ c. $y = -90$
   d. $y = (-14/5)x - 62$ e. $y = 6$ f. $x = 3$

4. The area is $90 \cdot 450 = \underline{40,500 \text{ square units}}$

Horizontal and Vertical Lines, cont.

Page 47

5. a. - (i)   b. - (iv)   c. - (ii)   d. - (iii)

6. This forms a right triangle. Its area is $4 \cdot 6/2 = 12$ square units.
   See the image on the right.

7. The horizontal line: $y = 15$. The vertical line: $x = -19$.
   The slanted line: $y = 3x + 81$

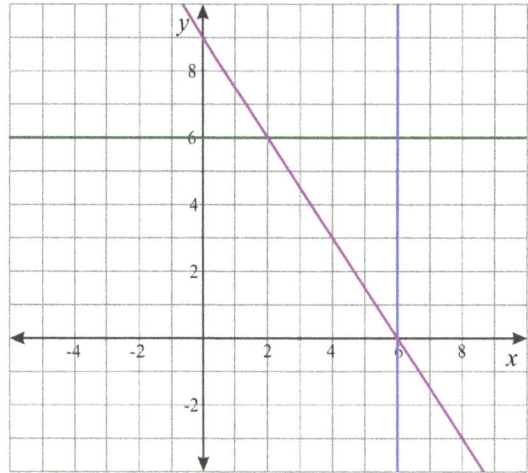

The Standard Form, pp. 48-50

Page 48

1. a. Yes.   b. No. $5x + 2y = -8$
   c. No. $x - 3y = 20$ d. Yes.
   e. Yes.   f. No. $x - 2y = 6$

2. a. $y = (1/2)x - 5/4$ b. $y = (-1/2)x - 5$
   c. $y = (-5/6)x - 1/2$ d. $y = -x + 7/3$

Page 49

3. See the graphs of the lines on the right.
   a. x-intercept: 2. y-intercept: -6

   b. x-intercept: -2. y-intercept: -5

   c. x-intercept: -3. y-intercept: 4

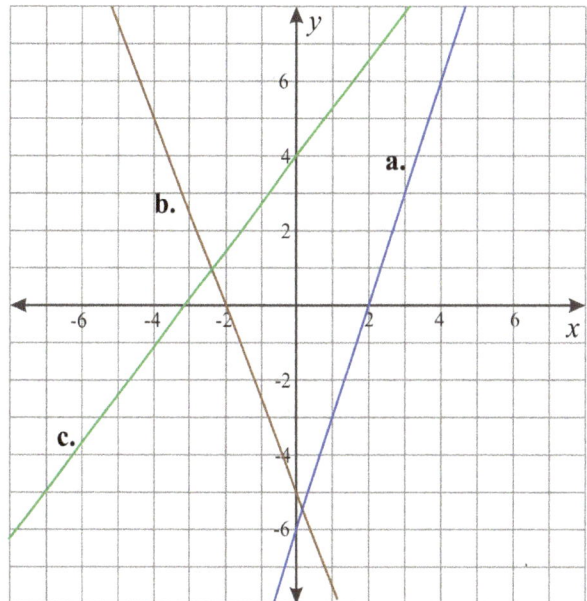

Page 50

4. a. Yes, it is in the standard form, because it is of the form
      $Ax + By = C$, with A= 0, B = 1, and C = 11.
      It is also in the slope-intercept form, with a slope of zero.

   b. Yes, it is in the standard form, because it is of the form
      $Ax + By = C$, with A= 1, B = 0, and C = -70.
      It is not in the slope-intercept form. This line has no slope.

5. a. $3x - y = -3$
   b. $x - 7y = 36$

6. See the graphs of the lines for (b) and (c) on the right.
   a. No, the point is not on the line.
   b. Yes. $3x - 2y = -12$
   c. Yes. $x - 4y = 0$

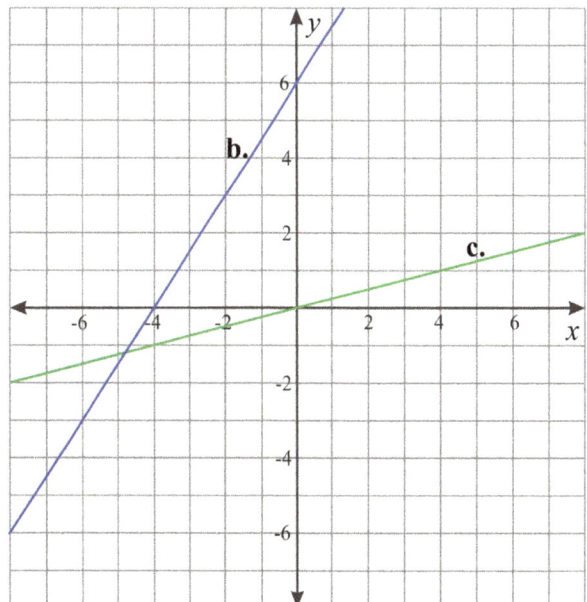

113

More Practice, pp. 51-52

Page 51

1. a. $5x - y = 3$ b. $x - 2y = -6$
   c. $2x + y = -6$ d. $8x - 15y = -20$

2. See the image on the right.

   a. x-intercept: 2. y-intercept: $-5/2$
   b. x-intercept: -2. y-intercept: -5
   c. x-intercept: -6. y-intercept: 2

Page 52

3. a. $4x - y = 18$ b. $5x + y = -1$
   c. $x = -3$ d. $y = -4$
   e. $x + 2y = 25$ f. $4x - y = -17$

4. a. $y = 2x$ b. $y = (-9/4)x + 7/4$

   c. $y = (-1/3)x - 2/9$ d. $y = -(4/3)x - 2$

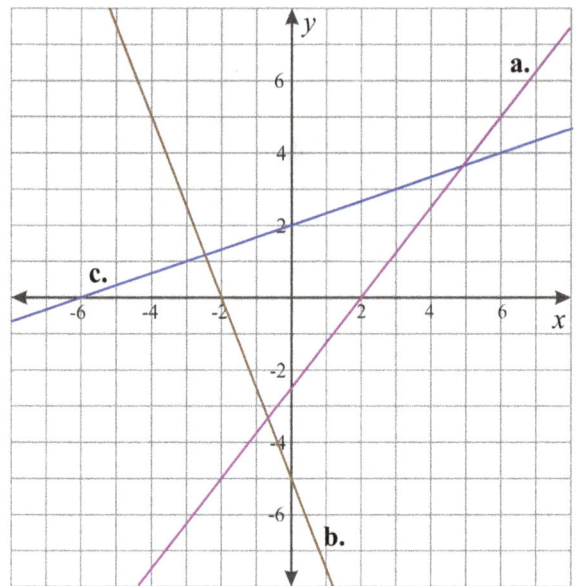

Parallel and Perpendicular Lines, pp. 53-55

Page 53

1. Lines a, d, and f are parallel. So are lines b and e. Line c is perpendicular to lines a, d, and f.

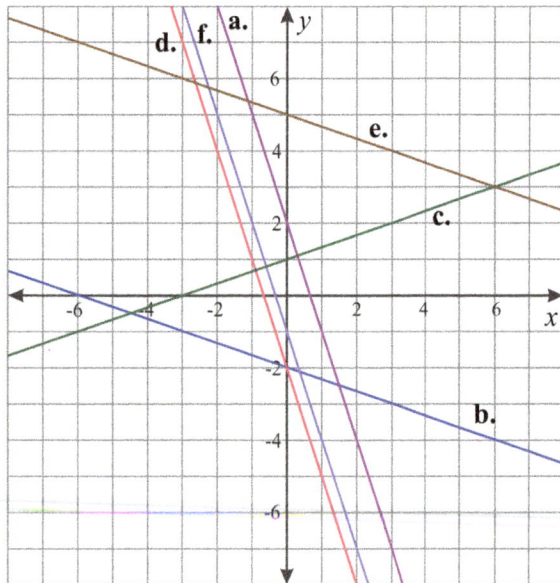

Lines that are parallel have the same slope.

Page 54

2. a. $-1/5$ b. $4/3$ c. $1/6$ d. -8

3. Lines a and c are perpendicular. Lines b and d are parallel.

4. a. Line L goes through the points $(-3, 5)$ and $(1, 2)$, so its slope is $-3/4$, and that is also the slope of line M.
   The equation of line M is therefore of the form $y = (-3/4)x + b$. To solve for b, we substitute $x = -1$ and $y = 0$
   to the equation: $0 = (-3/4)(-1) + b$, from which $b = -3/4$. The equation of Line M is therefore $y = (-3/4)x - 3/4$.

   b. The slope of Line N is $4/3$. Its equation is therefore of the form $y = (4/3) + b$. To solve for b, we substitute $x = -1$
   and $y = 0$ to the equation: $0 = (4/3)(-1) + b$, from which $b = 4/3$. The equation of Line M is therefore
   $y = (4/3)x + 4/3$.

114

Parallel and Perpendicular Lines, cont.

Page 54

5. a. The slope of the two parallel lines is $-1/4$, so the slope of the fourth line is <u>4</u>.

b.

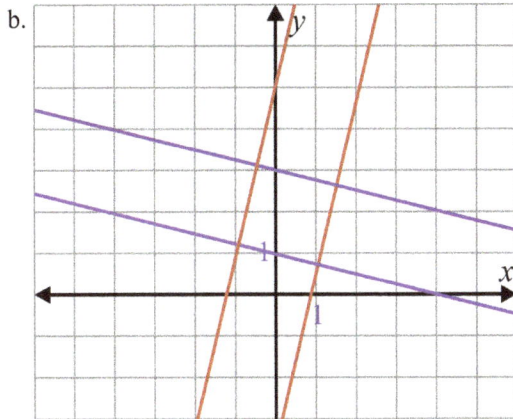

6. Yes. When we transform the equation of the first line into the slope-intercept form, we get $y = (-1/5)x + 2$. The slope of the first line $(-1/5)$ is the opposite of the reciprocal of the slope of the second line (5), so they are perpendicular.

Page 55

7. a. The slope of line t is $1/2$.

b. The rotated points are A$'(-4, 5)$ and B$'(-2, 1)$. Recall that to rotate a point, you take the absolute values of the x and y coordinates of the point, switch the two, and lastly figure out their sign based on in which quadrant the rotated points are located. So, for example, A$(-5, -4)$ becomes A$'(-4, 5)$.

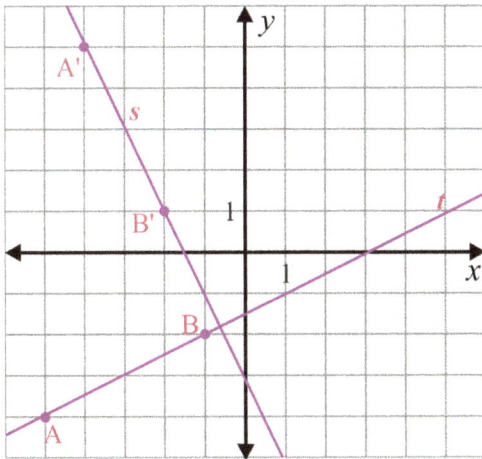

8. a. $10x + y = 5$ b. $x + y = -2$
   c. $y = 2$ d. $x + 2y = -8$
   e. $x = 4$ f. $3x - y = -12$

Page 56

1.

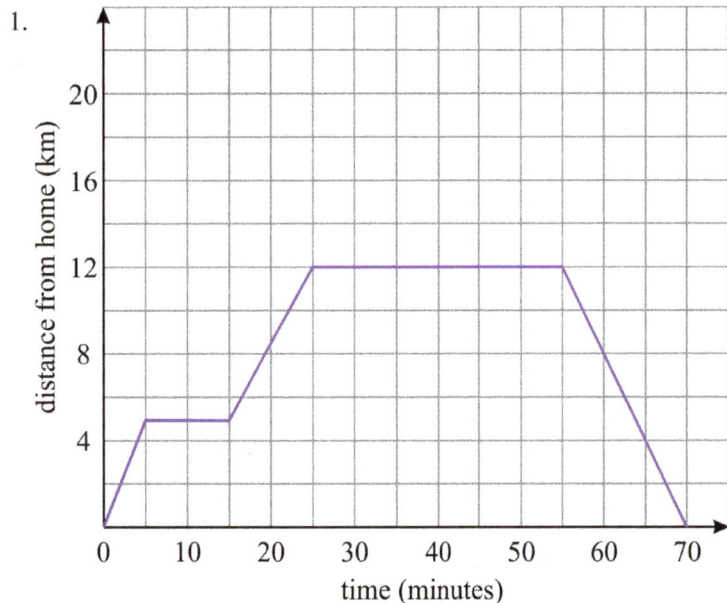

2. Answers will vary because there are many possible sequences of transformations that will map figure 3 onto figure 4. Check the student's answer. For example: Reflect Figure 3 in the vertical line at $x = 3$, and then translate it six units up and one unit to the right. Or: First, translate it six units up, then reflect it in the vertical line at $x = 3.5$.

Page 57

3. a. It has no solutions.
   b. Answers will vary; check the student's answer. For example, $3(y - 1) = 7 + 2y$ or $4(y - 1) = 7 + 3y$.
      After the change, if there is a term with y on both sides, those terms should not have the same coefficient.

4. a. Yes, they are. The two triangles have the same angles, so they are similar triangles. First, both triangles have one right angle. Secondly, the angles BCA and DCE are congruent because they are vertical angles.

   Lastly, the third angles of both triangles must be congruent. To see that, recall that the sum of the angles in any triangle is 180°. So, $\angle BAC = 180° - 90° - \angle BCA$ and $\angle EDC = 180° - 90° - \angle DCE$. But $\angle BCA$ and $\angle DCE$ are congruent, therefore $\angle BAC$ and $\angle EDC$ end up being congruent also.

   b. The side BC corresponds to the side CE = 9.6 ft in triangle CED. We can write a proportion: $x/5.3 = 9.6/10.3$, from which $x = 5.3 \cdot (9.6/10.3) \approx 4.939806$. So, $\underline{x \text{ is } 4.9 \text{ ft}}$.
      (The value of x can be found in other ways, too.)

5. a. <u>Function 3</u>. Its rate of change in the interval [1, 3] is $(6 - (-4)/2 = 5$. The rate of change of Function 2 is 2. The rate of change of Function 1 in this interval is negative.

   b. <u>Function 3</u>. In the interval [4, 5], the rate of change of Function 1 is $(2.5 - (-1.7)/1 = 4.2$. The rate of change of Function 2 is 2, and of Function 3 is 5.

6.

a. $10(x-3) + 2x - 5 = 6 - 3x$ $10x - 30 + 2x - 5 = 6 - 3x$ $12x - 35 = 6 - 3x$ $15x = 41$ $x = 41/15$	b. $\frac{1}{3}x - 5 = \frac{1}{4}x + 2$ $\Big| \cdot 12$ $4x - 60 = 3x + 24$ $x - 60 = 24$ $x = 84$
c. $\frac{1}{6}(x-7) = -\frac{7}{8}$ $\Big| \cdot 24$ $4(x-7) = -21$ $4x - 28 = -21$ $4x = 7$ $x = 7/4$	d. $\frac{5x-2}{10} - 2 = 3x$ $\Big| \cdot 10$ $(5x-2) - 20 = 30x$ $5x - 22 = 30x$ $-22 = 25x$ $x = -22/25$

7. Let M be Mary's age now, and R be Ryan's age now. Then, the equations we get are:

$$\begin{cases} M - 20 = (3/5)(R - 20) & \Big| \cdot 5 \\ M = (7/9)R & \Big| \cdot 9 \end{cases}$$

Substituting $M = (7/9)R$ into the first equation, we get:

$(7/9)R - 20 = (3/5)(R - 20)$ $\Big| \cdot 45$

$35R - 900 = 27(R - 20)$

$35R - 900 = 27R - 540$

$8R = 360$

$R = 45$

Then, $M = (7/9)R = (7/9)45 = 35$.

Solution: <u>Mary is 35 years old and Ryan is 45 years old.</u>

8. Let n be the number of nickels in her piggy bank. Then, the number of dimes is $2n$.
Since the value of her coins is $12.25, we can write the equation:

$5n + 10 \cdot 2n + 16 \cdot 25 = 1225$

$5n + 20n + 400 = 1225$

$25n + 400 = 1225$

$25n = 825$

$n = 33$

So, she has 33 nickels, 66 dimes, and 16 quarters.

Chapter 5 Review, pp. 59-62

1. a. Fridge 1. From the graph, we can see that Fridge 2 consumes 150 kWh in 5 months, which means it consumes 30 kWh/month. Fridge 1 consumes 37.5 kWh per month. So, Fridge 1 consumes <u>7.5 kWh more</u> in a month than Fridge 2.

   b. Fridge 1: $E = 37.5t$. Fridge 2: $E = 30t$.

   c. and d. See the image on the right.

2. a. The brown lines: $y = -2x$ and $y = -2x + 10$

   The green lines: $y = (1/2)x$ and $y = (1/2)x + 5/2$

   b. The easiest way is to draw an outer rectangle around it, calculate the areas of the triangles marked with 1, 2, 3, and 4, and lastly subtract the areas of those triangles from the area of the outer rectangle.

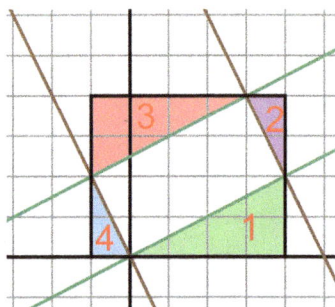

The area of the outer rectangle is 5 by 4 units = 20 square units.
The areas of triangles 1, 2, 3, and 4 are 4, 1, 4, and 1 square units, respectively.
The area of the inner rectangle is therefore $20 - 4 - 1 - 4 - 1 = $ <u>10 square units</u>.

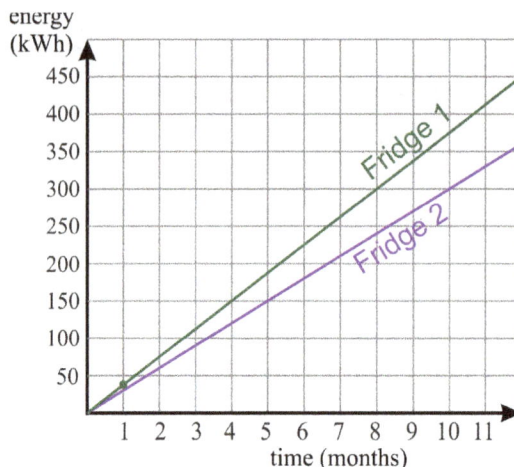

3. a. $y = (3/4)x + 9/2$ b. $y = -10$

4. a. This line passes through the points (5, 12) and (25, 3). The slope is therefore $-9/20$ or -0.45.
   b. This line passes through the points (10, 6) and (25, 9). The slope is therefore $3/15 = 1/5$.

Equations:

a. To find the y-intercept, we substitute (5, 12) into the equation $y = -9/20x + b$, and solve the resulting equation (see the solution on the right).

   The final equation is $y = -0.45x + 14.25$

$$12 = -9/20(5) + b$$
$$12 = -9/4 + b \qquad | +9/4$$
$$b = 12 + 9/4$$
$$b = 14\,\tfrac{1}{4} \text{ or } 14.25$$

b. To find the y-intercept, we substitute (10, 6) into the equation $y = (1/5)x + b$, and solve the resulting equation (see the solution on the right).

   The equation is $y = (1/5)x + 4$.

$$6 = (1/5)10 + b$$
$$6 = 2 + b$$
$$b = 4$$

5. We calculate the slope using any two of the three points, and check whether we get the same result.

   Using the first two points, the slope is $-5/2$. Using the last two points, the slope is $-4/2 = -2$.
   Using the first and the last point, the slope is $-9/4$. So, these three points do not fall on the same line.
   Encourage the student to also plot the points, and to see the slight difference in the slopes between the points.

6. Using the two known points, we can calculate the slope. It is $9/12 = 3/4$. The equation of the line is of the form $y = (3/4)x + b$. We substitute $(3, 9)$ in it: $9 = (3/4)(3) + b$. From that, $b = 27/4$. So, the equation is $y = (3/4)x + 27/4$.

Now, the point $(s, 12)$ also fulfills that equation, so we can write $12 = (3/4)s + 27/4$, and from that solve that $\underline{s = 7}$.

7. a. The equation of line S is $y = (6/5)x + 4$. The slope of line T is $-5/6$. Since it passes through $(1, 1)$, we can substitute those coordinates in the equation $y = (-5/6)x + b$ to get $1 = -5/6 + b$, from which $b = 11/6$.

So, the equation of line T is $y = (-5/6)x + 11/6$.

b. To transform this to the standard form, we multiply it by 6: $6y = -5x + 11$, from which we get $\underline{5x + 6y = 11}$.

8. It is a linear relationship. Equation: $C = 110d + 1{,}500$, where C is the cost of running the truck and d is the number of days.

9.

$y = (-4/3)x - 7$	Is parallel to $x = 9$ and passes through $(2, 7)$
$3x - y = -21$	Has y-intercept -4 and is perpendicular to $y = -2x$.
$y = -4$	Passes through $(-5, 6)$ and has slope 3.
$x - 2y = 8$	Passes through $(-9, 5)$ and $(-3, -3)$
$x = 2$	Passes through $(-3, 0)$ and $(0, 9)$
$y = 3x + 9$	Has y-intercept -4 and is parallel to $y = -2$.

10. a. $2x - y = -10$. x-intercept: -5, y-intercept: 10.
    b. $2x + 9y = -6$. x-intercept: -3, y-intercept: $-2/3$.

11. a. $T = (20/3)t - 5$, where T is temperature, and t is time in hours.
    b. 2:30 PM is 4.5 hours after 10 AM. Using our equation, $T = (20/3)(4.5) - 5 = 25°C$.

c.
$$
\begin{array}{rcl|l}
22 & = & (20/3)t - 5 & \cdot\, 3 \\
66 & = & 20t - 15 & +\, 15 \\
81 & = & 20t & \div\, 20 \\
81/20 & = & t & \\
t & = & 4.05 \text{ or } 4\ 1/20 &
\end{array}
$$

4 1/20 hours is 4 hours 3 minutes. The temperature reaches 22°C at 2:03 PM.

d. 1:45 is 3.75 hours after 10 AM. $T = (20/3)(3.75) - 5 = \underline{20°C}$

e. The equation for this changed situation is simply $T = (20/3)t - 12$. We solve the equation:

$$
\begin{array}{rcl|l}
22 & = & (20/3)t - 12 & \cdot\, 3 \\
66 & = & 20t - 36 & +\, 36 \\
102 & = & 20t & \div\, 20 \\
102/20 & = & t & \\
t & = & 5.1 \text{ or } 5\ 1/10 &
\end{array}
$$

5.1 hours is 5 hours 6 minutes. The house reaches the temperature of 22°C at 3:06 PM.

Chapter 6: Irrational Numbers and the Pythagorean Theorem

Square Roots, pp. 65-68

Page 65

1. a. 10   b. 8   c. 2   d. 0
   e. 9   f. 12   g. 1   h. 100

2. See the table →

x	x^2		x	x^2
1	1		11	121
2	4		12	144
3	9		13	169
4	16		14	196
5	25		15	225
6	36		16	256
7	49		17	289
8	64		18	324
9	81		19	361
10	100		20	400

3. a. 13   b. 30
   c. 15   d. 11
   e. 21   f. 90
   g. 18   h. 20
   i. 80   j. 160
   k. 130   l. 1000

Page 66

4. a. Between 2 and 3   b. Between 4 and 5
   c. Between 6 and 7   d. Between 9 and 10

5. a. 5 u   b. 40 u   c. $\sqrt{5}$ u   d. $\sqrt{11}$ u

6. a. 8 square units
   b. 7
   c. $\sqrt{130}$ meters

7. a. 0.4   b. 0.1   c. 1.1
   d. 4/5   e. 10/3 or 3 1/3   f. 7/6 or 1 1/6

Page 67

8. a. 8.367   b. 1.732   c. 38.079
   d. 0.671   e. 0.913   f. 2.104

9. a. 5   b. 11   c. 8
   d. 12   e. 6   f. 5

10. a. 5.191. If your calculator doesn't automatically perform the operations in order, you may need to write down the intermediate results (or enter them into the calculator's memory). If you write them down, keep at least 5 decimal digits. In other words, don't round the intermediate results to 3 decimal digits or your final answer may be off.

   b. 59.512

Page 68

11. a. 1,600 cm^2
    b. 37 sq. in.

12. a. Check the student's square. The side of the square is about $\sqrt{18}$ cm ≈ 4.2 cm.

    b. $4 \cdot \sqrt{18}$ cm ≈ 16.97 cm

13. a. Check the student's square. The side of the square is 4.5 cm.
    b. A = (4.5 cm)^2 = 20.25 cm^2

14. GIVES THEM SQUARE ROOTS

Puzzle corner: $19 = \sqrt{(2 \cdot 5 + 9 \cdot 6)} + 5 + 6$

How you enter this into a calculator depends. In some calculators, you would first calculate $(2 \cdot 5 + 9 \cdot 6) + 5 + 6$, and then press the square root button.

In others, you first press the square root button, then enter the opening parenthesis, then the rest.

Irrational Numbers, pp. 69-72

Page 69

1. The intermediate guesses by the student(s) will vary. For example:

Low Guess	(LG)^2	(HG)^2	High Guess
4.35	18.9225	19.0096	4.36
4.357	18.983449	18.992164	4.358
4.358	18.992164	19.000881	4.359
4.3587	18.99826569	18.99913744	4.3588

From this we can see that to three decimal digits, $\sqrt{19}$ is 4.359.

Irrational Numbers, cont.

2. The intermediate guesses by the student(s) will vary. The tables are just showing one possibility.

a. $\sqrt{7} \approx 2.65$

Low Guess	(LG)^2	(HG)^2	High Guess
2.6	6.76	7.29	2.7
2.64	6.9696	7.0225	2.65
2.645	6.996025	7.0225	2.65

Here is also an explanation in words of how this process could possibly go.
First we find two consecutive perfect squares so that 7 is between them: $4 < 7 < 9$.
From that fact we know that $2 < \sqrt{7} < 3$. Since 7 is closer to 9 than to 4.
Let's guess that $\sqrt{7} = 2.6$ and check: $2.6^2 = 6.76$ Too small.
Let's guess bigger: $2.7^2 = 7.29$

The above guesses show us that $\sqrt{7}$ is between 2.6 and 2.7.
 Now let's guess what the second decimal digit might be: $2.65^2 = 7.0225$ Too big.
 Let's guess smaller: $2.64^2 = 6.9696$ So $\sqrt{7}$ is between 2.64 and 2.65. Now we
 just need to know whether it would be rounded to 2.64 or 2.65. $2.645^2 = 6.996025$

This shows us that $\sqrt{7} > 2.645$, so when rounding to two decimal digits, $\sqrt{7} \approx 2.65$.

b. $\sqrt{51} \approx 7.14$

Low Guess	(LG)^2	(HG)^2	High Guess
7.1	50.41	51.84	7.2
7.14	50.9796	51.1225	7.15
7.141	50.993881	51.051025	7.145

Or, in words:

First we find two consecutive perfect squares so that 51 is between them: $49 < 51 < 64$.
From that fact we know that $7 < \sqrt{51} < 8$. Also, since 51 is much closer to 49 than to 64, $\sqrt{51}$ is much
closer to 7 than to 8. Let's first guess that $\sqrt{51} = 7.1$ and go on from there: $7.1^2 = 50.41$
Too small. Let's guess bigger. $7.2^2 = 51.84$
So $\sqrt{51}$ is between 7.1 and 7.2. Also, it is closer to 7.1 than to 7.2,
because 50.41 is closer to 51 than 51.84 is. $7.13^2 = 50.8369$
Too small. Let's guess bigger. $7.14^2 = 50.9796$
Still too small. Let's guess bigger. $7.15^2 = 51.1225$

So $\sqrt{51}$ is between 7.14 and 7.15. Now we just need to know whether it should be rounded to 7.14 or 7.15.

$7.145^2 = 51.051025$

This shows us that $\sqrt{51} < 7.145$, so when rounding to two decimal digits, $\sqrt{51} \approx 7.14$.

c. $\sqrt{99} \approx 9.95$

Low Guess	(LG)^2	(HG)^2	High Guess
9.9	98.01	99.0025	9.95
9.94	98.8036	99.0025	9.95
9.945	98.903025	99.0025	9.95

Irrational Numbers, cont.

3. No, she is not correct. If we square 3.317, we get $3.317^2 = 11.002489$. This is not exactly 11.

4. If we square 71/50, we will get the fraction $(71^2/50^2) = (71^2/2,500)$. Since 71^2 does not equal 5,000, the simplified form of this fraction does not equal 2. Thus, the square root of 2 cannot equal 71/50.

5. a. Rational; it can be written as the fraction 928/1000.
   b. Irrational, since 128 is not a perfect square.
   c. Rational, since it is a repeating decimal.
   d. Rational, since it is a terminating decimal.
   e. Irrational. Pi is irrational, and an irrational number divided by a rational number is irrational.
   f. Rational; this equals 10.
   g. Rational; it is a repeating decimal.
   h. Irrational. $\sqrt{3}$ is irrational, and an irrational number multiplied by a rational number is irrational.
   i. Irrational. $\sqrt{15}$ is irrational, and an irrational number divided by a rational number is irrational.
   j. Rational, as it equals 5/2.

6. a. Incorrect. 1.272727 is indeed rational, but the reason it is rational is because the decimal expansion terminates. As a fraction, this is 1,272,727/1,000,000.
   b. Incorrect. $\sqrt{49} = 7$ so the entire expression equals 21 and is rational.
   c. Incorrect. $\pi/3$ is irrational, because an irrational number (π) divided by a rational number (3) is irrational.
   d. Correct.

7.

Puzzle corner. Proof. Let x be an irrational and r a rational number. We need to prove that x/r is irrational. Suppose the contrary, that $x/r = s$ is a rational number. Then, $x = rs$, where both r and s are rational, thus are fractions. When two fractions are multiplied, the result is a fraction, or a rational number. So, x would be rational. This is a contradiction. Thus, the original statement is true: x/r is irrational.

Cube Roots and Approximations of Irrational Numbers, pp. 73-76

1. a. 3   b. 5   c. 4   d. 10
   e. 1   f. 6   g. 30   h. −2
   i. −1   j. −5   k. 0   l. −20

2. a. 6 cm
   b. 4
   c. 125,000 cm^3
   d. Its edge is 9 in. The surface area is
      $6 \cdot (9 \text{ in})^2 = 486 \text{ in}^2$.

3. a. 0.2   b. 0.5   c. −0.3
   d. 2/5   e. 4/3   f. −1/2

Page 74

4.

a. $5 < \sqrt{31} < 6$	b. $8 < \sqrt{65} < 9$	c. $9 < \sqrt{87} < 10$
d. $-3 < -\sqrt{5} < -2$	e. $-7 < -\sqrt{44} < -6$	f. $-8 < -\sqrt{50} < -7$
g. $1 < \sqrt[3]{7} < 2$	h. $3 < \sqrt[3]{37} < 4$	i. $4 < \sqrt[3]{101} < 5$

5.

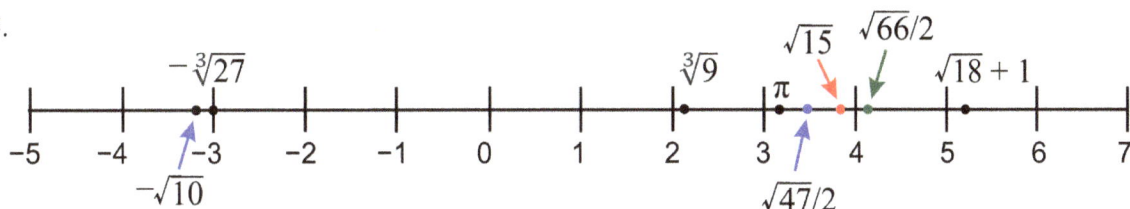

6.

a. $5 < \sqrt{27}$	b. $\sqrt{48} < 7$	c. $\sqrt{18} > 4$	d. $\sqrt[3]{9} > 2$
e. $2 < \sqrt{2} + 1$	f. $\sqrt{32} + 1 > 6$	g. $\sqrt{43} + 5 > 10$	h. $\sqrt{88} - 3 < 7$

7. a. $\sqrt{30}$ lies between 5 and 6, and $\sqrt{60}$ between 7 and 8.

   b. Therefore, $2\sqrt{30}$ is between 10 and 12, whereas $\sqrt{2 \cdot 30} = \sqrt{60}$ is between 7 and 8. So, $2\sqrt{30}$ is not equal to $\sqrt{2 \cdot 30}$.

Page 75

8. No, it is not. $\sqrt{50}$ is a little over 7, so $\dfrac{\sqrt{50}}{2}$ is a little over 3.5. On the other hand, $\sqrt{\dfrac{50}{2}} = \sqrt{25} = 5$. So, they are not equal.

9. a. 7.1   b. 9.9   c. 0.8   d. −2.6

10. a. The intermediate guesses by the student(s) will vary. The table is just showing one example.

$\sqrt{11} \approx 3.3$

Low Guess	$(LG)^2$	$(HG)^2$	High Guess
3.2	10.24	11.56	3.4
3.3	10.89	11.2225	3.35
3.31	10.9561	11.0224	3.32

10. b. $\sqrt{11} - \sqrt{2} \approx 3.3 - 1.4 = \underline{1.9}$. $3\sqrt{11} \approx \underline{9.9}$

11. She would take numbers between 6.7 and 6.8 and square those. A good starting point is the midpoint, 6.75. Since $6.75^2 = 45.5625$, a good point for the next guess is something quite a bit less than 6.75, such as 6.72. Let's continue:

$\sqrt{45} \approx 6.71$

Low Guess	$(LG)^2$	$(HG)^2$	High Guess
6.72	45.1584	45.5625	6.75
6.705	44.957025	45.0241	6.71

Page 76

12. Below each irrational number is an approximation for it.

$\sqrt[3]{1}$	$<$	$\sqrt{5} - 1$	$<$	$\sqrt[3]{8}$	$<$	$\sqrt{19}/2$	$<$	$\sqrt{9}$	$<$	$\sqrt{13}$	$<$	$\sqrt[3]{100}$	$<$	$\sqrt{22} + 1$	$<$	2π
1		between 1 and 2		2		between 2 and 3		3		between 3 and 4		between 4 and 5		between 5 and 6		6.28

Cube Roots and Approximations of Irrational Numbers, cont.

13.

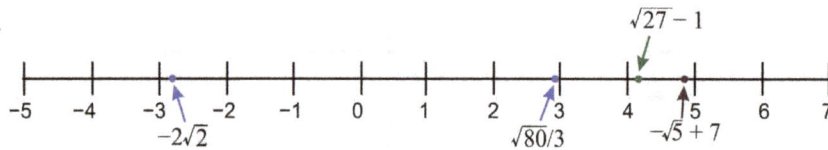

14.

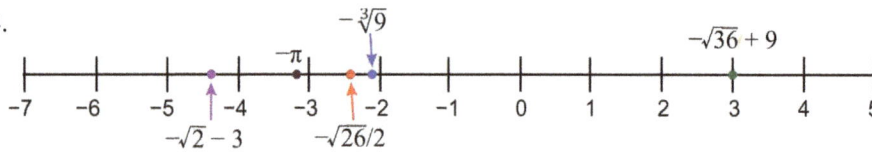

Fractions to Decimals, pp. 77-78

1. a. 0.02   b. 0.0278   c. 0.055073
   d. 4.508   e. 563.3   f. 4.0309

2. a. 0.4   b. 0.96   c. 0.27
   d. 1.75   e. 1.32   f. 0.056

3. a. $0.\overline{5}$ b. $18.8\overline{148}$ c. 0.37705

Decimals to Fractions, pp. 79-81

1. a. 9382/100 = 4691/50   b. 333/1000   c. 205,056/100,000 = 6408/3125
   d. 61,098/1000 = 30,549/500   e. 45/10,000,000 = 9/2,000,000   f. 4,932,048/1,000,000 = 308,253/62,500

2. No. Their difference is $0.000\overline{2}$.

3.

a. $10x = 4.444\ldots$	b. $10x = 2.11111\ldots$
$-\ x = 0.444\ldots$	$-\ x = 0.21111\ldots$
$9x = 4$	$9x = 1.9$
$x = \underline{4/9}$	$x = 1.9/9 = \underline{19/90}$
c. $1000x = 954.954954\ldots$	d. $100x = 253.2323232\ldots$
$-\ x = 0.954954\ldots$	$-\ x = 2.5323232\ldots$
$999x = 954$	$99x = 250.7$
$x = 954/999 = \underline{106/111}$	$x = 250.7/99 = \underline{2507/990}$

4.

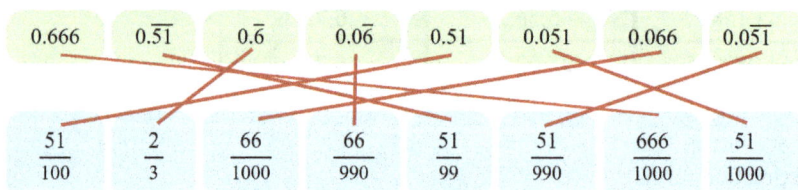

| 0.666 | $0.\overline{51}$ | $0.\overline{6}$ | $0.0\overline{6}$ | 0.51 | 0.051 | 0.066 | $0.0\overline{51}$ |

| $\frac{51}{100}$ | $\frac{2}{3}$ | $\frac{66}{1000}$ | $\frac{66}{990}$ | $\frac{51}{99}$ | $\frac{51}{990}$ | $\frac{666}{1000}$ | $\frac{51}{1000}$ |

Decimals to Fractions, cont.

Page 81

5. Both are correct. Erica's method is better since it provides a smaller numerator and denominator. Eric's answer is correct but would need simplified.

6. No. It is a fraction so it is rational. The decimal expansion of 305/55494 has 1541 repeating digits, starting at the hundredths place, and the calculator is not able to show that many.

7.

a. $1000x = 256.256256...$ $\underline{\quad -x = \quad\;\; 0.256256...}$ $999x = 256$ $\quad\quad x = 256/999$	b. $100x = 305.94949494...$ $\underline{\;-x = \quad\quad 3.05949494...}$ $99x = 302.89$ $\quad x = 302.89/99 = \underline{30,289/9900}$
c. $10,000x = 321.99\overline{2199}$ $\underline{-\quad\;\; x = \quad\; 0.03\overline{2199}}$ $9999x = 321.96$ $\quad\quad x = 321.96/9999$ $\quad\quad\quad = 32,196/999,900$ The student is not required to simplify this fraction, but it does simplify to 2,683/83,325.	d. $100,000x = 136,309.\overline{36309}$ $\underline{-\quad\quad\; x = \quad\quad\quad 1.36309}$ $99,999x = 136,308$ $\quad\quad x = 136,308/99,999$ $\quad\quad\quad = 45,436/33,333$

Square and Cube Roots as Solutions to Equations, pp. 82-84

Page 82

1.

a.	$x^2 = 25$ $x = 5$ or $x = -5$	b.	$y^2 = 3,600$ $y = 60$ or $y = -60$
c.	$x^2 = 500$ $x = \sqrt{500} \approx 22.36$ or $x = -\sqrt{500} \approx -22.36$	d.	$z^2 = 11$ $z = \sqrt{11} \approx 3.32$ or $z = -\sqrt{11} \approx -3.32$
e.	$w^2 = 287$ $w = \sqrt{287} \approx 16.94$ or $w = -\sqrt{287} \approx -16.94$	f.	$q^2 = 1,000,000$ $q = 1,000$ or $q = -1,000$

Page 83

2. a. $x = 4$ b. $n = 6$ c. $z = 30$
   d. $x = 1.91$ e. $b = 4.78$ f. $a = 2.62$

3. a. $s = 8.0$ m   b. $s = 29.0$ cm   c. $s = 1.80$ ft

4. The edge of that cube is about 2.4216 meters (we will keep a few extra decimals in this intermediate result). Its surface area is $6 \cdot (2.4216 \text{ m})^2 = 35.18487936 \text{ m}^2$. This last result needs rounded to three significant digits, since the volume was given with three. So, the surface are is 35.2 m^2.

Square and Cube Roots as Solutions to Equations, cont.

__Page 84__

5.

a. $a^2 - 8 = 37$ $a^2 = 45$ $a = \sqrt{45}$ or $a = -\sqrt{45}$ Check: $(\sqrt{45})^2 - 8 \overset{?}{=} 37$ $45 - 8 \overset{?}{=} 37$ $37 = 37$ ✓	b. $y^2 + 100 = 1{,}000$ $y^2 = 900$ $y = 30$ or $y = -30$ Check: $30^2 + 100 \overset{?}{=} 1{,}000$ $900 + 100 \overset{?}{=} 1{,}000$ $1{,}000 = 1{,}000$ ✓
c. $b^2 + 1.5 = 6.4$ $b^2 = 4.9$ $b = \sqrt{4.9}$ or $a = -\sqrt{4.9}$ Check: $(\sqrt{4.9})^2 + 1.5 \overset{?}{=} 6.4$ $4.9 + 1.5 = 6.4$ ✓	d. $x^2 - 26 = 709$ $x^2 = 735$ $x = \sqrt{735}$ or $y = -\sqrt{735}$ Check: $(\sqrt{735})^2 - 26 \overset{?}{=} 709$ $735 - 26 = 709$ ✓

6.

a. $x^3 - 5 = 59$ $x^3 = 64$ $x = 4$ Check: $4^3 - 5 \overset{?}{=} 59$ $64 - 5 = 59$ ✓	b. $x^3 + 78 = 437$ $x^3 = 359$ $x = \sqrt[3]{359}$ Check: $(\sqrt[3]{359})^3 + 78 \overset{?}{=} 437$ $359 + 78 = 437$ ✓

126

Page 85

1.

a. $\quad 5x^2 = 125$ $\quad\quad x^2 = 25$ $\quad\quad\ x = 5$ $\quad\ \text{or } x = -5$ Check: $\quad 5 \cdot 5^2 \overset{?}{=} 125$ $\quad\quad\ 5 \cdot 25 \overset{?}{=} 125$ $\quad\quad\quad\ 125 = 125\ \checkmark$	b. $\quad 8.2b^2 = 319$ $\quad\quad b^2 = 319/8.2$ $\quad\quad\ b = \sqrt{319/8.2} \approx 6.237$ $\quad\ \text{or } b = -\sqrt{319/8.2} \approx -6.237$ Check: $\ 8.2 \cdot 6.237^2 \overset{?}{=} 319$ $\quad 8.2 \cdot 38.900169 \overset{?}{=} 319$ $\quad\quad 318.9813858 \approx 319\ \checkmark$
c. $\quad a^2 + 4.5 = 10.7$ $\quad\quad\quad a^2 = 6.2$ $\quad\quad\quad\ a = \sqrt{6.2} \approx 2.490$ $\quad\ \text{or } a = -\sqrt{6.2} \approx -2.490$ Check: $(\sqrt{6.2})^2 + 4.5 \overset{?}{=} 10.7$ $\quad\quad 6.2 + 4.5 \overset{?}{=} 10.7$ $\quad\quad\quad\quad 10.7 = 10.7\ \checkmark$	d. $\quad 12b^2 = 36{,}000$ $\quad\quad b^2 = 3{,}000$ $\quad\quad\ b = \sqrt{3{,}000} \approx 54.772$ $\quad\ \text{or } b = -\sqrt{3{,}000} \approx -54.772$ Check: $12 \cdot (\sqrt{3{,}000})^2 \overset{?}{=} 36{,}000$ $\quad\quad 12 \cdot 3{,}000 \overset{?}{=} 36{,}000$ $\quad\quad\quad 36{,}000 \approx 36{,}000\ \checkmark$

Page 86

2.

a. $\quad a^2 + 3^2 = 7^2$ $\quad\quad a^2 + 9 = 49$ $\quad\quad\quad a^2 = 40$ $\quad\quad\quad\ a = \sqrt{40}$ $\quad\ \text{or } a = -\sqrt{40}$ Check: $(\sqrt{40})^2 + 3^2 \overset{?}{=} 7^2$ $\quad\quad 40 + 9 \overset{?}{=} 49$ $\quad\quad\quad 49 = 49\ \checkmark$	b. $\quad 43^2 + x^2 = 51^2$ $\quad\quad\quad x^2 = 51^2 - 43^2$ $\quad\quad\quad x^2 = 752$ $\quad\quad\quad\ x = \sqrt{752}$ $\quad\ \text{or } x = -\sqrt{752}$ Check: $43^2 + (\sqrt{752})^2 \overset{?}{=} 51^2$ $\quad 1{,}849 + 752 \overset{?}{=} 2{,}601$ $\quad\quad\ 2{,}601 = 2{,}601\ \checkmark$

More Equations that Involve Roots, cont.

3.

a. $\quad s^2 = 2.1^2 + 5.4^2$ $\quad\quad s^2 = 33.57$ $\quad\quad s = \sqrt{33.57} \approx 5.79$ $\quad$ or $s \approx -5.79$ Check: $\quad 5.79^2 \overset{?}{=} 2.1^2 + 5.4^2$ $\quad 33.5241 \approx 33.57$ ✓	b. $\quad 21^2 - w^2 = 15^2$ $\quad\quad -w^2 = 15^2 - 21^2$ $\quad\quad w^2 = 216$ $\quad\quad w = \sqrt{216} \approx 14.70$ $\quad$ or $w = -14.70$ Check: $\quad 21^2 - 14.70^2 \overset{?}{=} 15^2$ $\quad 224.91 \approx 225$ ✓
c. $\quad 121^2 - x^2 = 56$ $\quad\quad -x^2 = 56 - 14{,}641$ $\quad\quad x^2 = 14{,}585$ $\quad\quad x = \sqrt{14{,}585} \approx 120.77$ $\quad$ or $x = -120.77$ Check: $121^2 - 120.77^2 \overset{?}{=} 56$ $\quad 55.6071 \approx 56$ ✓	d. $\quad a^2 - 4.5^2 = 5.78$ $\quad\quad a^2 = 5.78 + 20.25$ $\quad\quad a^2 = 26.03$ $\quad\quad a = \sqrt{26.03} \approx 5.10$ $\quad$ or $a = -5.10$ Check: $\quad 5.10^2 - 4.5^2 \overset{?}{=} 5.78$ $\quad 5.76 \approx 5.78$ ✓

4.

a. $\quad 56 + s^3 = 542$ $\quad\quad s^3 = 486$ $\quad\quad s = \sqrt[3]{486} \approx 7.862$	b. $\quad 5x^3 = 180$ $\quad\quad x^3 = 36$ $\quad\quad x = \sqrt[3]{36} \approx 3.302$	c. $\quad 254 - z^3 = 46$ $\quad\quad -z^3 = -208$ $\quad\quad z = \sqrt[3]{208} \approx 5.925$

5.

a. $\quad x^3 = -27$ $\quad\quad x = -3$	b. $\quad w^3 = -343$ $\quad\quad w = -7$	c. $\quad 4t^3 = -4$ $\quad\quad t^3 - -1$ $\quad\quad t = -1$
d. $\quad 5x^3 + 3 = -27$ $\quad\quad 5x^3 = -30$ $\quad\quad x^3 = -6$ $\quad\quad x = -\sqrt[3]{6}$	e. $\quad -10r^3 = 10{,}000$ $\quad\quad r^3 = -1{,}000$ $\quad\quad r = -10$	f. $\quad 54 - x^3 = -1$ $\quad\quad -x^3 = -55$ $\quad\quad x = \sqrt[3]{55}$

Page 87

6.

a. $45 - x^2 = 20$	b. $112^2 + s^2 = 18{,}200$	c. $6{,}650 - y^2 = 70^2$
$-x^2 = -25$	$s^2 = 18{,}200 - 112^2$	$6{,}650 - 70^2 = y^2$
$x^2 = 25$	$s^2 = 5{,}656$	$1{,}750 = y^2$
$x = 5$	$s = \sqrt{5{,}656} \approx 75.206$	$y^2 = 1{,}750$
or $x = -5$	or $s = -\sqrt{5{,}656} \approx -75.206$	$y = \sqrt{1{,}750} \approx 41.833$
Check:	Check:	or $y = -\sqrt{1{,}750} \approx -41.833$
$45 - 5^2 \overset{?}{=} 20$	$112^2 + (\sqrt{5{,}656})^2 \overset{?}{=} 18{,}200$	Check:
$45 - 25 \overset{?}{=} 20$	$12{,}544 + 5{,}656 \overset{?}{=} 18{,}200$	$6{,}650 - (\sqrt{1{,}750})^2 \overset{?}{=} 70^2$
$20 = 20$ ✓	$18{,}200 = 18{,}200$ ✓	$6{,}650 - 1{,}750 \overset{?}{=} 4{,}900$
		$4{,}900 = 4{,}900$ ✓

Puzzle corner: solve $x^2 - x = 0$. You can use guess and check: Zero fulfills the equation because $0^2 - 0 = 0$. One is also a solution because $1^2 - 1 = 0$.

A way to see the solutions without guessing is to write the equation in the form $x(x - 1) = 0$. The product of x and $x - 1$ can only be zero if either x is zero or $x - 1$ is zero, which means either $x = 0$ or $x = 1$. From this form of the equation we can also see that there are no other solutions.

(We also know that there are no other solutions because of this principle of algebra: an equation where the highest exponent of the variable is n can have at most n solutions within the real numbers. Therefore, our equation, which has 2 as the highest exponent of the variable, can have at most two solutions within the real numbers.)

So the solution is: $x = 0$ or $x = 1$.

The Pythagorean Theorem, pp. 88-92

Page 88

1. $3^2 + 4^2 \overset{?}{=} 5^2$

   $9 + 16 \overset{?}{=} 25$

   $25 = 25$

2. Check the student's triangle, so that it has one right angle and the correct measurements. It should have the same shape as this one:

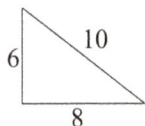

   Do the lengths 6, 8, and 10 fulfill the Pythagorean Theorem? <u>Yes</u>. See the calculation below:

   $6^2 + 8^2 \overset{?}{=} 10^2$

   $36 + 64 \overset{?}{=} 100$

   $100 = 100$

Page 89

3. Since we are solving for a length of side, we ignore the negative roots in these equations.

a. $5^2 + 3^2 = y^2$	b. $x^2 + 7^2 = 17^2$
$25 + 9 = y^2$	$x^2 + 49 = 289$
$34 = y^2$	$x^2 = 240$
$y = \sqrt{34}$	$x = \sqrt{240}$
c. $5^2 + 6^2 = s^2$	d. $w^2 + 11^2 = 12^2$
$25 + 36 = s^2$	$w^2 + 121 = 144$
$61 = s^2$	$w^2 = 23$
$s = \sqrt{61}$	$w = \sqrt{23}$

4. Let d be the length of the diagonal. The diagonal divides the square into two right triangles, each having sides 6, 6, and d. Using the Pythagorean Theorem, we get:

   $6^2 + 6^2 = d^2$, from which $d = \sqrt{72} \approx 8.49$. The diagonal measures <u>8.49 units</u>.

The Pythagorean Theorem, cont.

Page 90

5. Since we are solving for a length of side, we ignore the negative roots in these equations.

a. $\quad r^2 + 7^2 = (\sqrt{113})^2$ $r^2 + 49 = 113$ $r^2 = 64$ $r = 8$	b. $\quad x^2 + (\sqrt{52})^2 = 9^2$ $x^2 + 52 = 81$ $x^2 = 29$ $x = \sqrt{29}$
c. $\quad s^2 + (\sqrt{24})^2 = 8^2$ $s^2 + 24 = 64$ $s^2 = 40$ $s = \sqrt{40}$	d. $\quad x^2 + x^2 = (\sqrt{20})^2$ $2x^2 = 20$ $x^2 = 10$ $x = \sqrt{10}$

6. a. Let x be the hypotenuse. We get:

$$(\sqrt{7})^2 + (\sqrt{8})^2 = x^2$$
$$7 + 8 = x^2$$
$$x^2 = 15$$
$$x = \sqrt{15}$$

b. Let s be the unknown leg. We get:

$$s^2 + (\sqrt{41})^2 = (\sqrt{67})^2$$
$$s^2 + 41 = 67$$
$$s^2 = 26$$
$$s = \sqrt{26}$$

Page 91

7. a. $\quad w^2 + 37.0^2 = 42.1^2$

$$w^2 + 1{,}369 = 1{,}772.41$$
$$w^2 = 403.41$$
$$w = \sqrt{403.41} \approx 20.1 \text{ cm}$$

b. $\quad t^2 + 1044^2 = 1131^2$

$$t^2 + 1{,}089{,}936 = 1{,}279{,}161$$
$$t^2 = 189{,}225$$
$$t = \sqrt{189{,}225} = 435 \text{ ft}$$

c. $\quad 17^2 + 51^2 = x^2$

$$289 + 2{,}601 = x^2$$
$$x^2 = 2{,}890$$
$$x = \sqrt{2{,}890} \approx 54 \text{ m}$$

Page 91

8. To be able to use the Pythagorean Theorem, we need to convert the lengths of the sides into inches: 12 ft 5 in = 149 in and 7 ft 8 in = 92 in. Let x be the hypotenuse. Then:

$$149^2 + 92^2 = x^2$$
$$22{,}201 + 8{,}464 = x^2$$
$$30{,}665 = x^2$$
$$x = \sqrt{30{,}665} \approx 175.11$$

Since the legs were given to the accuracy of one inch, we will do the same with the hypotenuse. The hypotenuse measures about 175 in or 14 ft 7 in.

Page 92

9. a. Check the student's work. If you are using the digital version, the measurements below will not match your triangle unless the student page was printed at 100% (not "shrink to fit" or similar setting).

The sides measure 67 mm, 63 mm, and between 22 mm and 23 mm. Here, I used 63, 23, and 67:

$$63^2 + 23^2 \overset{?}{=} 67^2$$
$$3{,}969 + 529 \overset{?}{=} 4{,}489$$
$$4{,}498 \approx 4{,}489$$

While it may seem to you that 4,498 and 4,489 are quite different, they are actually very close to each other. To check how close they are, we must not simply look at their difference of 9, but at the *percentage* difference = (difference/reference). To calculate that, I will use the average value of 4,493.5 as reference.

Percentage difference = (difference/reference) = $9/4{,}493.5 \approx 0.0020 = 0.2\%$. This is an extremely small percentual difference.

10. Let x be the length of the diagonal. Applying the Pythagorean Theorem we get:

$$x^2 = 9.0^2 + 14.4^2$$
$$x^2 = 81 + 207.36$$
$$x^2 = 288.36$$
$$x = \sqrt{288.36} \text{ in} \approx \underline{17.0 \text{ in}}$$

130

The Pythagorean Theorem, cont.

Puzzle corner. The hypotenuse, 108 units, is shorter than one of the legs, 125 units. To fix it, the teacher could switch the two numbers so that the hypotenuse measures 125 units and the leg 108 units.

In that case, we get:

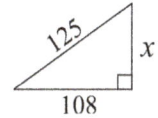

$$x^2 + 108^2 = 125^2$$
$$x^2 + 11{,}664 = 15{,}625$$
$$x^2 = 3{,}961$$
$$x = \sqrt{3{,}961} \text{ units} \approx 62.9 \text{ units}$$

Applications of The Pythagorean Theorem 1, pp. 93-95

1. The side of a square with an area of 100 m^2 is 10 m. The diagonal, d, is given by the Pythagorean Theorem:

$$d^2 = 10^2 + 10^2$$
$$d^2 = 100 + 100$$
$$d^2 = 200$$
$$d = \sqrt{200} \text{ m} \approx \underline{14.1 \text{ m}}$$

2. The length of the diagonal, d, is given by the Pythagorean Theorem:

$$d^2 = 48^2 + 30^2$$
$$d^2 = 2{,}304 + 900$$
$$d^2 = 3{,}204$$
$$d = \sqrt{3{,}204} \text{ m} \approx 56.6 \text{ m}$$

The walk around the park is 48 m + 30 m = 78 m. That route is therefore 78 m − 56.6 m = $\underline{21.4 \text{ m longer}}$.

3. We use the right triangle shown in the image.

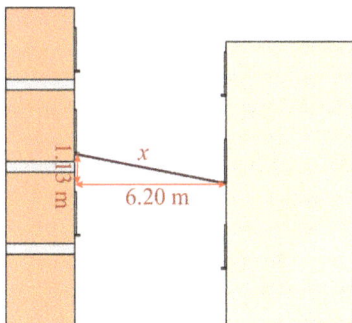

The side 1.13 m comes from subtracting 6.40 m − 5.27 m = 1.13 m. We get:

$$x^2 = 1.13^2 + 6.2^2$$
$$x^2 = 1.2769 + 38.44$$
$$x^2 = 39.7169$$
$$x = \sqrt{39.7169} \text{ m} \approx 6.30 \text{ m}$$

The clothes line is about $\underline{6.30 \text{ m long}}$.

4. Let x be the length of AC. Then:

$$x^2 = 6^2 + 3^2$$
$$x^2 = 36 + 9$$
$$x^2 = 45$$
$$x = \sqrt{45} \approx \underline{6.7}$$

Then, the perimeter is 6.7 + 3 + 6 = 15.7 units.

5. To calculate the height, we use the Pythagorean Theorem:

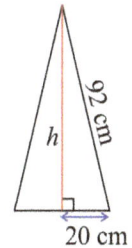

$$20^2 + h^2 = 92^2$$
$$h^2 = 92^2 - 20^2$$
$$h^2 = 8{,}064$$
$$h = \sqrt{8{,}064} \text{ cm} \approx 89.7998 \text{ cm}$$

Then, the area is A = bh/2 = 89.7998 cm · 40 cm / 2 = 1,795.996 cm^2 $\approx \underline{1{,}800 \text{ cm}^2}$.

6. First, we calculate the altitude with the Pythagorean Theorem:

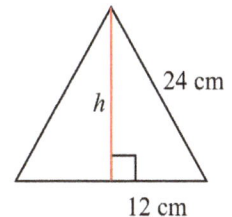

$$12^2 + h^2 = 24^2$$
$$h^2 = 24^2 - 12^2$$
$$h^2 = 432$$
$$h = \sqrt{432} \text{ cm} \approx 20.7846 \text{ cm}$$

Then, the area is A = bh/2 = 24 cm · 20.7846 cm / 2 = 249.4152 cm^2 $\approx \underline{249 \text{ cm}^2}$.

A Proof of the Pythagorean Theorem and of Its Converse, pp. 96-99

Page 96

1.

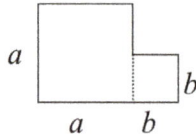

First, we have two squares
with areas a^2 and b^2.
The total area of the figure
is therefore $a^2 + b^2$.

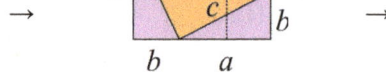

Two lines are drawn so that
two right triangles with legs
a and b are formed.

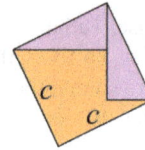

The two right triangles are moved
into new positions. Now we have
a square with sides c units long
and an area of c^2.

Since the total area of the figure is preserved through these changes, $a^2 + b^2 = c^2$.

Page 97

3. a. STATEMENT: If there are no clouds in the sky, then it is not raining.   <u>true</u> / false

    CONVERSE: If it is not raining, then there are no clouds in the sky.   true / <u>false</u>

  b. STATEMENT: If you eat spoiled food, you will get gastrointestinal symptoms.   <u>true</u> / false

    CONVERSE: If you get gastrointestinal symptoms, then you eat/ate spoiled food.   true / <u>false</u>

  c. STATEMENT: If you just had your 7th birthday, then you'll turn 8 in a year.   <u>true</u> / false

    CONVERSE: If you'll turn 8 in a year, then you just had your 7th birthday.   <u>true</u> / false

  d. STATEMENT: If $a + b = c$, then $c - b = a$.   <u>true</u> / false

    CONVERSE: If $c - b = a$, then $a + b = c$.   <u>true</u> / false

  e. STATEMENT: If you multiply two odd numbers, the product is also odd.   <u>true</u> / false

    CONVERSE: If the product of two numbers is odd, the numbers that were multiplied are odd.   <u>true</u> / false

Page 98

5. We check if the three numbers fulfill the Pythagorean Theorem:

$$40.2^2 + 36.4^2 \overset{?}{=} 49.1^2$$

$$1{,}616.04 + 1{,}324.96 \overset{?}{=} 2{,}410.81$$

$$2{,}941 \neq 2{,}410.81$$

No, the corner is not a right triangle, as 2,941 is very different from 2,410.81.
(You can check that by calculating the percent relative difference using the average of
the two numbers as the reference value. You should get $530.19/2675.905 \approx 19.81\%$.)

Page 99

6. They can measure the two diagonals and check that they are equal. If so, the
quadrilateral cannot be a (non-rectangular) parallelogram, so it must be a
rectangle. (In a parallelogram that is not a rectangle, the diagonals are of
different lengths.)

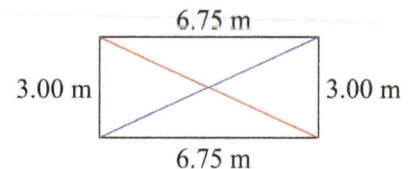

Another possibility is to actually calculate the length of the diagonal with the
Pythagorean Theorem, and then measure to check that the measurement agrees with the calculation.
The Pythagorean Theorem applied to the triangle gives us:

$$x^2 = 3^2 + 6.75^2$$

$$x^2 = 9 + 45.5625$$

$$x^2 = 54.5625$$

$$x = \sqrt{54.5625} \text{ m} \approx 7.39 \text{ m}$$

So if the diagonals measure 7.39 m, the shape is a rectangle.

A Proof of the Pythagorean Theorem and of Its Converse, cont.

Page 99

7.

a. $6^2 + 9^2 \overset{?}{=} 13^2$

$36 + 81 \overset{?}{=} 169$

$117 < 169$

The triangle is obtuse.

b. $12^2 + 5^2 \overset{?}{=} 13^2$

$144 + 25 \overset{?}{=} 169$

$169 = 169$

The triangle is right.

c. $4^2 + 5^2 \overset{?}{=} 7^2$

$16 + 25 \overset{?}{=} 49$

$41 < 49$

The triangle is obtuse.

d. $4^2 + 5^2 \overset{?}{=} 6^2$

$16 + 25 \overset{?}{=} 36$

$41 > 36$

The triangle is acute.

e. $11^2 + 10^2 \overset{?}{=} 13^2$

$121 + 100 \overset{?}{=} 169$

$221 > 169$

The triangle is acute.

f. $15^2 + 20^2 \overset{?}{=} 25^2$

$225 + 400 \overset{?}{=} 625$

$625 = 625$

The triangle is right.

133

Page 100

1. Using the Pythagorean Theorem:

$$(\sqrt{18})^2 + 5^2 = x^2$$
$$18 + 25 = x^2$$
$$43 = x^2$$
$$x = \sqrt{43} \text{ units}$$

2. Note the image on the right:

We will first solve for d using the right triangle on the bottom of the cube:

$$15^2 + 15^2 = d^2$$
$$450 = d^2$$
$$d = \sqrt{450}$$

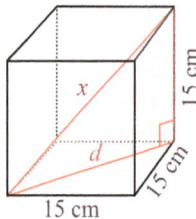

Next, we will use the Pythagorean Theorem again in the pink triangle:

$$(\sqrt{450})^2 + 15^2 = x^2$$
$$450 + 225 = x^2$$
$$x = \sqrt{675}$$
$$x \approx \underline{26 \text{ cm}}$$

Page 101

3. In order to calculate the surface area, we need to find out the height of each of the triangular faces (h in the image).

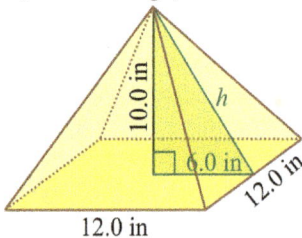

The height of the pyramid (10 in), half of its bottom edge (6 in), and h form a right triangle. Using the Pythagorean Theorem, we get:

$$6^2 + 10^2 = h^2$$
$$36 + 100 = h^2$$
$$h = \sqrt{136}$$

Now, the surface area is the area of the bottom square, plus four times the area of one of the triangular faces:

A = $(12 \text{ in})^2 + 4 \cdot 12 \text{ in} \cdot \sqrt{136} \text{ in} / 2$
$\approx 144 \text{ in}^2 + 279.88569 \text{ in}^2 \approx \underline{424 \text{ in}^2}$.

4. a. $rafter^2 = 3^2 + 12^2$
$rafter^2 = 9 + 144$
$rafter^2 = 153$
$rafter = \sqrt{153} \approx 12.37 \text{ ft} \approx \underline{12 \text{ ft } 4 \text{ in}}$

b. The rise of 5 ft 3 in is 5 1/4 ft = 5.25 ft.

$rafter^2 = 12^2 + 5.25^2$
$rafter^2 = 144 + 27.5625$
$rafter^2 = 171.5625$
$rafter = \sqrt{171.5625} \text{ ft} \approx 13.10 \text{ ft} \approx \underline{13 \text{ ft } 1 \text{ in}}$

Alternatively, you could calculate everything in inches (instead of in feet) and lastly convert to feet and inches. The answers will be the same as listed above.

Page 102

5. The roof consists of two identical rectangles. One dimension of each rectangle is given as 5 m. We need to calculate the other using the Pythagorean Theorem in the below triangle:

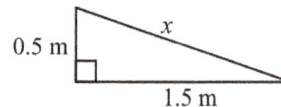

$$x^2 = 0.5^2 + 1.5^2$$
$$x^2 = 0.25 + 2.25$$
$$x^2 = 2.50$$
$$x = \sqrt{2.50} \text{ m} \approx 1.5811 \text{ m}$$

So the area of the roof is 2 · 5 m · 1.5811 m
= 15.811 m$^2 \approx \underline{15.8 \text{ m}^2}$.

6. The roof consists of four identical isosceles triangles. To calculate the area of those triangles, we need to find the altitude, h, of the triangles. You see one of the triangles below.

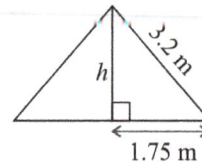

Applying the Pythagorean Theorem, we get:

$$1.75^2 + h^2 = 3.2^2$$
$$h^2 = 3.2^2 - 1.75^2$$
$$h^2 = 7.1775$$
$$h = \sqrt{7.1775} \text{ m} \approx 2.679 \text{ m}$$

The total surface area is then
4 · 3.5 m · 2.679 m / 2 $\approx \underline{18.8 \text{ m}^2}$.

Applications of the Pythagorean Theorem 2, cont.

7. a. We can calculate the length of the creek by applying the Pythagorean Theorem to the right triangle in the image:

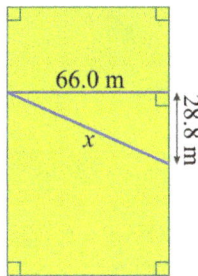

$$x^2 = 66^2 + 28.8^2$$
$$x^2 = 4{,}356 + 829.44$$
$$x^2 = 5{,}185.44$$
$$x = \sqrt{5{,}185.44} \text{ m} \approx 72.0 \text{ m}$$

There is also another way to draw a right triangle into the picture, but its dimensions are the same.

7. b. The two areas are trapezoids:

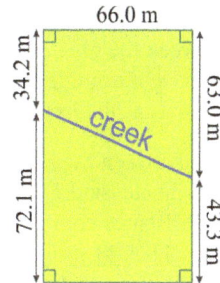

The northern one has a height of 66.0 m, and the two parallel sides measure 34.2 m and 63.0 m. The area is then:

(34.2 m + 63.0 m)/2 · 66.0 m = 3,207.6 m² ≈ 3,210 m²

Similarly, the area of the southern part is

(72.1 m + 43.3 m)/2 · 66.0 m = 3,808.2 m² ≈ 3,810 m²

Puzzle corner. Let x be the length of the one leg. Then the other leg is $2x$, and we can write and solve the equation:

$$x^2 + (2x)^2 = (12.0 \text{ m})^2$$
$$x^2 + 4x^2 = 144 \text{ m}^2$$
$$5x^2 = 144 \text{ m}^2$$
$$x^2 = 144/5 \text{ m}^2$$
$$x = \sqrt{144/5} \text{ m} \approx 5.37 \text{ m}$$

The other leg is $2\sqrt{144/5}$ m ≈ 10.73 m.

The sides of the triangle measure 5.37 m, 10.73 m, and 12.0 m.

Distance Between Points, pp. 104-106

1. a. When we connect the points and then draw a right triangle like in Example 1, the sides of the right triangle are 9 units and 5 units long. Using the Pythagorean Theorem, the desired distance is

$$9^2 + 5^2 = x^2$$
$$81 + 25 = x^2$$
$$106 = x^2$$
$$x = \sqrt{106} \approx 10.3 \text{ units}$$

b. The two sides of the right triangle are now 11 units and 3 units, and the equation is:

$$11^2 + 3^2 = x^2$$
$$121 + 9 = x^2$$
$$130 = x^2$$
$$x = \sqrt{130} \approx 11.4 \text{ units}$$

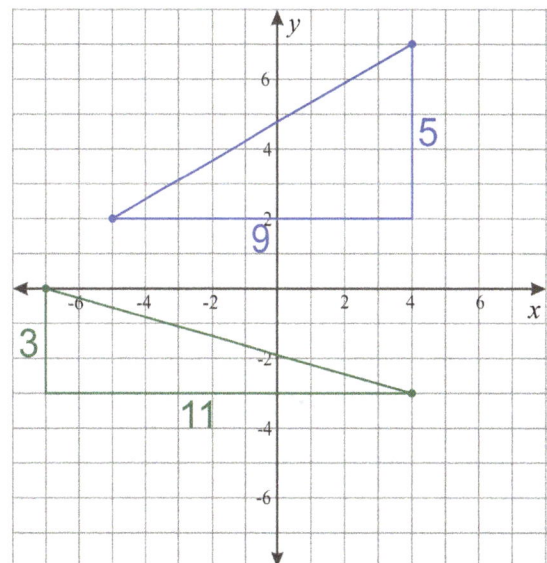

Distance Between Points, cont.

2. We need to find the vertical distance and the horizontal distance between the given points. Those will be the two legs of a right triangle that has the line between the two points as the hypotenuse.

   The horizontal distance between the two points comes from their x-coordinates and is $|-2 - 8|$ = 10 units. The vertical distance comes from the y-coordinates and is $|0 - 11|$ = 11 units. Using the Pythagorean Theorem, the desired distance is:

   $$10^2 + 11^2 = x^2$$
   $$100 + 121 = x^2$$
   $$221 = x^2$$
   $$x = \sqrt{221} \approx 14.9 \text{ units}$$

Page 105

3. First find the horizontal distance between the two points, which is found by taking the absolute value of the difference of their x-coordinates. Then find the vertical distance between the two points by taking the absolute value of the difference of their y-coordinates. Use those two distances as the legs of a right triangle, and solve for the hypotenuse using the Pythagorean Theorem.

   Here, the horizontal distance is $|30 - 5|$ = 25 units. The vertical distance is $|45 - 2|$ = 43 units. Using the Pythagorean Theorem, the desired distance between the two points is

   $$25^2 + 43^2 = x^2$$
   $$625 + 1{,}849 = x^2$$
   $$2{,}474 = x^2$$
   $$x = \sqrt{2{,}474} \text{ units}$$

4.

---

a. $(-10, 9)$ and $(22, 15)$

The horizontal distance is $|22 - (-10)| = 32$ units.
The vertical distance is $|15 - 9| = 6$ units.

$$32^2 + 6^2 = x^2$$
$$1024 + 36 = x^2$$
$$x = \sqrt{1{,}060} \text{ units}$$

---

b. $(30, -25)$ and $(-7, -32)$

The horizontal distance is $|-7 - 30| = 37$ units.
The vertical distance is $|-32 - (-25)| = 7$ units.

$$37^2 + 7^2 = x^2$$
$$1369 + 49 = x^2$$
$$x = \sqrt{1{,}418} \text{ units}$$

---

5. The distance between $(-4, -4)$ and $(-4, 5)$ is simply 9 units (it's their vertical distance). The distance between $(-4, 5)$ and $(2, 5)$ is simply 6 units (it's their horizontal distance). Let x be the distance between $(-4, -4)$ and $(2, 5)$. Then,

   $$x^2 = (-4 - 2)^2 + (-4 - 5)^2$$
   $$x^2 = (-6)^2 + (-9)^2$$
   $$x^2 = 36 + 81$$
   $$x = \sqrt{117} \text{ units} \approx 10.8 \text{ units}$$

   The perimeter is then $10.8 + 9 + 6 = 25.8$ units.

6. First we find the side of the square. Let s be the distance between $(5, 0)$ and $(0, 5)$. Then,

   $$s^2 = 5^2 + 5^2$$
   $$s^2 = 25 + 25$$
   $$s = \sqrt{50}$$

   The area of the square is then $s^2 = (\sqrt{50})^2$ = 50 square units.

Page 106

7. In triangle ABC, the distance from A to C, or AC, is $\sqrt{4^2 + 5^2}$ units = $\sqrt{41}$ units.
   The distance from B to C, or BC is $\sqrt{2^2 + 4^2}$ units = $\sqrt{20}$ units. And AB is 3 units.

   The perimeter is $\sqrt{41} + \sqrt{20} + 3 \approx 13.9$ units

   In trapezoid DEGF, DE = 2 units, GF = 5 units, EF = $\sqrt{2^2 + 3^2} = \sqrt{13}$ units, and DG = $\sqrt{1^2 + 3^2} = \sqrt{10}$ units. The perimeter is $\sqrt{13} + \sqrt{10} + 2 + 5 \approx 13.8$ units.

   The trapezoid has a shorter perimeter, by $13.875 - 13.768 \approx 0.1$ units.

8. If AB = 3 units is the base of the triangle, then its height is 4 units, and its area is $3 \cdot 4 \div 2$ = 6 square units.

   The area of the trapezoid is the average of its two parallel sides times its height. The two parallel sides measure 2 and 5 units, so their average is 3.5 units. Its height is 3 units. So, its area is $3 \cdot 3.5 = 10.5$ square units.

   The area of the trapezoid is 4.5 square units bigger than the area of the triangle.

9. First, let's calculate the distances using units on the map. Sarah has to travel $50 + 32 = 82$ map units. The crow flies a distance of $\sqrt{50^2 + 32^2} = \sqrt{3524} \approx 59.363$ map units. The difference between what Sarah travels and what the crow flies is about 22.6 map units, which is about 226 meters.

Puzzle corner. We divide the octagon into nine areas as in the image on the right. In the middle is a 2 by 2 square. Then there are four rectangles, with sides 2 and x units long. And in the corners we have four right triangles with base x and height x.

We can solve x using one of the corner triangles and the Pythagorean Theorem:

$$x^2 + x^2 = 2^2$$
$$2x^2 = 4$$
$$x^2 = 2$$
$$x = \sqrt{2}$$

Thus, the total area is: $4 + 4 \cdot 2 \cdot \sqrt{2} + 4 \cdot \sqrt{2} \cdot \sqrt{2} \div 2 = 4 + 8\sqrt{2} + 4 = 8 + 8\sqrt{2}$ which is about 19.3 square units.

Mixed Review Chapter 6, pp. 107-109

1.

a. $3x^4 y^5 y^2 \cdot 6x^6 = 18x^{10}y^7$	b. $(3x)^{-3} = \dfrac{1}{27x^3}$	c. $(3yz)^2 = 9y^2z^2$	d. $(b^{-2})^4 = \dfrac{1}{b^8}$
e. $\dfrac{8x^5}{28x^8} = \dfrac{2}{7x^3}$	f. $\dfrac{x^{-5}}{x^2} = \dfrac{1}{x^7}$	g. $\left(\dfrac{-2}{5y}\right)^2 = \dfrac{4}{25y^2}$	h. $\left(\dfrac{3s}{t^2}\right)^4 = \dfrac{81s^4}{t^8}$

2.

a. from point A with scale factor 1/3	b. from point C with scale factor 1/2

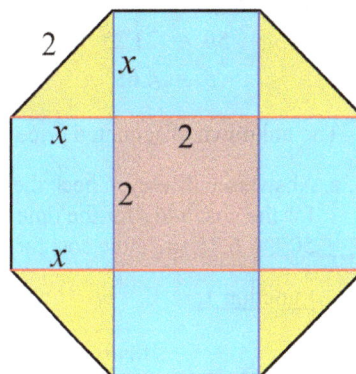

3. The ratio of the volume of the tennis balls to the volume of the cube is

$$\frac{8 \cdot (4/3) \cdot \pi \cdot (3\ \text{cm})^3}{(12\ \text{cm})^3} = \frac{(32/3) \cdot \pi \cdot 27\ \text{cm}^3}{12^3\ \text{cm}^3} = \frac{32 \cdot \pi \cdot \overset{9}{27}\ \text{cm}^3}{\underset{1}{3} \cdot 12^3\ \text{cm}^3} = \frac{32 \cdot \pi \cdot 9}{3 \cdot 4 \cdot 3 \cdot 4 \cdot 3 \cdot 4} = \frac{1 \cdot \pi}{3 \cdot 2} = \frac{\pi}{6} \approx 52.4\%.$$

Page 108

4. The equation is $5 \cdot 5.95 + 5p = 53$. Here is the solution:

$$
\begin{aligned}
5 \cdot 5.95 + 5p &= 53 \\
29.75 + 5p &= 53 \\
5p &= 23.25 \\
p &= 4.65
\end{aligned}
$$

The unknown, discounted price was $4.65 per yard.

5. a. Answers will vary. Check the student's answers. The first function should be of the form $y = mx$, or if using C for the cost and t for the time, of the form $C = mt$ for some constant m. Additionally, m can be a maximum of $50/8 = 6.25$ since the cost for eight hours has to be $50 or less. For example:

Function 1: $C = 6t$

Function 2:

time (hours)	0	1	2	3	4	5	6	7	8
Cost ($)	0	6	12	17	22	27	31	35	39

   b. Answers will vary. Check the student's answers. Using the functions above, both functions give the same price for renting a surfboard for 2 hours ($12). For 6 hours, Function 2 gives the better deal, for $31 (versus $36).

6. a. We can start with the equation $y = -2x + b$. To find b, we substitute -2 and 6 for x and y: $6 = -2(-2) + b$, from which $b = 6 - 4 = 2$. So, the equation is $y = -2x + 2$.

   b. We can start with the equation $y = (2/3)x + b$. To find b, we substitute 4 and -4 for x and y: $-4 = (2/3)4 + b$, from which $b = -4 - (2/3)4 = -4 - 8/3 = -20/3$. So, the equation is $y = (2/3)x - 20/3$.

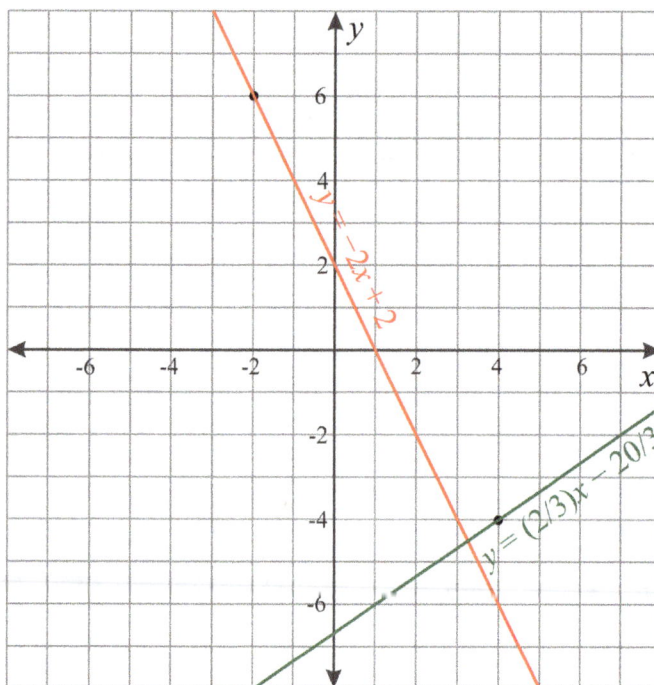

Page 109

7. a. $y = -5$
   b. $x = 9$
   c. $y = 5x - 3$

8. a. 28.0 in   b. 5,300 lb
   c. 71.6 m   d. 38.0 kg
   e. 51.48 ft   f. 11 cm

9. The volume of the square prism involved is found by simply multiplying the area of the bottom square by its height: 1 sq. mi $\cdot$ 1 inch. The first task is to convert one square mile to square inches.

One mile = 5,280 ft = 5,280 ft $\cdot$ 12 in/ft = 63,360 inches.

Then, a square mile is $(63,360 \text{ in})^2$, and the volume of our rainwater is then 1 in $\cdot$ $(63,360 \text{ in})^2 = 4,014,489,600 \text{ in}^3$.

Now, converting this to gallons, we get: V = $4,014,489,600 \text{ in}^3 \cdot 1 \text{ gal}/(231 \text{ in}^3)$ = or in a simpler format, $4,014,489,600/231$ gallons.

If you calculate this intermediate result, keep a few more than four significant digits: $4,014,489,600/231 \approx 17,378,743$ gallons.

And each of these gallons has 10^5 raindrops. So, we take the previous result and multiply it by 10^5: The number of raindrops = $10^5 \cdot 17,378,743 \approx 1.738 \cdot 10^{12}$ drops.

Page 110

1. a. 8   b. 13   c. 50   d. 0.9
   e. 6/10 = 3/5   f. −7   g. 5   h. 30

2. a. between 2 and 3   b. between 8 and 9   c. between 11 and 12   d. between 6 and 7

3. Here is one possible way the process could go. The intermediate guesses by the student(s) will vary. The table is just showing one example.

$\sqrt{13} \approx 3.6$

Low Guess	$(LG)^2$	$(HG)^2$	High Guess
3.6	12.96	13.3225	3.65
3.6	12.96	13.0321	3.61

4.

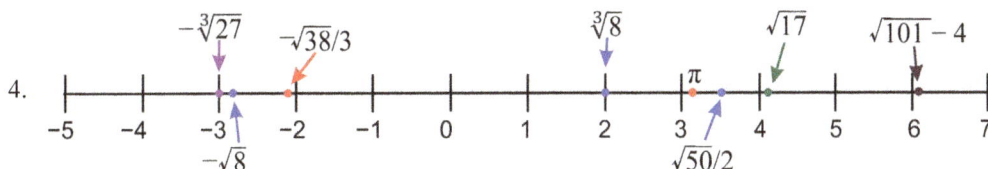

5.

a. $11 < \sqrt{150}$	b. $\sqrt{76} < 9$	c. $\sqrt{20} > 4$	d. $\sqrt[3]{10} > 2$
e. $4 < \pi + 1$	f. $\sqrt{85}/3 > 3$	g. $\sqrt{27} + 2 > 6$	h. $\sqrt{68} - 3 < 6$

Page 111

6. a. 12   b. −9   c. 40
   d. 8   e. 49   f. 20

7. a. 7 square units
   b. $4\sqrt{20}$ (or this can also be written as) $8\sqrt{5}$)

8. a. rational; it is a decimal that terminates
   b. rational; the value is 50
   c. irrational; π is irrational, 5 is rational, and an irrational number times a rational is irrational
   d. irrational; 56 is not a perfect square, so $\sqrt{56}$ is irrational and so is its opposite
   e. rational; this is a ratio of two integers
   f. rational; the value is 1/6
   g. rational; this is a repeating decimal
   h. rational; this is a terminating decimal
   i. rational; this is a repeating decimal
   j. irrational; Seven is not a perfect square so $\sqrt{7}$ is irrational. Four is rational, and an irrational number times a rational number ($4\sqrt{7}$) is irrational.

Page 112

9. a.
$$100x = 61.6161...$$
$$- \ x = \ \ 0.6161...$$
$$99x = 61$$
$$x = \underline{61/99}$$

   b.
$$10x = 41.7777...$$
$$- \ x = \ \ 4.1777...$$
$$9x = 37.6$$
$$x = 37.6/9 = 376/90 = \underline{188/45}$$

10. a. $x = \sqrt{147}$ or $x = -\sqrt{147}$
    b. $a = 13$ or $a = -13$
    c. $w = \sqrt[3]{0.36}$
    d. $x = \sqrt[3]{7}$
    e. $b = 5$
    f. $a = -2$

Page 112

11. a. $y^2 + 18 = 35$

$\qquad y^2 = 17$

$\qquad y = \sqrt{17} \approx 4.123$

$\qquad$ or $y = -\sqrt{17} \approx -4.123$

Check: $(\sqrt{17})^2 + 18 \overset{?}{=} 35$

$\qquad\qquad 17 + 18 = 35$ ✓

b. $0.6h^2 = 4$

$\qquad h^2 = 4/0.6 = 40/6 = 20/3$

$\qquad h = \sqrt{20/3} \approx 2.582$

$\qquad$ or $h = -\sqrt{20/3} \approx -2.582$

Check: $0.6 \cdot (\sqrt{20/3})^2 \overset{?}{=} 4$

$\qquad\qquad 0.6 \cdot (20/3) \overset{?}{=} 4$

$\qquad\qquad (6/10) \cdot (20/3) \overset{?}{=} 4$

$\qquad\qquad 120/30 = 4$ ✓

Page 113

12. a. $20^2 + 24^2 \overset{?}{=} 30^2$

$\qquad 400 + 576 \overset{?}{=} 900$

$\qquad\qquad 976 > 900$

No, they don't form a right triangle. (They would form an acute triangle.)

b. $1^2 + 2.4^2 \overset{?}{=} 2.6^2$

$\qquad 1 + 5.76 \overset{?}{=} 6.76$

$\qquad\qquad 6.76 = 6.76$

Yes, they form a right triangle.

13. We can ignore the negative answers because a side cannot have a negative length.

a. $s^2 = 3^2 + 5^2$

$\qquad s^2 = 9 + 25$

$\qquad s^2 = 34$

$\qquad s = \sqrt{34}$

b. $y^2 + 12^2 = 14^2$

$\qquad y^2 + 144 = 196$

$\qquad\qquad y^2 = 52$

$\qquad\qquad y = \sqrt{52}$

14. $\quad x^2 + 21.1^2 = 22.5^2$

$\qquad x^2 + 445.21 = 506.25$

$\qquad\qquad x^2 = 61.04$

$\qquad\qquad x = \sqrt{61.04} \approx \underline{7.8 \text{ m}}$

Page 114

15. Let x be the hypotenuse of this triangle.

$\qquad (\sqrt{7})^2 + (\sqrt{8})^2 = x^2$

$\qquad\qquad 7 + 8 = x^2$

$\qquad\qquad x^2 = 15$

$\qquad\qquad x = \sqrt{15}$

The hypotenuse is $\sqrt{15}$ units long.

16. The pennant is an isosceles triangle.

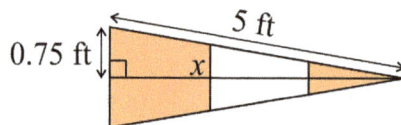

We calculate its altitude using the Pythagorean Theorem. From the right triangle in the image, we get:

$\qquad 0.75^2 + x^2 = 5^2$

$\qquad 0.5625 + x^2 = 25$

$\qquad\qquad x^2 = 24.4375$

$\qquad\qquad x = \sqrt{24.4375} \approx 4.94343... \text{ ft}$

So the area is A = bh/2 ≈ 1.5 ft · 4.94343 ft /2 = 3.70757 ft² ≈ $\underline{3.7 \text{ ft}^2}$.

17. The figure below has four right triangles, each with sides a, b and c. The sides of the outside square are $a + b$. The triangles enclose a square with sides c units long.

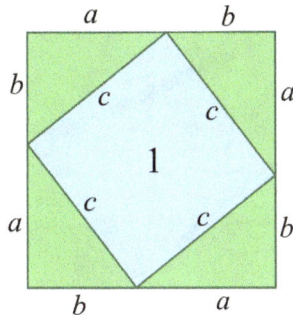

Below, the sides of the large square are still $a + b$, but the four right triangles have been rearranged so that two smaller squares are formed, with sides a and b.

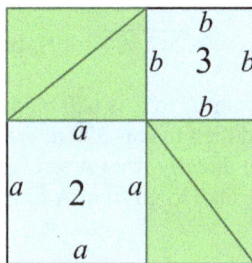

Since the areas of both large squares are equal, and the areas of the four right triangles are equal, it follows that the remaining (blue) areas are also equal. In other words, the area of square 1, which is c^2, equals the area of square 2 (which is a^2) plus the area of square 3 (which is b^2). In symbols, $c^2 = a^2 + b^2$. ☺

18. This is a trapezoid. Its area is the average length of the two parallel sides, times its altitude. The altitude is 5 units. The two parallel sides measure 5 and 9 units. So, the area is $(5 + 9)/2 \cdot 5 = 7 \cdot 5 = 35$ square units. Since each unit is 2 ft, then one square unit in the image corresponds to 2 ft $\cdot$ 2 ft = 4 ft^2. Then, the area is $35 \cdot 4$ ft^2 = $\underline{140 \text{ ft}^2}$.

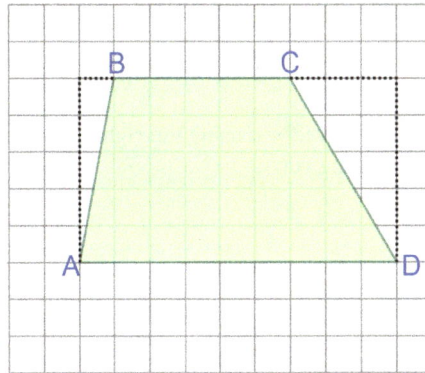

For the perimeter, we will find the side lengths AB and CD using the Pythagorean Theorem. The image shows with dashed lines the right triangles we will use.

$$\text{For AB: } \quad 5^2 + 1^2 \quad = \text{ AB}^2$$
$$26 = \text{ AB}^2$$
$$\text{AB} = \sqrt{26} \approx 5.099 \text{ units}$$

$$\text{For CD: } 3^2 + 5^2 \quad = \text{ CD}^2$$
$$34 = \text{ CD}^2$$
$$\text{CD} = \sqrt{34} \approx 5.831 \text{ units}$$

The perimeter is then the sum of all four sides:
P = 5.099 + 5 + 5.831 + 9 = 24.93 units.
Since each unit is 2.0 feet, the perimeter is 49.86 feet $\approx \underline{50 \text{ feet}}$.

19. a. The horizontal distance between the points is $|11 - (-3)| = 14$ and the vertical distance is $|-9 - 4| = 13$.

$$x^2 = 14^2 + 13^2$$
$$x^2 = 196 + 169$$
$$x = \sqrt{365} \approx 19.1 \text{ units}$$

b. The horizontal distance between the points is $|70 - 42| = 28$ and the vertical distance is $|100 - (-15)| = 115$.

$$x^2 = 28^2 + 115^2$$
$$x^2 = 14{,}009$$
$$x = \sqrt{14{,}009} \approx 118.4 \text{ units}$$

20. Please see the image below:

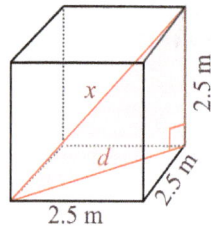

We will first solve for d using the right triangle on the bottom of the cube:

$$2.5^2 + 2.5^2 = d^2$$
$$12.5 = d^2$$
$$d = \sqrt{12.5}$$

Next, we will use the Pythagorean Theorem again in the pink triangle:

$$(\sqrt{12.5})^2 + 2.5^2 = x^2$$
$$12.5 + 6.25 = x^2$$
$$x = \sqrt{18.75}$$
$$x \approx \underline{4.3 \text{ m}}$$

21. Let x be the distance from B to C along 5th Avenue South.

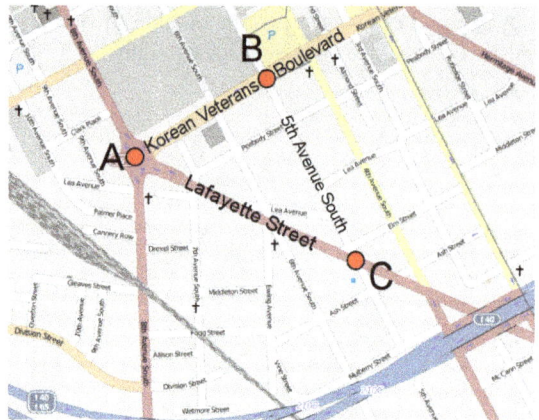

Then:

$$370^2 + x^2 = 620^2$$
$$136,900 + x^2 = 384,400$$
$$x^2 = 247,500$$
$$x = \sqrt{247,500} \approx 497.49 \text{ m} \approx 500 \text{ m}$$

To go directly from A to C is 620 m, and the distance from A to B and then to C is 370 m + 500 m = 870 m. Therefore, to go directly from A to C is 870 m − 620 m = $\underline{250 \text{ m shorter}}$ than to go from A to B and then to C.

Chapter 7: Systems of Linear Equations

Equations with Two Variables, pp. 119-122

Page 119

1. Answers will vary. Check the student's answers. For example: (0, 16), (2, 11), (4, 6). Or, (−2, 21), (−4, 26), (−6, 31).

2. Answers will vary. Check the student's answers. For example: (0, −6), (1, −2), (−1, −10).

Page 120

3. a. $y = (1/3)x + 2$. The student may write this same equation in a different format, such as $x − 3y = −6$.
   b. Answers will vary. For example: (0, 2), (−3, 1), (3, 3)

4. Using the two points (0, −5) and (2, 3), we calculate the slope of the line as $(3 − (−5))/(2 − 0) = 8/2 = 4$.
   The y-intercept is −5 (from the point (0, −5)). So, in the slope-intercept form, the equation is $y = 4x − 5$.
   It can also be given in other, equivalent forms, such as $4x − y = 5$.

5. Using the two given points (−1, −5) and (2, 8), we calculate the slope of the line as $(8 − (−5))/(2 − (−1)) = 13/3$.
   The generic form of the equation is $y = (13/3)x + b$. Substituting (2, 8) for x and y, we get $8 = (13/3)2 + b$, from
   which $b = 8 − 26/3 = −2/3$. So, the equation is $y = (13/3)x − 2/3$, or equivalently, $3y = 13x − 2$, or $13x − 3y = 2$.

6. a. $C = 2x + 3y$
   b. $2x + 3y = 48$. This equation has several possible solutions. For example, Randy could have bought 6 hats and
      12 whistles, because $2 \cdot 6 + 3 \cdot 12 = 48$.
   c. Here is a list of all the other solutions: 0 hats/16 whistles; 3 hats/14 whistles; 9 hats/10 whistles; 12 hats/8 whistles,
      15 hats/6 whistles; 18 hats/4 whistles, 21 hats/2 whistles; or 24 hats/0 whistles.

Page 121

7. a. 1/10 miles per minute and 1/5 miles per minute.

   b. $d = x/10 + y/5$

   c. 20/10 miles + 10/5 miles = 2 miles + 2 miles = 4 miles

   d. $x/10 + y/5 = 20$. Answers will vary. Check the student's answers.
      For example, she could have...
      jogged for 20 minutes (2 mi) and bicycled for 90 minutes (18 mi).
      jogged for 30 minutes (3 mi) and bicycled for 85 minutes (17 mi).
      jogged for 40 minutes (4 mi) and bicycled for 80 minutes (16 mi).
      jogged for 50 minutes (5 mi) and bicycled for 75 minutes (15 mi).
      jogged for 60 minutes (6 mi) and bicycled for 70 minutes (14 mi).
      Etcetera.

   e. See the graph on the right. The equation $x/10 + y/5 = 20$ in
      slope-intercept form is $y = −x/2 + 100$.

8. Answers will vary. Check the student's answers. For example:
   20 non-seniors ($300) and 60 seniors ($600)
   10 non-seniors ($150) and 75 seniors ($750)
   30 non-seniors ($450) and 45 seniors ($450)

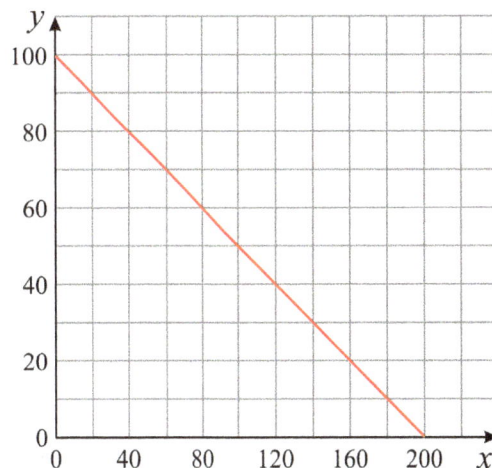

Equations with Two Variables, cont.

9. a. $4x + 0.5y = 20$

   b. Answers will vary. Check the student's answers. For example:
   2 cats (8 kg) and 24 kittens (12 kg), or 3 cats (12 kg)
   and 16 kittens (8 kg), or 4 cats (16 kg) and 8 kittens (4 kg)

   c. In slope-intercept form, the equation $4x + 0.5y = 20$ is
   $y = 40 - 8x$. See the graph on the right.

   d. See the graph on the right. If $x = 1.5$, then $y = 28$.
   However, we cannot have 1.5 adult cats. The number of cats
   (and of kittens) needs to be a whole number. The valid
   solutions are (0, 40), (1, 32), (2, 24), (3, 16), (4, 8), and (5, 0).

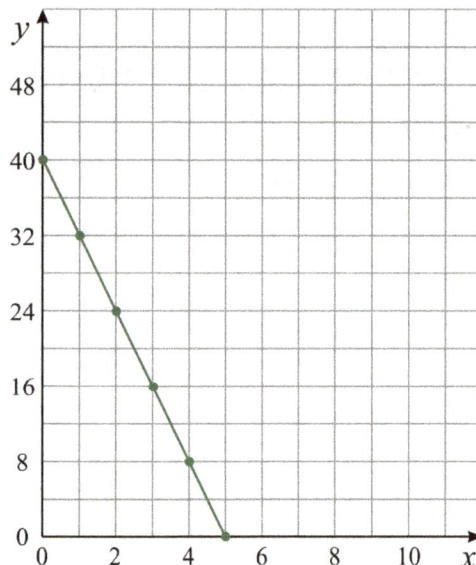

10. a. $120x + 20y$
    b. $120x + 20y = 760$. These are the possible answers,
    assuming they stayed at least one night and did at least
    one horse ride:

    1 night and 32 horse rides,
    2 nights and 26 horse rides,
    3 nights and 20 horse rides,
    4 nights and 14 horse rides,
    5 nights and 8 horse rides, or
    6 nights and 2 horse rides.
    It is easy to see the patterns in these amounts!

11. Answers will vary. Check the student's answers. For example: (1, −9), (−1, 3), (2, −9), (−2, 15).

Solving Systems of Equations by Graphing, pp. 123-127

1. a. (−3, −2)
   b. Answers will vary slightly since the student will
   have to estimate the coordinates. For example:
   (1.15, 3.7).
   c. (3, 0)

The image below shows the equations of the lines.

Question 2:

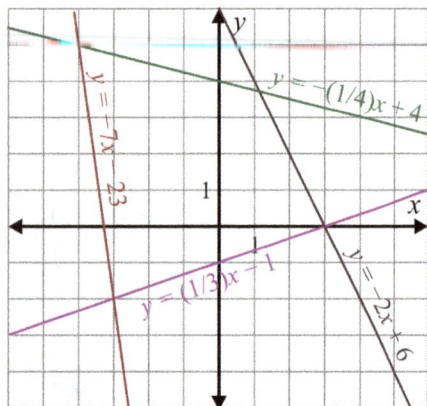

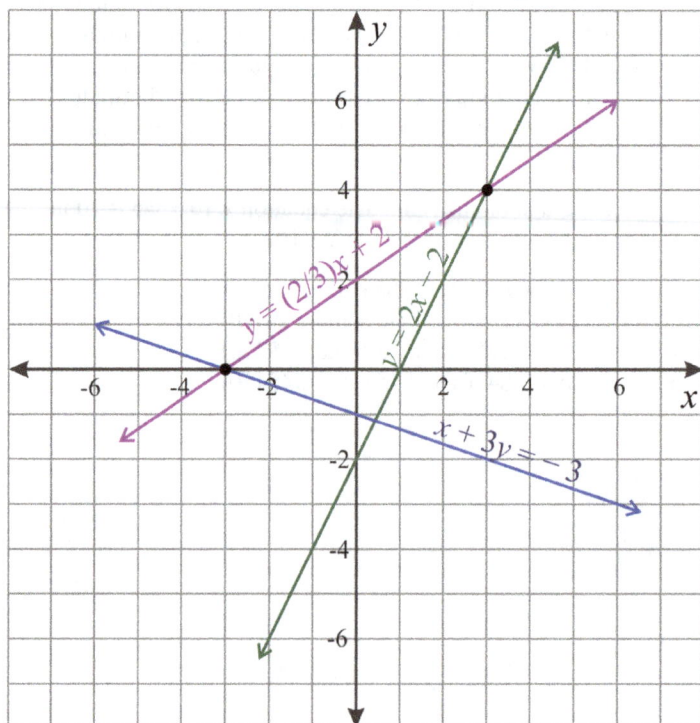

2. See the image on the right.
   a. (3, 4)   b. (−3, 0)

Solving Systems of Equations by Graphing, cont.

Page 123

3. a. x is about -0.6 or -0.7, y is about 0.1.

b. x is about 2.2, y is about -2.7.

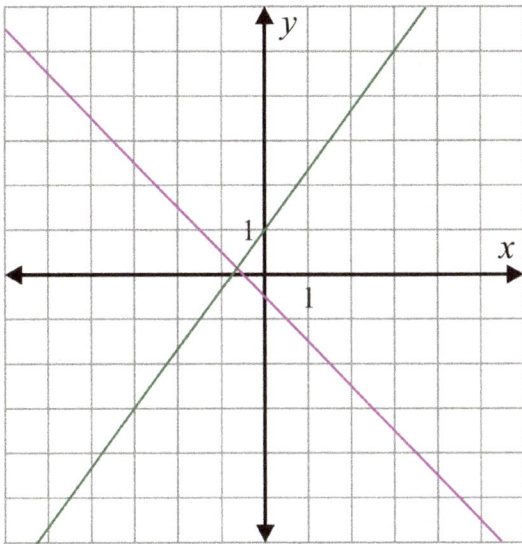

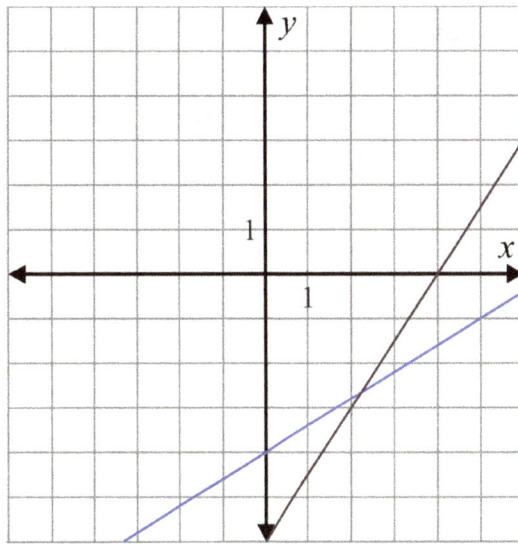

4. The exact coordinates of the solutions are:
   a. $(-9/14, 1/7)$ or about $(-0.643, 0.143)$
   b. $(20/9, -8/3)$ or about $(2.222, -2.667)$

Page 125

5. a. $2x + 3y = 24$. b. $x + y = 11$. For both (a) and (b), see the graph on the right.

   c. He got 9 hats and 2 whistles. Note that the point $(9, 2)$ is the point where the lines intersect.

6. To find the equation of each line, you can use the generic equation of a line, $y = mx + b$, and substitute to it the slope you get from the graph, plus the coordinates of one point on the line. Alternatively, the point-slope form of a linear equation could be used: $y - y_0 = m(x - x_0)$. where m is the slope and (x_0, y_0) is a known point.

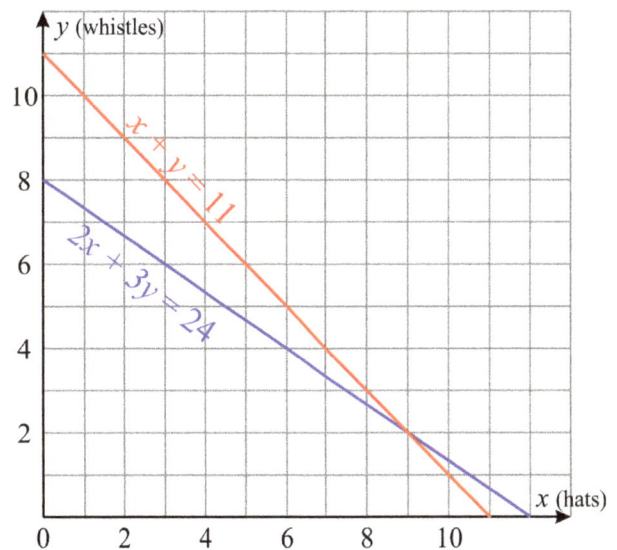

a.	b.
$\begin{cases} 2x - y = 6 \\ 2x - 5y = -10 \end{cases}$	$\begin{cases} x - 3y = 7 \\ 3x + 4y = -5 \end{cases}$
Solution: $(5, 4)$	Solution: $(1, -2)$

145

Solving Systems of Equations by Graphing, cont.

Page 126

7. See the graph on the right.

 a. $4x + 0.5y = 22$, or equivalently, $y = 44 - 8x$.
 b. $x + y = 16$, or equivalently, $y = 16 - x$.
 c. 4 adults cats and 12 kittens

8. a. One solution.
 b. Zero.
 c. An infinite number of solutions.

Puzzle corner: $m = -6$. Hint: substitute $(-1, 0)$ to the second equation, and solve for m.

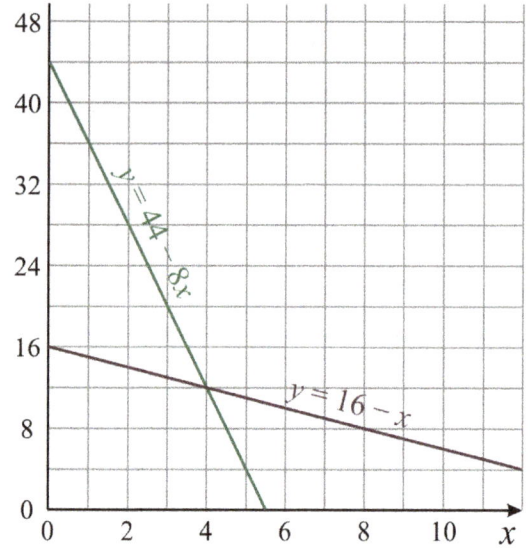

Page 127

9. a. $(6, -1)$ b. $(4, 6)$

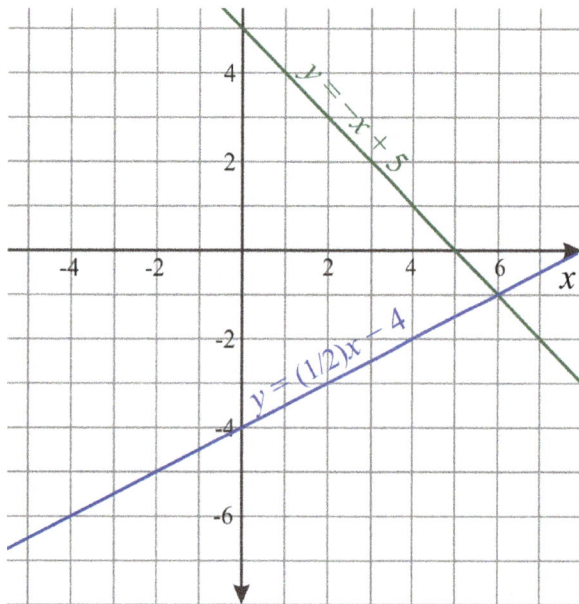
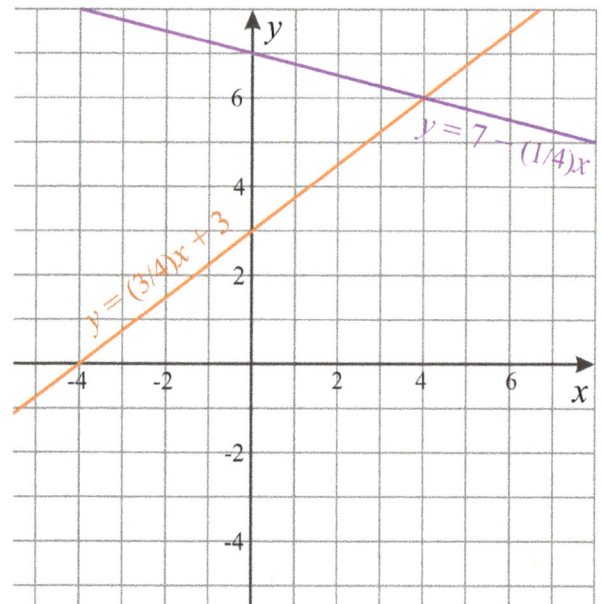

10.

a. $(0, -3)$ b. $(1.2, -2.9)$

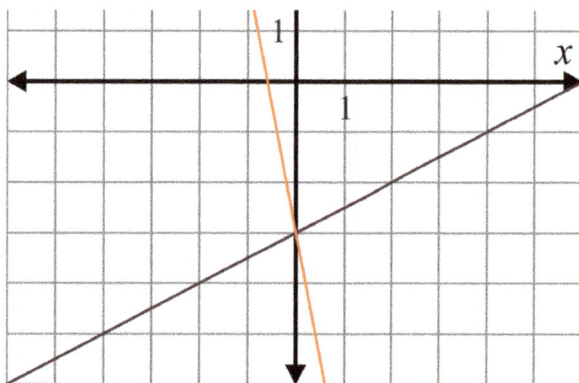
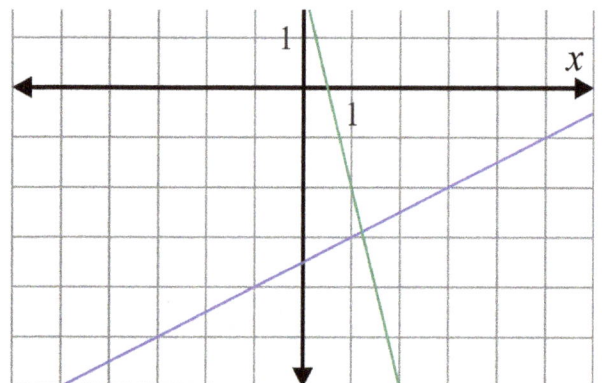

Page 128

1.

a. $\begin{cases} 2y = 4x - 1 \\ -2x + y = 3 \end{cases}$

The lines are parallel and do not intersect.
This means there is no solution.

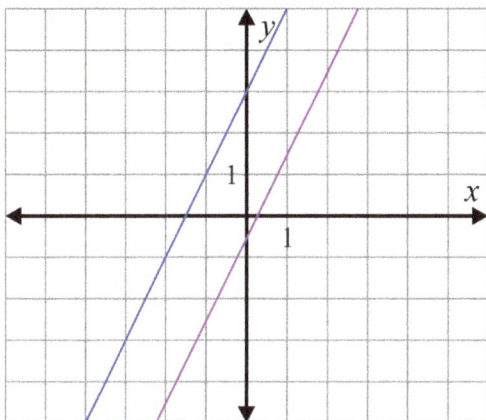

b. $\begin{cases} y = -2x + 2 \\ y = x - 4 \end{cases}$

The solution is (2, −2).

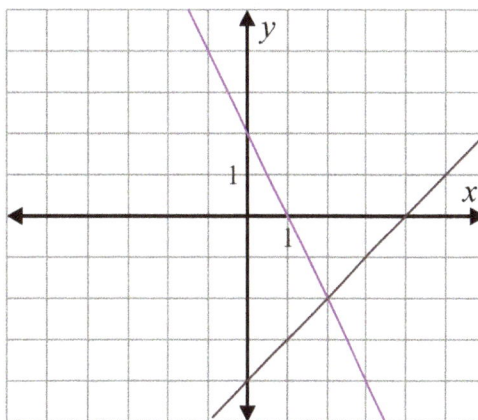

c. $\begin{cases} y = -(1/2)x + 1 \\ x + 2y = 2 \end{cases}$

The lines are the same; they meet at every point.

This means every point on the line is a solution.

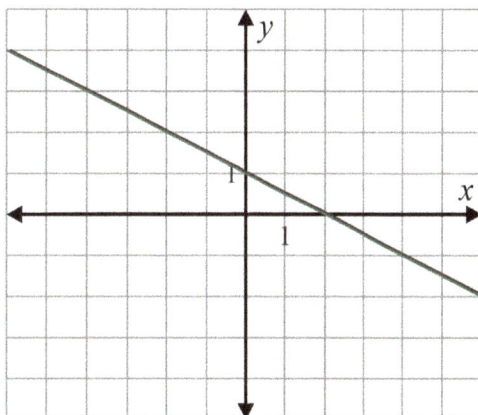

2. a. It has no solutions, since the lines are parallel and do not intersect.
   b. It has one solution. Since the lines have a different slope, they must intersect somewhere, in one point.

How many solutions can a system of two linear equations have? If you think about the system graphically, as two lines, and whether they intersect, it is easy to see that there are **three different cases**:

1. The two lines intersect in a single point. Therefore, the system has **one** solution (one ordered pair).
2. The two lines are _parallel_, and thus do not intersect at all. In this case, the system has _no_ solutions.
3. The two lines are identical. Here, the system has an _infinite_ number of solutions.
   It is as if the two lines "intersect in every point".

Number of Solutions, cont.

3. a. (2/5, 7). The two lines have different slopes, so they intersect in one point and there is one solution.
   b. There are no solutions. The lines have the same slope ($-1/3$) but different y-intercepts (-1 and 1). So, the lines are parallel.
   c. All points (x, y) that satisfy the equation $y = 4x + 1$. Here, the bottom equation is the top equation multiplied by 2. So, the lines are identical.

4. a. (0.7, 2.0)   b. All the points (x, y) that satisfy the equation $3y = -2x + 6$.   c. There are no solutions.

5. a. One solution. If we transform both to the slope-intercept form, the two lines have different slopes.
   b. An infinite number of solutions. If we divide the bottom equation by 4, we get $y = -(1/4)x$. The lines are identical.
   c. No solutions. If we transform the bottom equation to the slope-intercept form, we get $y = -4x$. Thus, the two equations have the same slope but different y-intercepts. The lines are parallel.

6. a. infinite number of solutions   b. no solutions   c. one solution
   d. no solutions   e. one solution   f. no solutions
   g. one solution   h. infinite number of solutions   i. no solutions

7. System (iii) has an infinite number of solutions. <u>Solution:</u> All points (x, y) that satisfy the equation $x/3 - y = 2$ (or the equation $x = 3y + 6$; either equation can be used.)

8. Answers will vary. Check the student's answer. For example:

a. the system has no solutions	b. the system has one solution	c. the system has an infinite number of solutions
$\begin{cases} y = 3x + 2 \\ y = 3x + 5 \end{cases}$	$\begin{cases} y = 3x + 2 \\ y = x + 2 \end{cases}$	$\begin{cases} y = 3x + 2 \\ 2y = 6x + 4 \end{cases}$

9. Answers will vary. Check the student's answer.

   a. In slope-intercept form, the two lines that intersect at $(2, -3)$ are $y = -(1/4)x - 5/2$ and $y = (4/3)x - 17/3$.

   But the equations can be given in other, equivalent forms, such as in the standard form or any other. So, the answers will vary but the student's equations need to be equivalent to the ones given here.

   For example: $\begin{cases} 4y = -x - 10 \\ 3y = 4x - 17 \end{cases}$ Or, in standard form: $\begin{cases} x + 4y = -10 \\ 4x - 3y = 17 \end{cases}$

   b. For the system to have no solutions, we need to use parallel lines. The two parallel lines are $y = -(1/4)x + 3$ and $y = -(1/4)x - 5/2$. The system of equations could have these two equations, or others equivalent to them, such as

   $\begin{cases} 4y = -x + 12 \\ 4y = -x - 10 \end{cases}$, or: $\begin{cases} x + 4y = 12 \\ x + 4y = -10 \end{cases}$

10. a. one solution   b. no solutions   c. no solutions
    d. infinite number of solutions   e. one solution   f. infinite number of solutions

Puzzle corner. Since System 1 has no solutions, we know that the two lines are parallel. Converting the first equation to slope-intercept form, we get $y = 4x$. Since the second line is parallel to the first, it will have the same slope, meaning $s = 4$. From the first equation in System 2, we can determine that $x = 2$. Substituting 4 for s and 2 for x, in the second equation we get: $y = 3(2) + 4 = 10$. So the solution is $(2, 10)$.

Solving Systems of Equations by Substitution, pp. 132-138

Page 132

1.

a. $\begin{cases} x + y = -9 \\ y = 3x - 1 \end{cases}$

Substituting $3x - 1$ to the first equation:
$$x + (3x - 1) = -9$$
$$4x - 1 = -9$$
$$4x = -8$$
$$x = -2$$

Now, we substitute $x = -2$ to the second equation:

$$y = 3x - 1 = 3(-2) - 1 = -7$$

Solution: $(-2, -7)$

b. $\begin{cases} x = 3y - 11 \\ 2x + 2y = 10 \end{cases}$

Substituting $3y - 11$ to the second equation:
$$2x + 2y = 10$$
$$2(3y - 11) + 2y = 10$$
$$6y - 22 + 2y = 10$$
$$8y - 22 = 10$$
$$8y = 32$$
$$y = 4$$

Now, we substitute $y = 4$ to the first equation:

$$x = 3y - 11 = 3(4) - 11 = 1$$

Solution: $(1, 4)$

Page 133

2.

a. $\begin{cases} x - 4y = -2 \\ y = 5 - 2x \end{cases}$

Substituting $y = 5 - 2x$ to the first equation:
$$x - 4(5 - 2x) = -2$$
$$x - 20 + 8x = -2$$
$$9x - 20 = -2$$
$$9x = 18$$
$$x = 2$$

Now, we substitute $x = 2$ in the second equation:

$$y = 5 - 2x = 5 - 2(2) = 1$$

Solution: $(2, 1)$

b. $\begin{cases} x = 10y + 1 \\ (1/2)x - y = 3 \end{cases}$

Substituting $x = 10y + 1$ to the second equation:
$$(1/2)x - y = 3$$
$$(1/2)(10y + 1) - y = 3$$
$$5y + 1/2 - y = 3$$
$$4y + 1/2 = 3$$
$$4y = 5/2$$
$$y = 5/8$$

Now, we substitute $y = 5/8$ in the first equation:

$$x = 10y + 1 = 10(5/8) + 1$$
$$= 25/4 + 1 = 29/4$$

Solution: $(29/4, 5/8)$

c. $\begin{cases} 3x = 3(y - 1) \\ y = 5x \end{cases}$

Substituting $5x$ in place of y in the first equation:
$$3x = 3(y - 1)$$
$$3x = 3(5x - 1)$$
$$3x = 15x - 3$$
$$3 = 12x$$
$$x = 1/4$$

Now, we substitute $1/4$ in place of x in the second equation:

$$y = 5x = 5(1/4) = 5/4$$

Solution: $(1/4, 5/4)$

3.

a.	b.

a.

$$\begin{cases} x + 3y = -5 & (1) \\ 2x + y = 3y - 1 & (2) \end{cases}$$

$$\downarrow$$

$$\begin{cases} x = -3y - 5 & (1) \\ 2x + y = 3y - 1 & (2) \end{cases}$$

$$\downarrow$$

(1)
$$\begin{aligned} (-3y - 5) + 3y &= -5 \\ -3y - 5 + 3y &= -5 \\ -5 &= -5 \end{aligned}$$

What now? Does this mean
y can be any number?

In general, arriving at a true equation, such as $-5 = -5$, would mean that the variable could take any value, but in this case there is an error instead.

In this bogus solution, the student first solves x from the first equation, and then substitutes the expression for x into the <u>same</u> equation (the first one).

To rectify the situation, the expression $-3y - 5$ needs substituted in the place of x in the <u>second</u> equation:

(2)
$$\begin{aligned} 2(-3y - 5) + y &= 3y - 1 \\ -6y - 10 + y &= 3y - 1 \\ -5y - 10 &= 3y - 1 \\ -8y &= 9 \\ y &= -9/8 \end{aligned}$$

Now, substitute $-9/8$ in place of y in either of the original equations, for example equation 1:

(1)
$$\begin{aligned} x + 3y &= -5 \\ x + 3(-9/8) &= -5 \\ x - 27/8 &= -5 \\ x &= 27/8 - 5 \\ x &= 27/8 - 40/8 = -13/8 \end{aligned}$$

Solution: $(-13/8, -9/8)$

b.

$$\begin{cases} -4x - y = 2(x - 3) & (1) \\ x + y = 3 & (2) \end{cases}$$

$$\downarrow$$

$$\begin{cases} -4x - y = 2(x - 3) & (1) \\ y = 3 - x & (2) \end{cases}$$

$$\downarrow$$

(1)
$$\begin{aligned} -4x - (3 - x) &= 2(x - 3) \\ -3x - 3 &= 2x - 6 \\ -5x - 3 &= -6 \\ -5x &= -3 \\ x &= 3/5 \end{aligned}$$

The solution is 3/5.

In this bogus solution, the student forgets to solve for y. Let's continue, by substituting 3/5 in place of x in the second equation:

(2)
$$\begin{aligned} x + y &= 3 \\ 3/5 + y &= 3 \\ y &= 3 - 3/5 \\ y &= 12/5 \end{aligned}$$

Solution: (3/5, 12/5)

Page 134

4.

$$\begin{cases} 5x - 2y = 10 & (1) \\ 3x - 8y = 4y - 48 & (2) \end{cases}$$

$$\downarrow$$

$$\begin{cases} 5x = 2y + 10 & (1) \\ 3x - 8y = 4y - 48 & (2) \end{cases}$$

$$\downarrow$$

$(2) \quad 3(2y + 10) - 8y = 4y - 48$

$6y + 30 - 8y = 4y - 48$

$-2y + 30 = 4y - 48$

$-6y = -78$

$y = 13$

The error is in the highlighted row above. In it, $2y + 10$ is substituted in place of x. The first equation was transformed into $5x = 2y + 10$, but x does not equal $2y + 10$. One has to continue one more step from that to solve for x: $x = (2/5)y + 2$.

(Continued in the second column on the right.)

Now, substituting $x = (2/5)y + 2$ to the second equation, we get:

$(2) \quad 3x - 8y = 4y - 48$

$3((2/5)y + 2) - 8y = 4y - 48$

$(6/5)y + 6 - 8y = 4y - 48$

$6y + 30 - 40y = 20y - 240$

$-34y + 30 = 20y - 240$

$-54y = -270$

$y = 5$

Then we substitute $y = 5$ to the first equation:

$(1) \quad 5x - 2y = 10$

$5x - 2(5) = 10$

$5x - 10 = 10$

$5x = 20$

$x = 4$

Solution: $(4, 5)$

Page 135

5.

a. $\begin{cases} 2x = 14 - 2y \\ x + y = 2 \end{cases}$

Substituting $x = 2 - y$ to the first equation:

$2(2 - y) = 14 - 2y$

$4 - 2y = 14 - 2y$

$4 = 14$

This is a false equation, so the original system has no solutions.

b. $\begin{cases} 2 - 4x = y + 8 \\ x - y = 5x + 6 \end{cases}$

Solving y from the first equation, we get $y = -6 - 4x$. Substituting that to the second equation:

$x - (-6 - 4x) = 5x + 6$

$x + 6 + 4x = 5x + 6$

$5x + 6 = 5x + 6$

At this point, we can see that x can be any number.

Solution: All points (x, y) that satisfy the equation $x - y = 5x + 6$.

(Note: the equation listed in the solution can be either of the original equations, or any equation equivalent to either, such as $y = -6 - 4x$.)

6.

---

a. $\begin{cases} 4x - 2y = 5 \\ y = 5x + 8 \end{cases}$

Substituting $y = 5x + 8$ to the first equation:

$$4x - 2y = 5$$
$$4x - 2(5x + 8) = 5$$
$$4x - 10x - 16 = 5$$
$$-6x - 16 = 5$$
$$-6x = 21$$
$$x = -21/6 = -7/2$$

Now, we substitute $x = -7/2$ in the second equation:

$y = 5x + 8 = 5(-7/2) + 8 = -35/2 + 16/2 = -19/2$.

Solution: $(-7/2, -19/2)$

---

b. $\begin{cases} 4x - 5y = 13 \\ y = 2(5 - x) \end{cases}$

Substituting $y = 2(5 - x)$ to the first equation:

$$4x - 5y = 13$$
$$4x - 5(2(5 - x)) = 13$$
$$4x - 5(10 - 2x) = 13$$
$$4x - 50 + 10x = 13$$
$$14x - 50 = 13$$
$$14x = 63$$
$$x = 63/14 = 9/2$$

Now, we substitute $x = 9/2$ in the second equation:

$y = 2(5 - 9/2) = 10 - 9 = 1$

Solution: $(9/2, 1)$

---

c. $\begin{cases} x - 2y = 5 - 3.5y \\ 2(5 - x) = 3y \end{cases}$

Solving x from the first equation, we get $x = 5 - 1.5y$. Substituting that to the second equation:

$$2[5 - (5 - 1.5y)] = 3y$$
$$2[5 - 5 + 1.5y] = 3y$$
$$2[1.5y] = 3y$$
$$3y = 3y$$

At this point, we can see that y can be any number.

Solution: All points (x, y) that satisfy the equation $2(5 - x) = 3y$.

(Note: the equation listed in the solution can be either of the original equations, or any equation equivalent to either, such as $x = 5 - 1.5y$.)

---

d. $\begin{cases} -4x - 3y = 2 \\ y + 4x = -6 \end{cases}$

We can solve y from the 2nd equation, to get that $y = -4x - 6$. Next, we substitute that in place of y in the first equation:

$$-4x - 3y = 2$$
$$-4x - 3(-4x - 6) = 2$$
$$-4x + 12x + 18 = 2$$
$$8x + 18 = 2$$
$$8x = -16$$
$$x = -2$$

Now, we substitute $x = -2$ in the second equation: $y + 4(-2) = -6$, from which $y = 2$.

Solution: $(-2, 2)$

---

e. $\begin{cases} 9x = 2y \\ 9x - 2y = 2 \end{cases}$

Rearranging the first equation, we get:

$$\begin{cases} -9x + 2y = 0 \\ 9x - 2y = 2 \end{cases}$$

Now, multiplying the top equation by -1, we get:

$$\begin{cases} 9x - 2y = 0 \\ 9x - 2y = 2 \end{cases}$$

This system has no solutions.

---

f. $\begin{cases} 2(2x - y) = 2 \\ x + 6y = 1 \end{cases}$

Solving x from the 2nd equation, $x = -6y + 1$. Next, we substitute that in place of x in the first equation:

$$(1) \quad 2[2(-6y + 1) - y] = 2$$
$$2[-12y + 2 - y] = 2$$
$$-12y + 2 - y = 1$$
$$-13y + 2 = 1$$
$$-13y = -1$$
$$y = 1/13$$

Now, we substitute $y = 1/13$ in the second equation: $x + 6(1/13) = 1$, from which $x = 7/13$.

Solution: $(7/13, 1/13)$

Solving Systems of Equations by Substitution, cont.

7.

a. $\begin{cases} 2x - 5y = 20 \\ -x - y = 3 \end{cases}$

Rearranging the second equation, we get $y = -x - 3$. Substituting that in the first equation, we get:

$$\begin{aligned} (2) \quad 2x - 5y &= 20 \\ 2x - 5(-x - 3) &= 20 \\ 2x + 5x + 15 &= 20 \\ 7x + 15 &= 20 \\ 7x &= 5 \\ x &= 5/7 \end{aligned}$$

Now, substituting $x = 5/7$ to the second equation, we get:
$-5/7 - y = 3$, from which $y = -5/7 - 3 = -26/7$

Solution: (5/7, −26/7) or about (0.7, −3.7), like the graph shows.

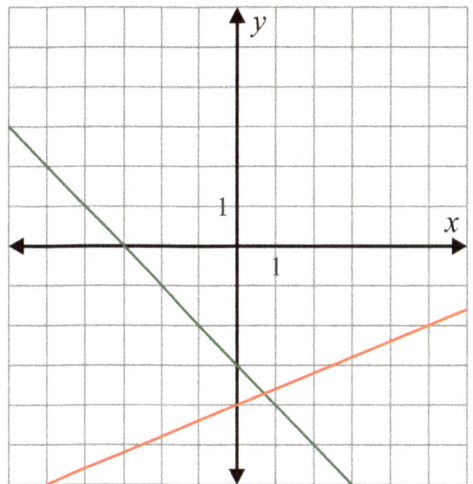

b. $\begin{cases} -x + 5y = 20 \\ y = 3x \end{cases}$

Substituting $y = 3x$ in the first equation, we get:

$$\begin{aligned} (1) \quad -x + 5y &= 20 \\ -x + 5(3x) &= 20 \\ 14x &= 20 \\ x &= 10/7 \text{ (about 1.4)} \end{aligned}$$

Now, substituting $x = 10/7$ to the second equation, we get:

$y = 3x = 3(10/7) = 30/7$ which is about 4.3. We can see in the graph that the intersection point is about (1.4, 4.3).

Solution: (10/7, 30/7)

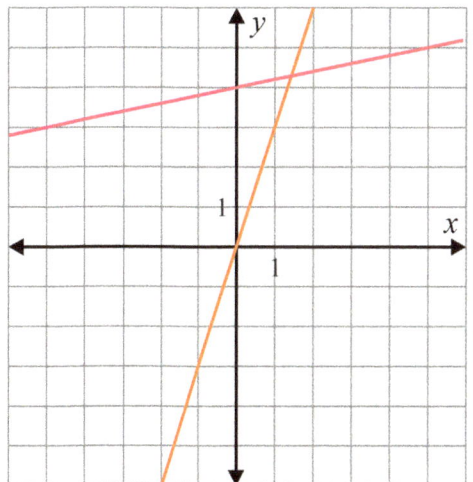

8.

a. $\begin{cases} 0.2x - 0.7y = 1 \quad (1) \\ x + y = 4 \quad (2) \end{cases}$

From (2), we can solve that $x = 4 - y$. Substituting $4 - y$ for x in the first equation, we get:

$$\begin{aligned} 0.2(4 - y) - 0.7y &= 1 \\ 0.8 - 0.2y - 0.7y &= 1 \\ 0.8 - 0.9y &= 1 \\ -0.9y &= 0.2 \\ y &= -2/9 \approx -0.22 \end{aligned}$$

Now, we substitute $y = -2/9$ in the second equation:

$$\begin{aligned} x + y &= 4 \\ x - 2/9 &= 4 \\ x &= 4\,2/9 = 38/9 \approx 4.22 \end{aligned}$$

Solution: (4.22, −0.22)

b. $\begin{cases} 1.2x - 3y = -2 \\ y = -0.6x + 5 \end{cases}$

Substituting $-0.6x + 5$ for y in the first equation:

$$\begin{aligned} 1.2x - 3y &= -2 \\ 1.2x - 3(-0.6x + 5) &= -2 \\ 1.2x + 1.8x - 15 &= -2 \\ 3x - 15 &= -2 \\ 3x &= 13 \\ x &= 13/3 \approx 4.33 \end{aligned}$$

Now, we substitute $x = 13/3$ in the second equation:

$$\begin{aligned} y &= -0.6(13/3) + 5 \\ y &= -0.6(13/3) + 5 \\ y &= -2.6 + 5 = 2.4 \end{aligned}$$

Solution: (4.33, 2.4)

9.

a. $\begin{cases} x = -(y-2) \\ 3y = 10(2-x) \end{cases}$

Substituting $-(y-2)$ in place of x in the second equation, we get:

$$3y = 10(2-x)$$
$$3y = 10(2-(-(y-2)))$$
$$3y = 10(2-(-y+2))$$
$$3y = 10(2+y-2)$$
$$3y = 20+10y-20$$
$$0 = 7y$$
$$y = 0$$

Now, we substitute $y = 0$ in the first equation:

$$x = -(y-2)$$
$$x = -(0-2)$$
$$x = 2$$

Solution: $(2, 0)$

b. $\begin{cases} 4x - 3y = -1 \\ 3x + y = -2 \end{cases}$

From the second equation, $y = -3x - 2$. Substituting $-3x - 2$ in place of y in the first equation, we get:

$$4x - 3y = -1$$
$$4x - 3(-3x - 2) = -1$$
$$4x + 9x + 6 = -1$$
$$13x + 6 = -1$$
$$13x = -7$$
$$x = -7/13$$

Now, we substitute $x = -7/13$ in the second equation:

$$3x + y = -2$$
$$3(-7/13) + y = -2$$
$$-21/13 + y = -2$$
$$y = 21/13 - 2 = 21/13 - 26/13 = -5/13$$

Solution: $(-7/13, -5/13)$

c. $\begin{cases} x = -3(y+8) \\ -x + 3y = 0 \end{cases}$

Substituting $-3(y+8)$ in place of x in the second equation, we get:

$$-x + 3y = 0$$
$$-(-3(y+8)) + 3y = 0$$
$$3(y+8) + 3y = 0$$
$$3y + 24 + 3y = 0$$
$$6y = -24$$
$$y = -4$$

Now, we substitute $y = -4$ in the first equation:

$$x = -3(y+8)$$
$$x = -3(-4+8)$$
$$x = -12$$

Solution: $(-12, -4)$

d. $\begin{cases} 3(x-2) + y = 12 \\ y = 4x - 3 \end{cases}$

Substituting $4x - 3$ in place of y in the first equation, we get:

$$3(x-2) + y = 12$$
$$3(x-2) + (4x-3) = 12$$
$$3x - 6 + 4x - 3 = 12$$
$$7x - 9 = 12$$
$$7x = 21$$
$$x = 3$$

Now, we substitute $x = 3$ in the second equation:

$$y = 4x - 3$$
$$y = 4(3) - 3$$
$$y = 9$$

Solution: $(3, 9)$

9.

e. $\begin{cases} 2x - 8y = 15 \\ -x + y = 20 \end{cases}$

From the second equation, we can solve that $y = x + 20$. Substituting $x + 20$ in place of y in the first equation, we get:

$$\begin{aligned} 2x - 8y &= 15 \\ 2x - 8(x + 20) &= 15 \\ 2x - 8x - 160 &= 15 \\ -6x - 160 &= 15 \\ -6x &= 175 \\ x &= -175/6 \end{aligned}$$

Now, we substitute $x = -175/6$ in the second equation:

$$\begin{aligned} -x + y &= 20 \\ -(175/6) = y &= 20 \\ y &= 20 - 175/6 = 120/6 - 175/6 = -55/6 \end{aligned}$$

Solution: $(-175/6, -55/6)$

f. $\begin{cases} x - 3(3y + 2) = 1 \\ 3x + 5y = -1 \end{cases}$

From the first equation, we can solve that $x = 3(3y + 2) + 1$. Substituting $3(3y + 2) + 1$ in place of x in the second equation, we get:

$$\begin{aligned} 3x + 5y &= -1 \\ 3(3(3y + 2) + 1) + 5y &= -1 \\ 3(9y + 6 + 1) + 5y &= -1 \\ 27y + 18 + 3 + 5y &= -1 \\ 32y + 21 &= -1 \\ 32y &= -22 \\ y &= -11/16 \end{aligned}$$

Now, we substitute $y = -11/16$ in the second equation:

$$\begin{aligned} 3x + 5y &= -1 \\ 3x + 5(-11/16) &= -1 \\ 3x - 55/16 &= -1 \\ 3x &= 55/16 - 1 \\ 3x &= 39/16 \\ x &= 13/16 \end{aligned}$$

Solution: $(13/16, -11/16)$

Applications, Part 1, pp. 139-142

1. Let a be the number of adults and c be the number of children that went to the park. We get these two equations:

$$\begin{cases} a + c = 15 \\ 70a + 45c = 900 \end{cases}$$

Solving for a from the first equation, we get $a = 15 - c$. Now substituting that to the second equation, we get:

$$\begin{aligned} 70a + 45c &= 900 \\ 70(15 - c) + 45c &= 900 \\ 1050 - 70c + 45c &= 900 \\ 1050 - 25c &= 900 \\ -25c &= -150 \\ c &= 6 \end{aligned}$$

If c is 6, then a is $15 - 6 = 9$. So, there were 6 children and 9 adults.

2. Let x be the number of hours Suzie skated during the week, and y be the number of hours she skated during the weekends. We get these two equations:

$$\begin{cases} x + y = 26 \\ 5x + 7y = 142 \end{cases}$$

Solving for x from the first equation, we get $x = 26 - y$. Now substituting that to the second equation, we get:

$$\begin{aligned} 5x + 7y &= 142 \\ 5(26 - y) + 7y &= 142 \\ 130 - 5y + 7y &= 142 \\ 130 + 2y &= 142 \\ 2y &= 12 \\ y &= 6 \end{aligned}$$

If y is 6, then x is $26 - 6 = 20$. So, she skated 20 hours during the weekdays and 6 hours during the weekends.

Page 140

3. No, the equations are not correct. Almost though! The second equation needs corrected to be $10d + 5n = 500$ (because we're dealing with cents, not dollars).

We can solve $d = 76 - n$ from the first equation. Substituting that in place of d in the second, we get:

$$
\begin{aligned}
10d + 5n &= 500 \\
10(76 - n) + 5n &= 500 \\
760 - 10n + 5n &= 500 \\
760 - 5n &= 500 \\
-5n &= -260 \\
n &= 52
\end{aligned}
$$

Then $d = 76 - 52 = 24$. Jamie has 24 dimes and 52 nickels.

4. Let q be the number of quarters and d be the number of dimes Lily has. Then:

$$
\begin{cases}
q + d = 44 \\
25q + 10d = 665
\end{cases}
$$

We can solve $q = 44 - d$ from the first equation. Substituting that in place of q in the second, we get:

$$
\begin{aligned}
25q + 10d &= 665 \\
25(44 - d) + 10d &= 665 \\
1100 - 25d + 10d &= 665 \\
1100 - 15d &= 665 \\
-15d &= -435 \\
d &= 29
\end{aligned}
$$

Then $q = 44 - 29 = 15$. Lily has 15 quarters and 29 dimes.

Page 141

5. Let s be the number of sheep and d be the number of ducks. Then, $s + d = 57$ and $4s + 2d = 158$. We can solve that $s = 57 - d$ from the first equation. Substituting $57 - d$ in place of s in the second equation, we get:

$$
\begin{aligned}
4s + 2d &= 158 \\
4(57 - d) + 2d &= 158 \\
228 - 4d + 2d &= 158 \\
228 - 2d &= 158 \\
-2d &= -70 \\
d &= 35
\end{aligned}
$$

There were 35 ducks and $57 - 35 = 22$ sheep.

Page 141

6. Let x be the length of the longer route and y the length of the shorter. Then:

$$
\begin{cases}
x = 2y - 9 \\
y = (4/5)x
\end{cases}
$$

Substituting $(4/5)x$ in place of y in the top equation:

$$
\begin{aligned}
x &= 2y - 9 \\
x &= 2[(4/5)x] - 9 \\
x &= (8/5)x - 9 \\
9 &= (8/5)x - x \\
9 &= (3/5)x \\
x &= (5/3) \cdot 9 = 15
\end{aligned}
$$

Then, $y = (4/5)x = (4/5) \cdot 15 = 12$. So, the longer route is 15 miles and the shorter route is 12 miles.

7. Let A be Ava's age and E be Eva's age. Then:

$$
\begin{cases}
A = 2E - 15 \\
E = (6/7)A
\end{cases}
$$

Substituting $(6/7)A$ in place of E in the top equation:

$$
\begin{aligned}
A &= 2E - 15 \\
A &= 2[(6/7)A] - 15 \\
A &= (12/7)A - 15 \\
15 &= (5/7)A \\
(7/5)15 &= A \\
A &= 21
\end{aligned}
$$

Then, $E = (6/7)A = (6/7) \cdot 21 = 18$. Ava is 21 years and Eva is 18 years.

Page 142

8. Let J be Juan's age now, and H be Henry's age now. Then:

$$
\begin{cases}
J + 10 = 2(H + 10) \\
J - 5 = 5(H - 5)
\end{cases}
$$

Solving J from the first equation, we get $J = 2(H + 10) - 10 = 2H + 10$. Let's substitute that for J in the second equation:

$$
\begin{aligned}
2H + 10 - 5 &= 5(H - 5) \\
2H + 5 &= 5H - 25 \\
30 &= 3H \\
10 &= H
\end{aligned}
$$

Then, $J = 2H + 10 = 2(10) + 10 = 30$. Juan is 30 years old and Henry is 10 years old.

Applications, Part 1, cont.

Page 142

9. Let f be the number of fruit bars and n be the number of nut bars that Amanda sold. Then,

$$\begin{cases} n = f + 5 \\ 3f + 5n = 121 \end{cases}$$

Substituting $f + 5$ in place of n in the second equation:

$$\begin{aligned} 3f + 5n &= 121 \\ 3f + 5(f + 5) &= 121 \\ 3f + 5f + 25 &= 121 \\ 8f + 25 &= 121 \\ 8f &= 96 \\ f &= 12 \end{aligned}$$

Then, $n = f + 5 = 17$. She sold 12 fruit bars and 17 nut bars.

Puzzle corner. Let q be the number of quarters Diane has. Then, she has $2q$ dimes, and $2q + 10$ nickels. The total worth of her coins is $25q + 10(2q) + 5(2q + 10)$ which equals 325 cents:

$$\begin{aligned} 25q + 10(2q) + 5(2q + 10) &= 325 \\ 25q + 20q + 10q + 50 &= 325 \\ 55q + 50 &= 325 \\ 55q &= 275 \\ q &= 5 \end{aligned}$$

She has 5 quarters, 10 dimes, and 20 nickels.

The Addition Method, Part 1, pp. 143–147

Page 143

1.

a.
$$\begin{cases} x + y = -7 \\ 3x - y = 3 \end{cases}$$
$$\begin{aligned} 4x &= -4 \\ x &= -1 \end{aligned}$$

Then, from the top equation, we can solve that $y = -x - 7 = -(-1) - 7 = -6$.

Solution: $(-1, -6)$

b.
$$\begin{cases} -x + 3y = 11 \\ x + 2y = 9 \end{cases}$$
$$\begin{aligned} 5y &= 20 \\ y &= 4 \end{aligned}$$

Then, from the bottom equation, we can solve that $x = -2y + 9 = -2(4) + 9 = 1$.

Solution: $(1, 4)$

Page 144

2.

a.
$$\begin{cases} 2x + 2y = 7 \\ -2x - 12y = 23 \end{cases}$$
$$\begin{aligned} -10y &= 30 \\ y &= -3 \end{aligned}$$

Then we can solve x from the top equation:

$$\begin{aligned} 2x + 2y &= 7 \\ 2x &= 7 - 2y \\ x &= (7 - 2y)/2 \end{aligned}$$

Substituting now $y = -3$, we get
$x = (7 - 2(-3))/2 = 13/2$

Solution: $(13/2, -3)$

b.
$$\begin{cases} 8x - 5y = 9 \\ -8x - y = 21 \end{cases}$$
$$\begin{aligned} -6y &= 30 \\ y &= -5 \end{aligned}$$

Then we can solve x from the bottom equation:

$$\begin{aligned} -8x - y &= 21 \\ -8x &= y + 21 \\ x &= (-y - 21)/8 \end{aligned}$$

Substituting now $y = -5$, we get
$x = (-(-5) - 21)/8 = -16/8 = -2$.

Solution: $(-2, -5)$

157

Page 144

2.

c.
$$\begin{cases} 2x - 3y &= 11 \\ -2x + 3y &= -11 \end{cases}$$
$$\overline{\ 0 \qquad = 0}$$

This means the two lines are the same line (which is also easy to see by simple inspection).

Solution: All points (x, y) that satisfy $2x - 3y = 11$.

This time, we will *subtract* the equations.

d.
$$-\begin{cases} 4x + 9y &= 16 \\ 4x + 2y &= 2 \end{cases}$$
$$\overline{\qquad\quad 7y = 14}$$
$$y = 2$$

Then we can solve x from the bottom equation:

$$4x + 2y = 2$$
$$4x = 2 - 2y$$
$$x = (2 - 2y)/4 = (1 - y)/2$$

Substituting now $y = 2$, we get
$x = (1 - 2)/2 = -1/2$.

Solution: $(-1/2, 2)$

Page 145

3.

a.
$$\begin{cases} 7x &= 10 - 4y \\ -x - 4y &= 26 \end{cases} \rightarrow \begin{cases} 7x + 4y = 10 \\ -x - 4y = 26 \end{cases}$$
$$\overline{\qquad 6x \qquad = 36}$$
$$x \qquad = 6$$

Then we can solve y from the top equation:

$$7x + 4y = 10$$
$$4y = 10 - 7x$$
$$y = (10 - 7x)/4$$

Substituting now $x = 6$, we get
$y = (10 - 7(6))/4 = -32/4 = -8$.

Solution: $(6, -8)$

b.
$$\begin{cases} -2x + 6y &= 4 \\ 4 + 5x &= 2y + 9 + 4y \end{cases} \rightarrow \begin{cases} -2x + 6y = 4 \\ 5x - 6y = 5 \end{cases}$$
$$\overline{\qquad 3x \qquad = 9}$$
$$x \qquad = 3$$

Then we can solve y from the top equation:

$$-2x + 6y = 4$$
$$6y = 4 + 2x$$
$$y = (4 + 2x)/6 = (x + 2)/3$$

Substituting now $x = 3$, we get
$y = (3 + 2)/3 = 5/3$.

Solution: $(3, 5/3)$

c.
$$\begin{cases} 10 &= 2x - 4y - 2 \\ -2x + 7y &= 18 + 2y \end{cases} \rightarrow \begin{cases} 2x - 4y = 12 \\ -2x + 5y = 18 \end{cases}$$
$$\overline{\qquad\qquad y = 30}$$

Then we can solve x from the top equation:

$$10 = 2x - 4y - 2$$
$$12 + 4y = 2x$$
$$x = 2y + 6$$

Substituting now $y = 30$, we get
$x = 2(30) + 6 = 66$.

Solution: $(66, 30)$

d.
$$\begin{cases} 50x + 120 &= 20y \\ 35y - 50x &= 180 \end{cases} \rightarrow \begin{cases} 50x - 20y = -120 \\ -50x + 35y = 180 \end{cases}$$
$$\overline{\qquad\qquad 15y = 60}$$
$$y = 4$$

Then we can solve x from the top equation:

$$50x + 120 = 20y$$
$$50x = 20y - 120$$
$$x = (2y - 12)/5$$

Substituting now $y = 4$, we get
$x = (2(4) - 12)/5 = -4/5$.

Solution: $(-4/5, 4)$

The Addition Method, Part 1, cont.

4. Since as per the top diagram, $2x + y = 22$, then in the bottom diagram, top line, we can replace $2x + y$ with 22, which leaves only one x and 22 to equal 31. So, $x = 9$. Then since $2x + y = 22$, and $x = 9$, it follows that $y = 4$.

5. a. From the bottom diagram, we can see that $x + 2y = 34$. In the top diagram, top line, we have $3x$ and $2y$. One of those x's plus two of the y's equal 34, leaving the top line to equal $2x + 34$. This $2x + 34$ equals 62, from which we can solve that $x = 14$. Looking at the bottom diagram again, if x is 14, then two y's will equal 20, and $y = 10$.

b. Here, the bottom diagram tells us that $x = 2y + 1$. If we substitute $2y + 1$ in place of each of the three x's in the top diagram, top line, the top line becomes $6y + 3$. But this is equal to $3y + 15$. Solving $6y + 3 = 3y + 15$, we get:

$$6y + 3 = 3y + 15$$
$$3y + 3 = 15$$
$$3y = 12$$
$$y = 4$$

Then, $x = 2y + 1 = 2(4) + 1 = 9$. Solution: $(9, 4)$.

6.

a. $\begin{cases} 2x - 10y = -36 \\ x + 10y = 24 \end{cases}$ $\ \ 3x \qquad\quad = -12$ $\qquad x = -4$ Substituting $x = -4$ in the second equation, we get: $-4 + 10y = 24$ $10y = 28$ $y = 14/5$ or 2.8 Solution: $(-4, 14/5)$	b. After applying the distributive property, since both equations have the term $5x$, we will subtract the equations, instead of adding. $\begin{cases} 5(x + y) = 20 \\ 5x + 4y = 34 \end{cases} \rightarrow \begin{array}{r} \begin{cases} 5x + 5y = 20 \\ 5x + 4y = 34 \end{cases} \\ \hline y = -14 \end{array}$ Substituting $y = -14$ in the second equation, we get: $5x + 4(-14) = 34$ $5x - 56 = 34$ $5x = 90$ $x = 18$ Solution: $(18, -14)$
c. Since both equations have the term $5x$, we will subtract the equations, instead of adding. $\begin{cases} 5x - 11 = -2y \\ 5x + 2y = 14 \end{cases} \rightarrow \begin{array}{r} \begin{cases} 5x + 2y = 11 \\ 5x + 2y = 14 \end{cases} \\ \hline 0 = -3 \end{array}$ We get a false equation, so the system has no solutions.	d. $\begin{cases} 4(x - 1) = 2y \\ 3y = -4x + 8 \end{cases} \rightarrow \begin{cases} 4x - 4 = 2y \\ 4x + 3y = 8 \end{cases}$ (In the next step, we will subtract the equations.) $\begin{array}{r} \begin{cases} 4x - 2y = 4 \\ 4x + 3y = 8 \end{cases} \\ \hline -5y = -4 \\ y = 4/5 \end{array}$ Substituting $y = 4/5$ in the second equation, we get: $3(4/5) = -4x + 8$ $12/5 = -4x + 8$ $12 = -20x + 40$ $20x = 28$ $x = 28/20 = 7/5$ Solution: $(7/5, 4/5)$

The Addition Method, Part 1, cont.

6.

e.
$\begin{cases} 0.2x - 2 = 0.1y \\ 0.1y = -0.2x + 2 \end{cases}$ $\rightarrow$ $\begin{cases} 0.2x - 0.1y = 2 \\ 0.2x + 0.1y = 2 \end{cases}$
$$0.4x = 4$$
$$x = 10$$

Substituting $x = 10$ in the first equation, we get:
$$0.2(\mathbf{10}) - 2 = 0.1y$$
$$2 - 2 = 0.1y$$
$$0 = 0.1y$$
$$y = 0$$

Solution: $(10, 0)$

f.
$\begin{cases} 200x - 120y = 200 \\ 120y + 400x = 600 \end{cases}$ $\rightarrow$ $\begin{cases} 200x - 120y = 200 \\ 400x + 120y = 600 \end{cases}$
$$600x = 800$$
$$x = 4/3$$

Substituting $x = 4/3$ in the first equation, we get:
$$200(\mathbf{4/3}) - 120y = 200$$
$$800/3 - 120y = 200$$
$$800 - 360y = 600$$
$$-360y = -200$$
$$y = 200/360 = 5/9$$

Solution: $(4/3, 5/9)$

The Addition Method, Part 2, pp. 148-152

1.

a.
$\begin{cases} 3x + y = -7 \\ -6x - 4y = 8 \end{cases}$ $\Big| \cdot 2$

$\downarrow$

$\begin{cases} 6x + 2y = -14 \\ -6x - 4y = 8 \end{cases}$
$$+ \qquad -2y = -6$$
$$y = 3$$

Substituting $y = 3$ in the first equation, we get:
$$3x + \mathbf{3} = -7$$
$$3x = -10$$
$$x = -10/3$$

Solution: $(-10/3, 3)$

b.
$\begin{cases} -5x + 3y = -8 \\ 3x - y = 4 \end{cases}$ $\Big| \cdot 3$

$\downarrow$

$\begin{cases} -5x + 3y = -8 \\ 9x - 3y = 12 \end{cases}$
$$+ \qquad 4x = 4$$
$$x = 1$$

Substituting $x = 1$ in the second equation, we get:
$$3(\mathbf{1}) - y = 4$$
$$3 - y = 4$$
$$-y = 1$$
$$y = 1$$

Solution: $(1, -1)$

2. a. Multiply the top equation by 3 and the bottom equation by 7.
Or, multiply the top equation by −5 and the bottom equation by 4.
Or, multiply the top equation by 5 and the bottom equation by −4.

b. Multiply the top equation by −4, and the bottom equation by 3.
Or, multiply the top equation by 4, and the bottom equation by −3.
Or, multiply the top equation by −9 and the bottom equation by 2.
Or, multiply the top equation by 9 and the bottom equation by −2.

Page 149

3.

a. $\begin{cases} 2x + 7y = -3 \\ 3x - 2y = 3 \end{cases} \quad \begin{matrix} \cdot\ 3 \\ \cdot\ (-2) \end{matrix}$

$\downarrow$

$+ \begin{cases} 6x + 21y = -9 \\ -6x + 4y = -6 \end{cases}$
$\overline{\qquad\qquad 25y = -15}$
$\qquad\qquad\quad y = -15/25 = -3/5$

Substituting $y = -3/5$ in the first equation, we get:

$2x + 7(-3/5) = -3$
$2x - 21/5 = -3$
$2x = 21/5 - 15/5$
$2x = 6/5$
$x = 3/5$

Solution: $(3/5, -3/5)$

b. $\begin{cases} 6x - 2y = -38 \\ -10x + 5y = 70 \end{cases} \quad \begin{matrix} \cdot\ 5 \\ \cdot\ 2 \end{matrix}$

$\downarrow$

$+ \begin{cases} 30x - 10y = -190 \\ -20x + 10y = 140 \end{cases}$
$\overline{\qquad\quad 10x \qquad\quad = -50}$
$\qquad\qquad\quad x = -5$

Substituting $x = -5$ in the first equation, we get:

$6(-5) - 2y = -38$
$-30 - 2y = -38$
$-2y = -8$
$y = 4$

Solution: $(-5, 4)$

Page 150

4.

a. $\begin{cases} 6x - 2y = -20 \\ 12x + 6y = 30 \end{cases} \quad \begin{matrix} \cdot\ 3 \end{matrix}$

$\downarrow$

$+ \begin{cases} 18x - 6y = -60 \\ 12x + 6y = 30 \end{cases}$
$\overline{\quad 30x \qquad = -30}$
$\qquad\quad x = -1$

Substituting $x = -1$ in the first equation, we get:

$6(-1) - 2y = -20$
$-6 - 2y = -20$
$-2y = -14$
$y = 7$

Solution: $(-1, 7)$

b. $\begin{cases} 10x - 6y = 48 \\ (5/3)x - y = 8 \end{cases} \quad \begin{matrix} \cdot\ (-1) \\ \cdot\ 6 \end{matrix}$

$\downarrow$

$+ \begin{cases} -10x + 6y = -48 \\ 10x - 6y = 48 \end{cases}$
$\overline{\qquad\quad 0 = 0}$

This means the two lines are the same line.

Solution: All points (x, y) that satisfy $10x - 6y = 48$.

Page 150

4.

c.
$$\begin{cases} 2x + 3y = -19 \\ -3x + 5y = 95 \end{cases} \quad \begin{array}{l} \cdot (-5) \\ \cdot 3 \end{array}$$

$\downarrow$

$$+ \begin{cases} -10x - 15y = 95 \\ -9x + 15y = 285 \end{cases}$$
$$\overline{\quad -19x \qquad\quad = 380}$$
$$x = -20$$

Substituting $x = -20$ in the first equation, we get:

$$2(-20) + 3y = -19$$
$$-40 + 3y = -19$$
$$3y = 21$$
$$y = 7$$

Solution: $(-20, 7)$

d.
$$\begin{cases} 2x - 5y = -11 \\ 7x + 3y = -100 \end{cases} \quad \begin{array}{l} \cdot (-7) \\ \cdot 2 \end{array}$$

$\downarrow$

$$+ \begin{cases} -14x + 35y = 77 \\ 14x + 6y = -200 \end{cases}$$
$$\overline{\qquad\quad 41y = -123}$$
$$y = -3$$

Substituting $y = -3$ in the first equation, we get:

$$2x - 5(-3) = -11$$
$$2x + 15 = -11$$
$$2x = -26$$
$$x = -13$$

Solution: $(-13, -3)$

Page 151

5. The error is in the multiplication by 8. When the top equation is multiplied by 8, the term $-9y$ should become $-72y$, not $-70y$. The corrected version is on the right.

$$\begin{cases} -7x - 9y = -5 \\ 6x + 8y = 4 \end{cases} \quad \begin{array}{l} \cdot 8 \\ \cdot 9 \end{array} \quad \rightarrow \quad + \begin{cases} -56x - 72y = -40 \\ 54x + 72y = 36 \end{cases}$$
$$\overline{\quad -2x \qquad\quad = -4}$$
$$x = 2$$

Substituting $x = 2$ in the second equation, we get:

$$6(2) + 8y = 4$$
$$12 + 8y = 4$$
$$8y = -8$$
$$y = -1$$

Solution: $(2, -1)$

6. The error can be noticed in the addition of the two equations. The addition should produce $20x + 38y = 20$, not just $38y = 20$. The $10x$ and $10x$ do not cancel each other in addition since they're not a zero pair. The two equations should be *subtracted*, not added. Alternatively, in the beginning, the bottom equation could have been multiplied by -5 instead of by 5, and then the equations could have been added.

The solution $(13/32, 5/16)$ checks:

$$2(13/32) + 7(5/16) \overset{?}{=} 3$$
$$26/32 + 35/16 \overset{?}{=} 3$$
$$26/32 + 70/32 \overset{?}{=} 3$$
$$96/32 = 3$$

The faulty solution:

(1)
(2)
$$\begin{cases} 10x + 3y = 5 \\ 2x + 7y = 3 \end{cases} \quad \cdot 5$$

$\downarrow$

$$\begin{cases} 10x + 3y = 5 \\ 10x + 35y = 15 \end{cases}$$
$$\overline{\quad 20x + 38y = 20}$$
$$y = 10/19$$

These should have been added, also.

The correct solution:

(1)
(2)
$$\begin{cases} 10x + 3y = 5 \\ 2x + 7y = 3 \end{cases} \quad \cdot (-5)$$

$\downarrow$

$$\begin{cases} 10x + 3y = 5 \\ -10x - 35y = -15 \end{cases}$$
$$\overline{\quad -32y = -10}$$
$$y = 10/32 = 5/16$$

$\downarrow$

(1) $10x + 3(\mathbf{5/16}) = 5$
$$10x + 15/16 = 5$$
$$10x = 5 - 15/16$$
$$10x = 65/16$$
$$x = 65/160 = 13/32$$

Page 151

7. Yes, dividing both sides of an equation by something is fine. Dividing by number n is actually the same as multiplying by $1/n$.

See the solution on the right.

$$\begin{cases} -x + 5y = 11 \\ 3x - 12y = 18 \end{cases} \div 3$$

$$\downarrow$$

$$\begin{array}{r} \begin{cases} -x + 5y = 11 \quad \text{(1)} \\ x - 4y = 6 \quad \text{(2)} \end{cases} \\ \hline y = 17 \end{array}$$

Substituting $y = 17$ in the second equation, we get:

$$3x - 12(17) = 18$$
$$3x - 204 = 18$$
$$3x = 222$$
$$x = 74$$

Solution: $(74, 17)$.

Page 152

8.

a.
$$\begin{cases} 3x + 8y = 10 \\ 15x + 20y = 4 \end{cases} \cdot (-5)$$

$$\downarrow$$

$$\begin{array}{r} \begin{cases} -15x - 40y = -50 \\ 15x + 20y = 4 \end{cases} \\ \hline -20y = -46 \\ y = 2.3 \end{array}$$

Substituting $y = 2.3$ in the first equation, we get:

$$3x + 8(2.3) = 10$$
$$3x + 18.4 = 10$$
$$3x = -8.4$$
$$x = -2.8$$

Solution: $(-2.8, 2.3)$

b.
$$\begin{cases} -30x + 40y = 50 \\ 9x - 12y = -15 \end{cases} \begin{array}{l} \div 10 \\ \div 3 \end{array}$$

$$\downarrow$$

$$\begin{array}{r} \begin{cases} -3x + 4y = 5 \\ 3x - 4y = -5 \end{cases} \\ \hline 0 = 0 \end{array}$$

This means the two lines are the same line.

Solution: All points (x, y) that satisfy $3x - 4y = -5$.

c.
$$\begin{cases} 0.4x - 0.5y = 0 \\ 2.5x + 0.3y = 1 \end{cases} \begin{array}{l} \cdot 3 \\ \cdot 5 \end{array}$$

$$\downarrow$$

$$\begin{array}{r} \begin{cases} 1.2x - 1.5y = 0 \\ 12.5x + 1.5y = 5 \end{cases} \\ \hline 13.7x = 5 \\ x \approx 0.36496 \end{array}$$

Substituting $x = 0.36496$ in the first equation, we get:

$$0.4(0.36496) - 0.5y = 0$$
$$0.145984 - 0.5y = 0$$
$$0.145984 = 0.5y$$
$$y \approx 0.291968$$

Solution: $(0.36, 0.29)$

d.
$$\begin{cases} 40x - 20y = 140 \\ -30x + 15y = 90 \end{cases} \begin{array}{l} \cdot 3 \\ \cdot 4 \end{array}$$

$$\downarrow$$

$$\begin{array}{r} \begin{cases} 120x - 60y = 420 \\ -120x + 60y = 360 \end{cases} \\ \hline 0 = 780 \end{array}$$

The system has no solutions.

The Addition Method, Part 2, cont.

Page 152

8.

e.
$$\begin{cases} -0.3x + 0.7y = -0.5 \\ x + 0.9y = 0 \quad\Big|\cdot 0.3 \end{cases}$$

↓

$$\begin{array}{r} \{\begin{array}{rcl} -0.3x + 0.7y &=& -0.5 \\ + \quad 0.3x + 0.27y &=& 0 \end{array} \\ \hline \begin{array}{rcl} 0.97y &=& -0.5 \\ y &\approx& -0.51546 \end{array}\end{array}$$

Substituting $y = -0.51546$ in the second equation, we get:

$$\begin{aligned} x + 0.9(-0.51546) &= 0 \\ x - 0.463914 &= 0 \\ x &= 0.463914 \end{aligned}$$

Solution: $(0.46, -0.52)$

f.
$$\begin{cases} 150x - 0.25y = 17.5 \quad\Big|\cdot 4 \\ 600x + y = 70 \end{cases}$$

↓

$$\begin{array}{r} \{\begin{array}{rcl} 600x - y &=& 70 \\ + \quad 600x + y &=& 70 \end{array} \\ \hline \begin{array}{rcl} 1200x &=& 140 \\ x &=& 7/60 \approx 0.11667 \end{array}\end{array}$$

In this case, using the fraction works well because of the denominator of 60 and the 600 in the equation. When we substitute $x = 7/60$ in the second equation, we get:

$$\begin{aligned} 600(7/60) + y &= 70 \\ 70 + y &= 70 \\ y &= 0 \end{aligned}$$

Solution: $(7/60, 0)$ or $(0.12, 0)$

Puzzle corner. The system is this:

$$\begin{cases} 6x - by = 2 \\ ax + by = 3 \end{cases}$$

There are several ways to solve this. One is to immediately add the equations, arriving at the equation $(6 + a)x = 5$. Now substituting $x = 1$, we arrive at $6 + a = 5$, from which $a = -1$.

Then we can solve for b using the top equation and the facts that $x = 1$, $y = -3$, and $a = -1$. The top equation is $6x - by = 2$. Substituting the facts mentioned earlier, we get:

$$\begin{aligned} 6(1) - b(-3) &= 2 \\ 6 + 3b &= 2 \\ 3b &= -4 \\ b &= -4/3 \end{aligned}$$

So, $a = -1$ and $b = -4/3$.

More Practice, pp. 153-156

Page 153

1.
$$\begin{cases} -x - 6y = -10 \\ -x + 3y = 4 \quad\Big|\cdot 2 \end{cases} \rightarrow \begin{array}{r} \{\begin{array}{rcl} -x - 6y &=& -10 \\ + \quad -2x + 6y &=& 8 \end{array} \\ \hline \begin{array}{rcl} -3x &=& -2 \\ x &=& 2/3 \end{array}\end{array}$$

Substituting $x = 2/3$ in the first equation, we get:
$$\begin{aligned} -2/3 - 6y &= -10 \\ -6y &= -10 + 2/3 \\ -6y &= -30/3 + 2/3 \\ -6y &= -28/3 \\ y &= 28/18 = 14/9 \end{aligned}$$

Solution: $(2/3, 14/9)$

More Practice, cont.

Page 153

2.

a. $\begin{cases} 5y = -x + 7 \\ 4(y-1) = -3(x-2) \end{cases}$

$\downarrow$

$\begin{cases} x + 5y = 7 \\ 4y - 4 = -3x + 6 \end{cases}$

$\downarrow$

$\begin{cases} x + 5y = 7 \quad \Big| \cdot (-3) \\ 3x + 4y = 10 \end{cases}$

$\downarrow$

$\begin{aligned} &\begin{cases} -3x - 15y = -21 \\ 3x + 4y = 10 \end{cases} \\ +&\rule{5cm}{0.4pt} \\ &\phantom{+\{\ }-11y = -11 \\ &\phantom{+\{\ -11}y = 1 \end{aligned}$

Substituting $y = 1$ in the first equation, we get:

$5(1) = -x + 7$

$5 = -x + 7$

$-2 = -x$

$x = 2$

Solution: $(2, 1)$

b. $\begin{cases} 3y - 21 = 6(x + 1) \\ x + y + 3 = -4x + 5 \end{cases}$

$\downarrow$

$\begin{cases} 3y - 21 = 6x + 6 \\ 5x + y + 3 = 5 \end{cases}$

$\downarrow$

$\begin{cases} -6x + 3y = 27 \\ 5x + y = 2 \quad \Big| \cdot (-3) \end{cases}$

$\downarrow$

$\begin{aligned} &\begin{cases} -6x + 3y = 27 \\ -15x - 3y = -6 \end{cases} \\ +&\rule{5cm}{0.4pt} \\ &\phantom{+\{\ }-21x = 21 \\ &\phantom{+\{\ -21}x = -1 \end{aligned}$

Substituting $x = -1$ in the second equation, we get:

$(-1) + y + 3 = -4(-1) + 5$

$y + 2 = 4 + 5$

$y = 7$

Solution: $(-1, 7)$

Page 154

3.

a. $\begin{cases} 2y + 8x = 51 \\ x + 10y = 2(y + 10x) \end{cases}$

$\downarrow$

$\begin{cases} 8x + 2y = 51 \\ x + 10y = 2y + 20x \end{cases}$

$\downarrow$

$\begin{cases} 8x + 2y = 51 \quad \Big| \cdot (-4) \\ -19x + 8y = 0 \end{cases}$

$\downarrow$

$\begin{aligned} &\begin{cases} -32x - 8y = -204 \\ -19x + 8y = 0 \end{cases} \\ +&\rule{5cm}{0.4pt} \\ &\phantom{+\{\ }-51x = -204 \\ &\phantom{+\{\ -51}x = 4 \end{aligned}$

Substituting $x = 4$ in the first equation, we get:

$8(4) + 2y = 51$

$32 + 2y = 51$

$2y = 19$

$y = 9.5$

Solution: $(4, 9.5)$

b. $\begin{cases} 5y = 15x - 20 \\ -y - 12 + 13x = 2(5x - 4) \end{cases}$

$\downarrow$

$\begin{cases} -15x + 5y = -20 \\ -y - 12 + 13x = 10x - 8 \end{cases}$

$\downarrow$

$\begin{cases} -15x + 5y = -20 \\ 3x - y = 4 \quad \Big| \cdot 5 \end{cases}$

$\downarrow$

$\begin{aligned} &\begin{cases} -15x + 5y = -20 \\ 15x - 5y = 20 \end{cases} \\ +&\rule{5cm}{0.4pt} \\ &\phantom{+\{\ }0 = 0 \end{aligned}$

This means the two equations are equivalent.

Solution: All points (x, y) that satisfy $3x - y = 4$.

More Practice, cont.

Page 154

3.

c. $\begin{cases} x + y + 5 = -3x + 46 \\ -3(y+7) + 2x = 3(x+7) \end{cases}$

$\downarrow$

$\begin{cases} 4x + y = 41 \\ -3y - 21 + 2x = 3x + 21 \end{cases}$

$\downarrow$

$\begin{cases} 4x + y = 41 \\ -x - 3y = 42 \quad | \cdot 4 \end{cases}$

$\downarrow$

$\begin{aligned} &\begin{cases} 4x + y = 41 \\ -4x - 12y = 168 \end{cases} \\ +&\rule{4cm}{0.4pt} \\ & -11y = 209 \\ & y = -19 \end{aligned}$

Substituting $y = -19$ in the first equation, we get:

$x - 19 + 5 = -3x + 46$

$4x - 14 = 46$

$4x = 60$

$x = 15$

Solution: $(15, -19)$

---

d. $\begin{cases} 3(x+y) + 50 = 4x - 65 - 2y \\ 3(x-15) + 5y = 4y - 20 \end{cases}$

$\downarrow$

$\begin{cases} 3x + 3y + 50 = 4x - 65 - 2y \\ 3x - 45 + 5y = 4y - 20 \end{cases}$

$\downarrow$

$\begin{cases} -x + 5y = -115 \quad | \cdot 3 \\ 3x + y = 25 \end{cases}$

$\downarrow$

$\begin{aligned} &\begin{cases} -3x + 15y = -345 \\ 3x + y = 25 \end{cases} \\ +&\rule{4cm}{0.4pt} \\ & 16y = -320 \\ & y = -20 \end{aligned}$

Substituting $y = -20$ in the second equation, in the version $3x + y = 25$, we get:

$3x - 20 = 25$

$3x = 45$

$x = 15$

Solution: $(15, -20)$

Page 155

4.

a. $\begin{cases} 3x + y = 6 \\ -5x - \dfrac{y}{2} = 4 \quad | \cdot 2 \end{cases}$

$\downarrow$

$\begin{aligned} &\begin{cases} 3x + y = 6 \\ -10x - y = 8 \end{cases} \\ +&\rule{4cm}{0.4pt} \\ &-7x - 14 \\ & x = -2 \end{aligned}$

Substituting $x = -2$ in the first equation, we get:

$3(-2) + y = 6$

$-6 + y = 6$

$y = 12$

Solution: $(-2, 12)$

---

b. $\begin{cases} -x + 9y = 7 \\ \dfrac{x}{3} - 2y = 1 \quad | \cdot 3 \end{cases}$

$\downarrow$

$\begin{aligned} &\begin{cases} -x + 9y = 7 \\ x - 6y = 3 \end{cases} \\ +&\rule{4cm}{0.4pt} \\ & 3y = 10 \\ & y = 10/3 \end{aligned}$

Substituting $y = 10/3$ in the first equation, we get:

$-x + 9(10/3) = 7$

$-x + 30 = 7$

$-x = -23$

$x = 23$

Solution: $(23, 10/3)$

More Practice, cont.

5.

a. $\begin{cases} 2x + \dfrac{y}{2} = 1 & \bigg| \cdot 2 \\ \\ x - \dfrac{y}{3} = 4 & \bigg| \cdot 3 \end{cases}$

$\downarrow$

$\begin{aligned} & \begin{cases} 4x + y = 2 \\ 3x - y = 12 \end{cases} \\ \hline & 7x = 14 \\ & x = 2 \end{aligned}$

Substituting $x = 2$ in the first equation, we get:

$2(2) + y/2 = 1$

$4 + y/2 = 1$

$y/2 = -3$

$y = -6$

Solution: $(2, -6)$

b. $\begin{cases} \dfrac{x}{3} + y = 3 & \bigg| \cdot 3 \\ \\ \dfrac{2x}{3} - 2y = -1 & \bigg| \cdot 3 \end{cases}$

$\downarrow$

$\begin{cases} x + 3y = 9 & \big| \cdot (-2) \\ 2x - 6y = -3 \end{cases}$

$\downarrow$

$\begin{aligned} & \begin{cases} -2x - 6y = -18 \\ 2x - 6y = -3 \end{cases} \\ \hline & -12y = -21 \\ & y = 21/12 = 7/4 \end{aligned}$

Substituting $y = 7/4$ in the first equation, we get:

$\dfrac{x}{3} + \dfrac{7}{4} = 3 \qquad \Big| \cdot 12$

$4x + 21 = 36$

$4x = 15$

$x = 15/4$

Solution: $(15/4, 7/4)$

6.

a. $\begin{cases} 10 = \dfrac{1}{4}(x+y) & \bigg| \cdot 4 \\ \\ x - \dfrac{y}{3} = 4 & \bigg| \cdot 3 \end{cases}$

$\downarrow$

$\begin{cases} 40 = x + y \\ 3x - y = 12 \end{cases}$

$\downarrow$

$\begin{aligned} & \begin{cases} x + y = 40 \\ 3x - y = 12 \end{cases} \\ \hline & 4x = 52 \\ & x = 13 \end{aligned}$

Substituting $x = 13$ in the equation $40 = x + y$ (which is a version of the first equation), we get:

$40 = 13 + y$

$y = 27$

Solution: $(13, 27)$

b. $\begin{cases} 6x - y = \dfrac{4}{5} & \bigg| \cdot 5 \\ \\ \dfrac{y}{2} - \dfrac{3x}{4} = 1 - y & \bigg| \cdot 4 \end{cases}$

$\downarrow$

$\begin{cases} 30x - 5y = 4 \\ 2y - 3x = 4 - 4y \end{cases}$

$\downarrow$

$\begin{cases} 30x - 5y = 4 \\ -3x + 6y = 4 & \big| \cdot 10 \end{cases}$

$\downarrow$

$\begin{aligned} & \begin{cases} 30x - 5y = 4 \\ -30x + 60y = 40 \end{cases} \\ \hline & 55y = 44 \\ & y = 4/5 \end{aligned}$

Substituting $y = 4/5$ in the first equation, we get:

$6x - 4/5 = 4/5$

$6x = 8/5$

$x = 8/30 = 4/15$

Solution: $(4/15, 4/5)$

7.

a.
$$\begin{cases} x+y=2 \\ 5y-7=10x-6 \end{cases}$$

↓

$$\begin{cases} x+y=2 \quad \big| \cdot (-5) \\ -10x+5y=1 \end{cases}$$

↓

$$\begin{aligned} & \begin{cases} -5x-5y=-10 \\ -10x+5y=1 \end{cases} \\ \hline & \quad -15x \quad\quad = -9 \\ & \quad\quad\quad x = 9/15 = 3/5 \end{aligned}$$

Substituting $x = 3/5$ in the first equation, we get:

$$3/5 + y = 2$$
$$y = 7/5$$

Solution: $(3/5, 7/5)$

b.
$$\begin{cases} -45x+10y=-39 \\ 15x+30y=-27 \quad \big| \cdot 3 \end{cases}$$

↓

$$\begin{aligned} & \begin{cases} -45x+10y=-39 \\ 45x+90y=-81 \end{cases} \\ \hline & \quad\quad 100y = -120 \\ & \quad\quad\quad y = -120/100 = -6/5 \end{aligned}$$

Substituting $y = -6/5$ in the second equation, we get:

$$15x + 30(-6/5) = -27$$
$$15x - 36 = -27$$
$$15x = 9$$
$$x = 9/15 = 3/5$$

Solution: $(3/5, -6/5)$

c.
$$\begin{cases} -1=2x-y \\ 5y-15=2(5x+1) \end{cases}$$

↓

$$\begin{cases} 2x-y=-1 \\ 5y-15=10x+2 \end{cases}$$

↓

$$\begin{cases} 2x-y=-1 \quad \big| \cdot 5 \\ -10x+5y=17 \end{cases}$$

↓

$$\begin{aligned} & \begin{cases} 10x-5y=-5 \\ -10x+5y=17 \end{cases} \\ \hline & \quad\quad 0 = 12 \end{aligned}$$

This system has no solutions.

d.
$$\begin{cases} \dfrac{1}{2}x+y=70 \quad \big| \cdot 2 \\ -5x+y=-30 \end{cases}$$

↓

$$\begin{cases} x+2y=140 \\ -5x+y=-30 \quad \big| \cdot (-2) \end{cases}$$

↓

$$\begin{aligned} & \begin{cases} x+2y=140 \\ 10x-2y=60 \end{cases} \\ \hline & \quad 11x \quad\quad = 200 \\ & \quad\quad\quad x = 200/11 = 18.\overline{18} \end{aligned}$$

Substituting $x = 18.\overline{18}$ in the second equation, we get:

$$-5(18.\overline{18}) + y = -30$$
$$y = -30 + 90.\overline{90}$$
$$y = 60.\overline{90}$$

Solution: $(18.\overline{18}, 60.\overline{90})$, or, rounded to two decimal digits, $(18.18, 60.91)$.

More Practice, cont.

7.

e. $\begin{cases} 4(y+2) = 3(4x-1) \\ y + 3x + 2 = x + \dfrac{1}{2} \end{cases}$ $\Big| \cdot 2$

↓

$\begin{cases} 4y + 8 = 12x - 3 \\ 2y + 6x + 4 = 2x + 1 \end{cases}$

↓

$\begin{cases} -12x + 4y = -11 \\ 4x + 2y = -3 \end{cases}$ $\Big| \cdot (-2)$

↓

$\begin{array}{r} \begin{cases} -12x + 4y = -11 \\ -8x - 4y = 6 \end{cases} \\ \hline -20x = -5 \\ x = 1/4 \end{array}$

Substituting $x = 1/4$ in the first equation, we get:

$4(y+2) = 3(4(1/4) - 1)$
$4y + 8 = 0$
$4y = -8$
$y = -2$

Solution: $(1/4, -2)$

f. $\begin{cases} \dfrac{1}{3}(x+5) + y = 30 \\ -y - 30 = 2(3x - y + 15) \end{cases}$ $\Big| \cdot 3$

↓

$\begin{cases} x + 5 + 3y = 90 \\ -y - 30 = 6x - 2y + 30 \end{cases}$

↓

$\begin{cases} x + 3y = 85 \\ -6x + y = 60 \end{cases}$ $\Big| \cdot 6$

↓

$\begin{array}{r} \begin{cases} 6x + 18y = 510 \\ -6x + y = 60 \end{cases} \\ \hline 19y = 570 \\ y = 30 \end{array}$

Substituting $y = 30$ in the second equation, we get:

$-30 - 30 = 2(3x - 30 + 15)$
$-60 = 6x - 30$
$-30 = 6x$
$x = -5$

Solution: $(-5, 30)$

g. $\begin{cases} x + 2y - 16 = -1 \\ y - 6 = \dfrac{1}{4}(x+9) \end{cases}$ $\Big| \cdot 4$

↓

$\begin{cases} x + 2y = 15 \\ 4y - 24 = x + 9 \end{cases}$

↓

$\begin{array}{r} \begin{cases} x + 2y = 15 \\ -x + 4y = 33 \end{cases} \\ \hline 6y = 48 \\ y = 8 \end{array}$

Substituting $y = 8$ in the equation $x + 2y = 15$ (which is equivalent to the first equation), we get:

$x + 2(8) = 15$
$x + 16 = 15$
$x = -1$

Solution: $(-1, 8)$

h. $\begin{cases} y + 3x = -\dfrac{3}{2} \\ 2x + y = 12x + 5 \end{cases}$ $\Big| \cdot 2$

↓

$\begin{cases} 2y + 6x = -3 \\ -10x + y = 5 \end{cases}$

↓

$\begin{cases} 6x + 2y = -3 \\ -10x + y = 5 \end{cases}$ $\Big| \cdot (-2)$

↓

$\begin{array}{r} \begin{cases} 6x + 2y = -3 \\ 20x - 2y = -10 \end{cases} \\ \hline 26x = -13 \\ x = -1/2 \end{array}$

Substituting $x = -1/2$ in the equation $2y + 6x = -3$ (which is equivalent to the first equation), we get:

$2y + 6(-1/2) = -3$
$2y - 3 = -3$
$2y = 0$
$y = 0$

Solution: $(-1/2, 0)$

Applications, Part 2, pp. 157-159

1. The system of equations is this:

$$\begin{cases} t + u = 8 \\ 10t + u = 7u \end{cases}$$

Solving t from the top one, we get $t = 8 - u$.
Substituting $8 - u$ in the second equation, in place of t, we get:

$$10(8 - u) + u = 7u$$
$$80 - 10u + u = 7u$$
$$-16u = -80$$
$$u = 5$$

Then, $t = 8 - u = 8 - 5 = 3$. Solution: <u>The number is 35</u>.

2. The system of equations is this:

$$\begin{cases} t + u = 11 \\ t = 3u - 1 \end{cases}$$

Substituting $3u - 1$ in the first equation, in place of t, we get:

$$3u - 1 + u = 11$$
$$4u = 12$$
$$u = 3$$

Then, $t = 3u - 1 = 9 - 1 = 8$. Solution: <u>The number is 83</u>.

3. Reversing the digits means forming the number $10u + t$. The system of equations is this:

$$\begin{cases} t + u = 9 \\ 10u + t = 10t + u + 9 \end{cases}$$
$$\downarrow$$
$$\begin{cases} t + u = 9 \quad \bigg| \cdot 9 \\ -9t + 9u = 9 \end{cases}$$
$$\downarrow$$
$$\begin{array}{r} \begin{cases} 9t + 9u = 81 \\ -9t + 9u = 9 \end{cases} \\ \hline 18u = 90 \\ u = 5 \end{array}$$

Since the digit sum is 9, then $t = 4$ and <u>the number is 45</u>.

4. The system of equations is this:

$$\begin{cases} t = u + 4 \\ 10t + u = 7(t + u) \end{cases}$$

Substituting $u + 4$ in place of t in the second equation, we get:

$$10(u + 4) + u = 7(u + 4 + u)$$
$$10u + 40 + u = 14u + 28$$
$$-3u = -12$$
$$u = 4$$

Then, $t = u + 4 = 4 + 4 = 8$. Solution: <u>The number is 84</u>.

5. The system of equations is this:

$$\begin{cases} u = t + 3 \\ 10u + t = 2(10t + u) + 2 \end{cases}$$

Substituting $t + 3$ in place of u in the second equation, we get:

$$10(t + 3) + t = 2(10t + t + 3) + 2$$
$$10t + 30 + t = 20t + 2t + 6 + 2$$
$$-11t = -22$$
$$t = 2$$

Then, $u = t + 3 = 5$. Solution: <u>The number is 25</u>.

6. Answers will vary. Check the student's answer.

7. a. It signifies that the total sales from the two kinds of energy bars was $132.
   b. The second equation states that three more nut bars than fruit bars were sold on that day.
   c. The solution will tell us the number of fruit bars and of nut bars that were sold.

$$\begin{cases} 4f + 5n = 132 \\ n = f + 3 \end{cases}$$

Substituting $f + 3$ in place of n in the first equation, we get:

$$4f + 5(f + 3) = 132$$
$$4f + 5f + 15 = 132$$
$$9f = 117$$
$$f = 13$$

Then, $n = f + 3 = 16$. Solution: <u>The store sold 13 fruit bars and 16 nut bars</u>.

Applications, Part 2, cont.

8. Let R be the price of one rose and L be the price of one lily. Then:

$$\left\{\begin{array}{l} 12R + 10L = 39 \\ 8R + 16L = 40 \end{array}\right. \quad \begin{array}{l} \cdot(-2) \\ \cdot 3 \end{array}$$

$$\downarrow$$

$$+ \left\{\begin{array}{l} -24R - 20L = -78 \\ 24R + 48L = 120 \end{array}\right.$$

$$\begin{aligned} 28L &= 42 \\ L &= 42/28 = 6/4 = \$1.50 \end{aligned}$$

Substituting L = 1.50 into the second equation, we get:

$$\begin{aligned} 8R + 16(1.5) &= 40 \\ 8R + 24 &= 40 \\ 8R &= 16 \\ R &= 2 \end{aligned}$$

Solution: One rose costs $2 and one lily costs $1.50.

9. Let x and y be the sides of the rectangle. Then:

$$\left\{\begin{array}{l} 2x + 2y = 472 \\ x = y + 56 \end{array}\right.$$

Substituting $y + 56$ for x in the first equation, we get:

$$\begin{aligned} 2(y + 56) + 2y &= 472 \\ 2y + 112 + 2y &= 472 \\ 4y &= 360 \\ y &= 90 \end{aligned}$$

Then, $x = y + 56 = 90 + 56 = 146$ feet.

Solution: the sides are 146 ft and 90 ft.

10. Let x be the price of the cheaper kind and y be the price of the more expensive kind of solar panel. Then:

$$\left\{\begin{array}{l} 20x + 10y = 3,300 \\ 10x + 20y = 3,900 \end{array}\right. \quad \cdot(-2)$$

$$\downarrow$$

$$+ \left\{\begin{array}{l} 20x + 10y = 3,300 \\ -20x - 40y = -7,800 \end{array}\right.$$

$$\begin{aligned} -30y &= -4,500 \\ y &= 150 \end{aligned}$$

Substituting $y = 150$ into the first equation, we get:

$$\begin{aligned} 20x + 10(150) &= 3,300 \\ 20x + 1,500 &= 3,300 \\ 20x &= 1,800 \\ x &= 90 \end{aligned}$$

Solution: The cheaper solar panels cost $90 each and the more expensive ones $150 each.

Puzzle corner. Let the sides of the rectangle be s and $4s$. Then, its area is $s \cdot 4s$, which is $4s^2$. Since the area is 3,364 square units, we get the equation:

$$\begin{aligned} 4s^2 &= 3,364 \\ s^2 &= 841 \\ s &= 29 \end{aligned}$$

Then, the perimeter is $s + 4s + s + 4s = 10s = \underline{290 \text{ units}}$.

Speed, Time, and Distance Problems, pp. 160-165

1. $$\left\{\begin{array}{l} d = 90t \\ d = 100(t - 1/6) \end{array}\right.$$

Since we are given two expressions for d, we can set those to be equal, and thus solve for t:

$$\begin{aligned} 90t &= 100(t - 1/6) \\ 90t &= 100t - 100/6 \quad \cdot 6 \\ 540t &= 600t - 100 \\ 100 &= 60t \\ t &= 10/6 = 1.\overline{6} \text{ hr} = 1 \text{ hr } 40 \text{ min} \end{aligned}$$

Since the first train left at 9:00 AM and traveled $t = 1$ hr 40 min, the second train caught up to it at 10:40 AM.

Speed, Time, and Distance Problems, cont.

Page 161

2. The distance you've covered when you reach your friend will be the same that your friend has covered, so we will denote that with d. The two times are different, but since they differ by 15 minutes, or 1/4 hour, we can use t for the time your friend takes, and $t - 1/4$ for the time you take.

	distance	velocity	time
Your friend	d	6 km/h	t
You	d	18 km/h	$t - 1/4$

Using $d = vt$, the two equations we get are:

$$\begin{cases} d = 6t \\ d = 18(t - 1/4) \end{cases}$$

Since we are given two expressions for d, we can set those to be equal, and thus solve for t:

$$\begin{aligned} 6t &= 18(t - 1/4) \\ 6t &= 18t - 9/2 \qquad |\cdot 2 \\ 12t &= 36t - 9 \\ 9 &= 24t \\ t &= 9/24 \text{ hr} = 3/8 \text{ hr} = 22.5 \text{ min} \end{aligned}$$

To calculate the distance, since the speed is in km/h, the time has to be in hours. The distance then is $d = 6t$ = 6 km/h $\cdot$ (3/8) hr = 18/8 km = 9/4 km = <u>2.25 km</u>.

3. Let t be the time that the hare takes to run the race. Then, the tortoise takes $t + 16$ minutes to run it. However, since the given speed is in meters per second, we need to use seconds as a time unit. So, instead of 16 minutes, we will have 960 seconds.

	distance	velocity	time
Tortoise	100	v	$t + 960$
Hare	100	15	t

Using $d = vt$, the two equations we get are:

$$\begin{cases} 100 = v(t + 960) \\ 100 = 15t \end{cases}$$

From the bottom equation, we can solve that $t = 100/15$ = $20/3 = 6.\overline{6}$ seconds. Substituting that to the first equation, we get:

$$\begin{aligned} 100 &= v(6.\overline{6} + 960) \\ 100 &= 966.\overline{6}v \\ v &= 100/966.\overline{6} \approx 0.103 \text{ m/s} \end{aligned}$$

The tortoise's speed is <u>0.10 m/s</u> and it takes about 967 seconds, or <u>16 minutes 7 seconds</u> to run the race.

Page 162

4. Here, the amount of time is the same for both trains. Since they are 50 km apart from each other, the two distances they travel add up to 50: $d_1 + d_2 = 50$, from which $d_2 = 50 - d_1$. So, we only need to use one variable for distance.

	distance	velocity	time
Train 1	d_1	120	t
Train 2	$50 - d_1$	100	t

Using $d = vt$, the two equations we get are:

$$\begin{cases} d_1 = 120t \\ 50 - d_1 = 100t \end{cases}$$

Substituting $120t$ in place of d_1 in the second equation, we get:

$$\begin{aligned} 50 - 120t &= 100t \\ 50 &= 220t \\ t &= 50/220 \approx 0.2272 \text{ hr} \end{aligned}$$

The trains are 50 km apart from each other after about 0.2272 hours $\approx$ 13.6 minutes = 13 minutes 36 seconds.

Page 163

5. Since the trains are 75 mi apart from each other, $d_1 + d_2 = 75$, from which $d_2 = 75 - d_1$. So, we only need to use one variable for distance.

	distance	velocity	time
Train 1	d_1	70	1/2
Train 2	$75 - d_1$	v	1/2

Using $d = vt$, the two equations we get are:

$$\begin{cases} d_1 = 70(1/2) \\ 75 - d_1 = (1/2)v \end{cases}$$

The distance d_1 is 35 miles. Substituting that in the second equation, we get:

$$\begin{aligned} 75 - 35 &= (1/2)v \\ 40 &= (1/2)v \\ v &= 80 \end{aligned}$$

Train 2 is traveling at a speed of 80 mph.

Speed, Time, and Distance Problems, cont.

Page 163

6. a. We will denote the distance Chip has run with d, and then, Ranger has run $d + 600$ meters.

	distance	velocity	time
Ranger	$d + 600$	16 m/s	100 s
Chip	d	v	100 s

The equations are:
$$\begin{cases} d + 600 = 16(100) \\ d = 100v \end{cases}$$

Substituting $100v$ for d in the first equation, we get:

$$100v + 600 = 16(100)$$
$$100v + 600 = 1600$$
$$100v = 1000$$
$$v = 10$$

Chip is running at the speed of 10 m/s — quite a bit less than Ranger.

b. If the horses are only 50 m apart, the equations are:

$$\begin{cases} d + 50 = 16(100) \\ d = 100v \end{cases}$$

Substituting $100v$ for d in the first equation, we get:

$$100v + 50 = 16(100)$$
$$100v + 50 = 1600$$
$$100v = 1550$$
$$v = 15.5$$

Chip is running at the speed of 15.5 m/s.

Page 164

7. $$\begin{cases} 5 = 20(v_b + v_w) \\ 5 = 24(v_b - v_w) \end{cases}$$

$\downarrow$

$$\begin{cases} 5 = 20v_b + 20v_w & \cdot (-6) \\ 5 = 24v_b - 24v_w & \cdot 5 \end{cases}$$

$\downarrow$

$$+\begin{cases} -30 = -120v_b - 120v_w \\ 25 = 120v_b - 120v_w \end{cases}$$
$$\overline{}$$
$$-5 = -240v_w$$
$$v_w = 5/240 = 0.020\overline{83}$$

Substituting $5/240$ for v_w in the first equation, we get:
$$5 = 20v_b + 20(5/240)$$
$$5 = 20v_b + 5/12$$
$$20v_b = 5 - 5/12 = 55/12$$
$$v_b = 55/240 = 0.2291\overline{6}$$

(Continued in the next column.)

(7 continued)

Both of these speeds are in miles per minute, but it will make more sense to give them in miles per hour, so we multiply them by 60. The speed of the boat in still water is $0.2291\overline{6}$ mi/min $\cdot$ 60 min = <u>13.75 mph</u> and the speed of the water is $0.020\overline{83}$ mi/min $\cdot$ 60 min = <u>1.25 mph</u>.

8. Let v_w be the speed of the wind and d the distance between the cities.

	distance	velocity	time
From A to B	d	$900 + v_w$	2.5
From B to A	d	$900 - v_w$	2.75

Using $d = vt$, the two equations we get are:

$$\begin{cases} d = 2.5(900 + v_w) \\ d = 2.75(900 - v_w) \end{cases}$$

Since we are given two expressions for d, we can set those to be equal, and thus solve for v_w, first of all:

$$2.5(900 + v_w) = 2.75(900 - v_w)$$
$$2250 + 2.5v_w = 2475 - 2.75v_w$$
$$5.25v_w = 225$$
$$5.25v_w = 225$$
$$v_w \approx 42.85714 \text{ km/h}$$

Now we can find the distance: $d = 2.5(900 + v_w)$ = $2.5(900 + 42.85714) \approx 2360$ km.

173

Speed, Time, and Distance Problems, cont.

9. Let v_p be the speed of the plane in still air. Since the speed 30 km/h is in kilometers per hour, we need to express the time of 40 minutes in hours.

	distance	velocity	time
With tailwind	650 km	$v_p + 30$	2/3 hrs
With headwind	650 km	$v_p - 30$	t

Using $d = vt$, the two equations we get are:

$$\begin{cases} 650 = 2/3(v_p + 30) \\ 650 = t(v_p - 30) \end{cases}$$

Since in the bottom equation we would be multiplying the two unknowns, t and v_p, this is best solved by solving v_p from the top equation first.

$$\begin{aligned} 650 &= 2/3(v_p + 30) \quad | \cdot 3 \\ 1{,}950 &= 2(v_p + 30) \\ 1{,}950 &= 2v_p + 60 \\ 2v_p &= 1{,}890 \\ v_p &= 945 \text{ km/h} \end{aligned}$$

Now we can find t using the bottom equation:

$$\begin{aligned} 650 &= t(v_p - 30) \\ 650 &= t(945 - 30) \\ 650 &= 915t \\ t &= 0.71038 \text{ hr} \end{aligned}$$

It takes the plane 0.71038 hours, or 42.6 minutes to fly the same distance against the wind. This wind speed is not great enough to make much difference in the flying time.

10. Let v_w be the speed of the water. We need to be careful with the time units. The speed 1 mph is in miles per *hour*. The swimming time is in seconds. We either need to convert the speed to miles per second or the time to hours.

If we do the former, 1 mph = 1 mi/(60 min) = 1 mi/(3,600 sec) = (1/3,600) mi/sec.
If the latter, 43 seconds = 43/60 min = 43/360 hrs, and 53 seconds = 53/360 hours. The choice is yours, but here we will use the time converted to hours.

	distance	velocity	time
Downstream	d	$1 + v_w$	43/360
Upstream	d	$1 - v_w$	53/360

Using $d = vt$, the two equations we get are:

$$\begin{cases} d = 43/360(1 + v_w) \\ d = 53/360(1 - v_w) \end{cases}$$

Since we have two expressions for d, we can set them to be equal, and then multiply to get rid of the fractions:

$$\begin{aligned} 43/360(1 + v_w) &= 53/360(1 - v_w) \quad | \cdot 360 \\ 43(1 + v_w) &= 53(1 - v_w) \\ 43 + 43v_w &= 53 - 53v_w \\ 96v_w &= 10 \\ v_w &\approx 0.1041667 \text{ miles/hour} \end{aligned}$$

The speed of the water is about 0.10 mph.

Puzzle corner. (1) Neither. The trains are in the exact same spot, thus at the same distance from Atlanta.
(2) Never. The locations and directions are such that you will not meet each other (unless you go past the poles...).
(3) We don't know. There is not enough information to find that out.

Mixtures and Comparisons, pp. 166-170

1.
$$\begin{cases} r + a = 1 \\ 3.34r + 6.23a = 4.50 \end{cases}$$

Solving r from the first equation, we get $r = 1 - a$. Substituting that in place of r in the second equation, we get:

$$\begin{aligned} 3.34(1 - a) + 6.23a &= 4.50 \\ 3.34 - 3.34a + 6.23a &= 4.50 \\ 2.89a &= 1.16 \\ a &\approx 0.40138 \end{aligned}$$

The weight of almonds in the mixture is 0.4 lb and the weight of raisins is 0.6 lb.

2.

	weight (kg)	protein (%)	protein (kg)
Corn	w_1	12	$0.12w_1$
Chickpeas	w_2	22	$0.22w_2$
Mixture	2	16	$0.16(2)$

Equations:
$$\begin{cases} w_1 + w_2 = 2 \\ 0.12w_1 + 0.22w_2 = 0.16(2) \end{cases}$$

Mixtures and Comparisons, cont.

3. We can solve that $w_1 = 2 - w_2$ from the first equation. Substituting that into the second:

$$0.12(2 - w_2) + 0.22w_2 = 0.16(2)$$
$$0.24 - 0.12w_2 + 0.22w_2 = 0.32$$
$$0.1w_2 = 0.08$$
$$w_2 = 0.8$$

So, the weight of chickpeas in the mixture is 0.8 kg. The weight of corn is $w_1 = 2 - w_2 = 2$ kg $- 0.8$ kg $= 1.2$ kg.

4.

	weight (kg)	cost per kg	cost ($)
Wildflower honey	w_1	50	$50w_1$
Regular honey	w_2	20	$20w_2$
Mixture	2	25	50

Equations: $\begin{cases} w_1 + w_2 = 2 \\ 50w_1 + 20w_2 = 50 \end{cases}$

Substituting $2 - w_2$ for w_1 in the second equation:

$$50(2 - w_2) + 20w_2 = 50$$
$$100 - 50w_2 + 20w_2 = 50$$
$$-30w_2 = -50$$
$$w_2 = 5/3 = 1.667 \text{ kg}$$

So, the weight of the regular honey in the mixture is 1.667 kg. The weight of expensive honey is 0.333 kg.

5. Notice that we can look at the amount of copper and the amount of zinc in the mixture separately. The chart actually gives rise to three equations but we only need two.

	weight	copper	amount of copper	zinc	amount of zinc
Alloy 1	w_1	90%	$0.9w_1$	10%	$0.1w_1$
Alloy 2	w_2	65%	$0.65w_2$	35%	$0.35w_2$
Mixture	1000 kg	80%	800 kg	20%	200 kg

Equations: $\begin{cases} w_1 + w_2 = 1000 \\ 0.9w_1 + 0.65w_2 = 800 \end{cases}$

We can solve that $w_1 = 1000 - w_2$, from the first equation. Substituting that into the second:

$$0.9(1000 - w_2) + 0.65w_2 = 800$$
$$900 - 0.9w_2 + 0.65w_2 = 800$$
$$-0.25w_2 = -100$$
$$w_2 = 400$$

So, the weight of the alloy 2 in the mixture is 400 kg and the the weight of the alloy 1 is 600 kg.

Lastly, we might check that these values will fulfill the constraint for zinc: $0.1(600$ kg$) + 0.35(400$ kg$)$ $= 60$ kg $+ 140$ kg $= 200$ kg. Yes, they do.

6.

	volume	H$_2$O$_2$ %	amount of H$_2$O$_2$
10% solution	V_1	10%	$0.1V_1$
3% solution	V_2	3%	$0.03V_2$
Mixture (8% solution)	1,000 mL	8%	80 mL

Equations: $\begin{cases} V_1 + V_2 = 1000 \\ 0.1V_1 + 0.03V_2 = 80 \end{cases}$

Solving V_2 from the first equation, $V_2 = 1000 - V_1$. Substituting that into the second:

$$0.1V_1 + 0.03(1000 - V_1) = 80$$
$$0.1V_1 + 30 - 0.03V_1 = 80$$
$$0.07V_1 = 50$$
$$V_1 \approx 714 \text{ mL}$$

So, the volume of the 10% solution is 714 mL and the volume of the 3% one is 286 mL.

Page 169

7. a. Plan 1: $C = 20 + 3.6x$ Plan 2: $C = 50 + 3x$

b. $\begin{cases} C = 20 + 3.6x \\ C = 50 + 3x \end{cases}$

Since we have two expressions for C, we can set them to be equal, and solve for x:

$$20 + 3.6x = 50 + 3x$$
$$0.6x = 30$$
$$x = 50$$

Then, $C = 50 + 3x = 50 + 3(50) = 200$.

c. It signifies that if you were buying 50 energy bars, the cost will be $200 in *either* plan. In other words, 50 is the amount of bars for which the two plans have the same cost, which is $200.

d. See the graph on the right.

e. When you are buying less than 50 bars per year, plan 1 is more cost-effective.
When you are buying more than 50 bars per year, plan 2 is more cost-effective.

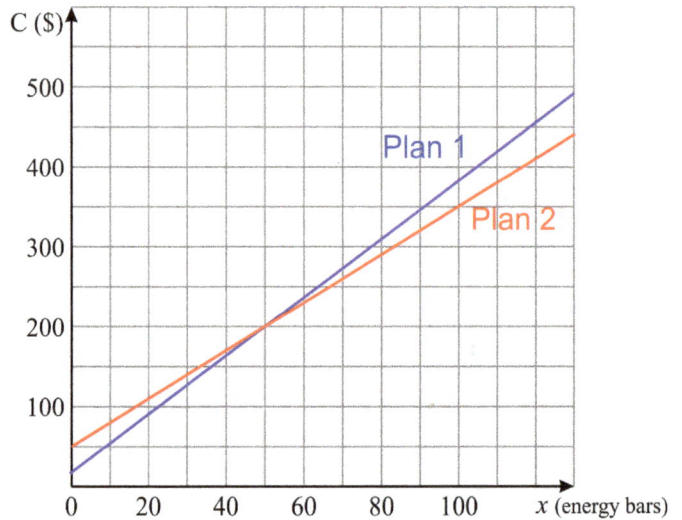

Page 170

8. a. Plan 1: $C = 45t$ Plan 2: $C = 100 + 20t$

b. $\begin{cases} C = 45t \\ C = 100 + 20t \end{cases}$

Since we have two expressions for C, we can set them to be equal, and solve for t:

$$45t = 100 + 20t$$
$$25t = 100$$
$$t = 4$$

So, t is four months. Then, $C = 45(4) = \$180$.

c. Four months is the amount of time for which both plans cost the same, $180.

d. Answers will vary since the student can choose the scaling for the vertical axis. For one possibility, see the graph on the right.

e. When you will be using the gym for less than four months.

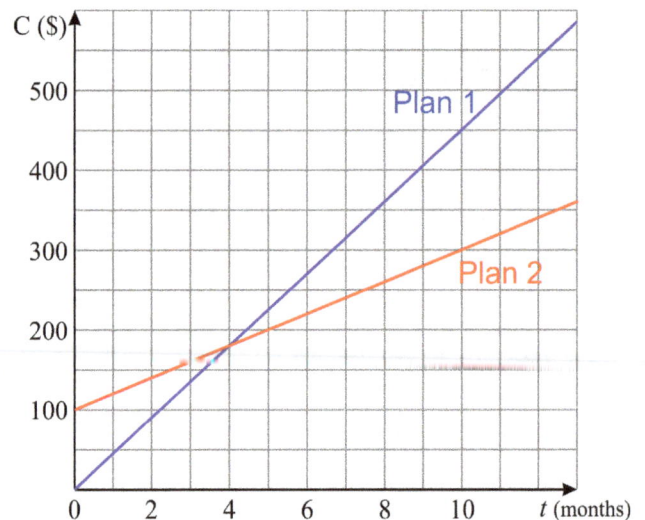

176

Page 171

1. Answers will vary; check the student's answer. For example:

First, reflect the trapezoid in the x-axis, so it becomes the trapezoid A'B'C'D'. Then, dilate it from point C' and with scale factor 1/2. Lastly, translate it 6.5 units to the right and one unit down.

Another possibility:

First, dilate the trapezoid from point D and with scale factor 1/2. Then, reflect it in the x-axis. Lastly, translate it six units to the right and two units down.

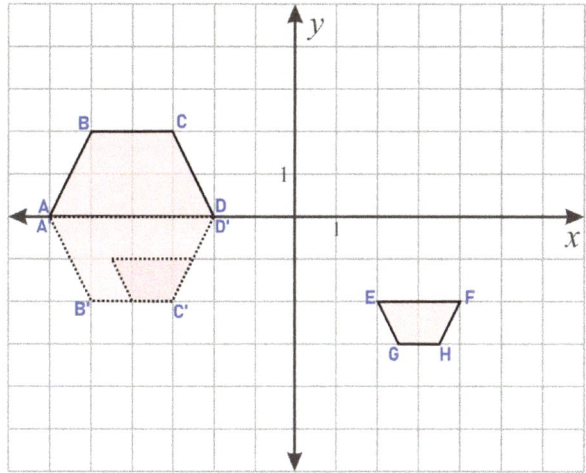

2.

a.
$$\begin{aligned} s + 2 - 5s &= 3 - 8(s - 1) - 6s \\ 2 - 4s &= 3 - 8s + 8 - 6s \\ 2 - 4s &= 11 - 14s \\ 2 - 4s &= 11 - 14s \\ 10s &= 9 \\ s &= 9/10 \end{aligned}$$

b.
$$\begin{aligned} 6 + \frac{2x - 5}{3} &= x - 2 \qquad \cdot 3 \\ 18 + (2x - 5) &= 3x - 6 \\ 13 + 2x &= 3x - 6 \\ -x &= -19 \\ x &= 19 \end{aligned}$$

c.
$$\begin{aligned} 0 &= \frac{x - 5}{3} + \frac{x + 5}{4} \qquad \cdot 12 \\ 0 &= 4(x - 5) + 3(x + 5) \\ 0 &= 4x - 20 + 3x + 15 \\ 0 &= 7x - 5 \\ 7x &= 5 \\ x &= 5/7 \end{aligned}$$

d.
$$\begin{aligned} 10 &= y + \frac{2 - 3y}{5} \qquad \cdot 5 \\ 50 &= 5y + (2 - 3y) \\ 50 &= 5y + 2 - 3y \\ 48 &= 2y \\ y &= 24 \end{aligned}$$

Page 172

3. Answers will vary. Check the student's answers.

a. Here, the right side should have $2x$ and some constant that is not -5. For example: $2x - 5 = 2x + 2$. Or, $2x - 5 = 2(x - 3)$.

b. Here, the coefficient of the x term on the right side should be different from 2. For example: $2x - 5 = 3x + 1$, or $2x - 5 = x + 10 + 5x$.

c. Here, the right side should equal $2x - 5$. But we might make it look a little different, so that only after combining like terms does it equal $2x - 5$. For example: $2x - 5 = 2(x - 3) + 1$ or $2x - 5 = 6x - 4x - 7 + 2$.

4. a. $y = (3/4)x - 3$ b. $y = -(1/6)x + 3$ c. $y = -(1/2)x - 7/2$

Page 172

5.

---

a.
$$\begin{cases} y = -2(x + 10) \\ 2(y - x) = 8 \end{cases}$$

↓

$$\begin{cases} y = -2x - 20 \\ 2y - 2x = 8 \end{cases}$$

↓

$$\begin{cases} 2x + y = -20 \quad \Big| \cdot (-2) \\ -2x + 2y = 8 \end{cases}$$

↓

$$\begin{aligned} &\begin{cases} -4x - 2y = 40 \\ -2x + 2y = 8 \end{cases} \\ + \\ \hline &\quad -6x = 48 \\ &\quad\quad x = -8 \end{aligned}$$

Substituting $x = -8$ in the first equation, we get:

$$y = -2(-8 + 10)$$
$$y = -2(2)$$
$$y = -4$$

Solution: $(-8, -4)$

---

b.
$$\begin{cases} 2x - 4y = -3 \quad \Big| \cdot 3 \\ 5x + 3y = 1 \quad \Big| \cdot 4 \end{cases}$$

↓

$$\begin{aligned} &\begin{cases} 6x - 12y = -9 \\ 20x + 12y = 4 \end{cases} \\ + \\ \hline &\quad 26x = -5 \\ &\quad\quad x = -5/26 \end{aligned}$$

Substituting $x = -5/26$ in the first equation, we get:

$$2(-5/26) - 4y = -3$$
$$-10/26 - 4y = -3$$
$$-4y = 10/26 - 3$$
$$-4y = 5/13 - 3$$
$$-4y = 5/13 - 39/13$$
$$-4y = -34/13$$
$$y = 34/52 = 17/26$$

Solution: $(-5/26, 17/26)$

Page 173

6. a. Let d be the distance Charlie has run in 20 seconds. Then:

	distance	*velocity*	*time*
Rocky	$d + 8$	10 m/s	20 sec
Charlie	d	v	20 sec

Using $d = vt$, the two equations we get are:

$$\begin{cases} d + 8 = 10(20) \\ d = 20v \end{cases}$$

We will substitute $20v$ for d in the first equation:

$$20v + 8 = 200$$
$$20v = 192$$
$$v = 192/20 = 9.6$$

So, Charlie's speed is 9.6 m/s.

6. b. We will use the same type of chart but just change Rocky's distance to $d + 2$:

	distance	velocity	time
Rocky	$d + 2$	10 m/s	20 sec
Charlie	d	v	20 sec

Using $d = vt$, the two equations we get are:

$$\begin{cases} d + 2 = 10(20) \\ d = 20v \end{cases}$$

We will substitute $20v$ for d in the first equation:

$$\begin{aligned} 20v + 2 &= 200 \\ 20v &= 198 \\ v &= 198/20 = 9.9 \end{aligned}$$

Charlie's speed would need to be 9.9 m/s so that he'd only be 2 m behind Rocky after 20 seconds.

7. We can look at the amount of copper and the amount of zinc in the mixture separately. The chart actually gives rise to three equations but we only need two.

	weight	copper	amount of copper	zinc	amount of zinc
Alloy 1	w_1	90%	$0.9w_1$	10%	$0.1w_1$
Alloy 2	w_2	65%	$0.65w_2$	35%	$0.35w_2$
Mixture	5000 kg	75%	3,750 kg	25%	1,250 kg

Equations: $$\begin{cases} w_1 + w_2 = 5,000 \\ 0.9w_1 + 0.65w_2 = 3,750 \end{cases}$$

(The third equation we could use comes from the amount of zinc: $0.1w_1 + 0.35w_2 = 1,250$.)

From the first equation, we can solve that $w_1 = 5,000 - w_2$. Substituting that into the second:

$$\begin{aligned} 0.9(5,000 - w_2) + 0.65w_2 &= 3,750 \\ 4,500 - 0.9w_2 + 0.65w_2 &= 3,750 \\ -0.25w_2 &= -750 \\ w_2 &= 3,000 \end{aligned}$$

So, the weight of the alloy 2 in the mixture is 3,000 kg and the weight of the alloy 1 is 2,000 kg.

Lastly, we might check that these values will fulfill the constraint for zinc:
$0.1(2,000 \text{ kg}) + 0.35(3,000 \text{ kg}) = 200 \text{ kg} + 1,050 \text{ kg} = 1,250 \text{ kg}$. Yes, they do.

8. a. In a minute, the elephant will cover $60 \cdot 6$ m $= 360$ m. So, his speed is 360 m/min.

   b. $d = 360t$

   c. Student graphs will vary because the student chooses the scaling for the vertical axis. For example:

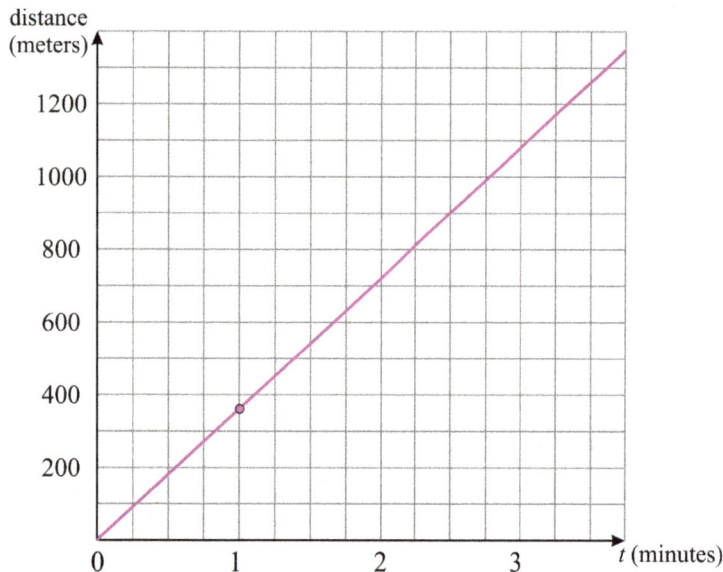

   d. The slope is 360 (or, 360 m/min).

   e. Elizabeth, because she goes at the speed of 400 meters per minute.

   f. Each minute, Elizabeth covers 40 m more than the elephant. So, in 3 minutes, she will cover 120 m more than the elephant.

Chapter 7 Review, pp. 175-179

Page 175

1. a. An infinite number of solutions, because the second equation is two times the first, so they're equivalent equations.

   b. No solutions, because $7x - 2y$ cannot equal both 3 and -1.

   c. One solution. Considering these as lines, they have different slopes (-1 and 3/2).

2. a. $(-1, 2)$ b. All points (x, y) that satisfy $2y - x = 1$ (or that satisfy $y = (1/2)x + 1/2$).

Page 175

3. (b) has a single solution. $\begin{cases} y = (1/2)x + 6 \\ x + 6y = 0 \end{cases}$

Substituting $y = (1/2)x + 6$ to the second equation:

$$x + 6((1/2)x + 6) = 0$$
$$x + 3x + 36 = 0$$
$$4x = -36$$
$$x = -36/4 = -9$$

Now, we substitute $x = -9$ in the first equation: $y = (1/2)(-9) + 6 = -9/2 + 6 = -9/2 + 12/2 = 3/2$.
Solution: $(-9, 3/2)$.

Page 176

4. $\begin{cases} x - 2y = 3 \\ 2x + y = -8 \end{cases}$

We can solve from the first equation that $x = 2y + 3$. Substituting $2y + 3$ for x in the second equation:

$$2(2y + 3) + y = -8$$
$$4y + 6 + y = -8$$
$$5y + 6 = -8$$
$$5y = -14$$
$$y = -14/5$$

Now, we substitute $y = -14/5$ in the first equation:

$$x - 2(-14/5) = 3$$
$$x + 28/5 = 3$$
$$x = 3 - 28/5 = 15/5 - 28/5 = -13/5$$

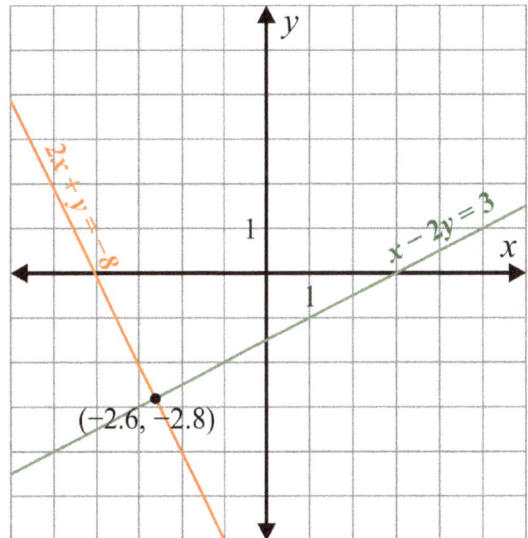

$(-2.6, -2.8)$

Solution: $(-13/5, -14/5)$ or $(-2.6, -2.8)$. This is verified by the graph.

5.

a. $\begin{cases} x = -(y - 5) \\ 2y = 12(1 - x) \end{cases}$

$\downarrow$

$\begin{cases} x = -y + 5 \\ 2y = 12 - 12x \end{cases}$

$\downarrow$

$\begin{cases} x + y = 5 \\ 12x + 2y = 12 \end{cases}$ $\bigg| \cdot (-12)$

$\downarrow$

$+ \begin{cases} -12x - 12y = -60 \\ 12x + 2y = 12 \end{cases}$

$$\overline{\qquad -10y = -48}$$
$$y = 4.8$$

Substituting $y = 4.8$ in the first equation, we get:

$$x = -(4.8 - 5)$$
$$x = -(-0.2) = 0.2$$

Solution: $(0.2, 4.8)$ or $(1/5, 24/5)$

b. $\begin{cases} 4.5x - 3y = 0 \\ 7x + 2y = -2 \end{cases}$ $\begin{matrix} \cdot 2 \\ \cdot 3 \end{matrix}$

$\downarrow$

$+ \begin{cases} 9x - 6y = 0 \\ 21x + 6y = -6 \end{cases}$

$$\overline{\qquad 30x = -6}$$
$$x = -6/30 = -1/5$$

Substituting $x = -1/5$ (or -0.2 if you prefer) in the second equation, we get:

$$7(-1/5) + 2y = -2$$
$$-7/5 + 2y = -2$$
$$2y = 7/5 - 2$$
$$2y = -3/5$$
$$y = -3/10$$

Solution: $(-1/5, -3/10)$ or $(-0.2, -0.3)$

Page 176

5.

c. $\begin{cases} 3x = -9(y + 1/3) \\ -2x - 6y = 2 \end{cases}$

$\downarrow$

$\begin{cases} 3x = -9y - 3 \\ -2x - 6y = 2 \end{cases}$

$\downarrow$

$\begin{cases} 3x + 9y = -3 \quad | \cdot 2 \\ -2x - 6y = 2 \quad | \cdot 3 \end{cases}$

$\downarrow$

$\begin{array}{r} \begin{cases} 6x + 18y = -6 \\ -6x - 18y = 6 \end{cases} \\ \hline 0 = 0 \end{array}$ $+$

Solution: all points (x, y) that satisfy the equation $-2x - 6y = 2$.

(The student can list some other, equivalent equation instead.)

d. $\begin{cases} 5x - 3(y + 2) = 0 \\ -5x + 6y = 7 \end{cases}$

$\downarrow$

$\begin{cases} 5x - 3y - 6 = 0 \\ -5x + 6y = 7 \end{cases}$

$\downarrow$

$+ \begin{array}{r} \begin{cases} 5x - 3y = 6 \\ -5x + 6y = 7 \end{cases} \\ \hline 3y = 13 \\ y = 13/3 \end{array}$

Substituting $y = 13/3$ in the second equation, we get:

$$-5x + 6(13/3) = 7$$
$$-5x + 26 = 7$$
$$-5x = -19$$
$$x = 19/5$$

Solution: (19/5, 13/3)

Page 177

6. The error is in the highlighted portion. Instead, $y = 36/30 = 6/5$. The corrected solution is on the right.

(1) $\begin{cases} 6x + 3y = 3 \\ -2x + 9y = 11 \quad | \cdot 3 \end{cases}$
(2)

$\downarrow$

$\begin{array}{r} \begin{cases} 6x + 3y = 3 \\ -6x + 27y = 33 \end{cases} \\ \hline 30y = 36 \\ \boxed{y = 5/6} \end{array}$

(1) $\begin{cases} 6x + 3y = 3 \\ -2x + 9y = 11 \quad | \cdot 3 \end{cases}$
(2)

$\downarrow$

$\begin{array}{r} \begin{cases} 6x + 3y = 3 \\ -6x + 27y = 33 \end{cases} \\ \hline 30y = 36 \\ y = 6/5 \end{array}$

$\downarrow$

$$(1) \quad 6x + 3(6/5) = 3$$
$$6x + 18/5 = 3$$
$$6x = 3 - 18/5$$
$$6x = -3/5$$
$$x = -3/30 = -1/10$$

The solution $(-1/10, 6/5)$ does fulfill the 2nd equation:

$$-2(-1/10) + 9(6/5) \overset{?}{=} 11$$
$$1/5 + 54/5 \overset{?}{=} 11$$
$$55/5 = 11$$

Chapter 7 Review, cont.

7.

a. $\begin{cases} 7x + 6y = -1 \\ 11x + 2y = 3 \end{cases} \Big| \cdot (-3)$

$\downarrow$

$+ \begin{cases} 7x + 6y = -1 \\ -33x - 6y = -9 \end{cases}$

$-26x = -10$

$x = 10/26 \approx 0.38462$

Substituting $x = 0.38462$ in the first equation, we get:

$7(0.38462) + 6y = -1$

$6y = -1 - 7(0.38462)$

$y = -0.61539$

Solution: $(0.385, -0.615)$

b. $\begin{cases} 3.4x + 0.7y = 5 \\ 0.5x - 0.2y = -2 \end{cases} \begin{matrix} \cdot 2 \\ \cdot 7 \end{matrix}$

$\downarrow$

$+ \begin{cases} 6.8x + 1.4y = 10 \\ 3.5x - 1.4y = -14 \end{cases}$

$10.3x = -4$

$x \approx -0.38835$

Substituting $x = -0.38835$ in the first equation, we get:

$3.4(-0.38835) + 0.7y = 5$

$-1.32039 + 0.7y = 5$

$0.7y = 5 + 1.32039$

$y = 9.02913$

Solution: $(-0.388, 9.029)$

8. Let b be the number of bikes, and t be the number of trikes. Then:

$\begin{cases} b + t = 31 \\ 2b + 3t = 80 \end{cases}$

From the top equation, we can solve that $b = 31 - t$. We will then substitute that in place of b in the second equation:

$2(31 - t) + 3t = 80$

$62 - 2t + 3t = 80$

$t = 18$

Then, $b = 31 - 18 = 13$. There are 13 bikes and 18 trikes.

9. Let D be Denny's age, and S be Sam's age. Then:

$\begin{cases} S + 4 = (2/3)(D + 4) \\ D = 2S - 14 \end{cases}$

Substituting $2S - 14$ for D in the first equation:

$S + 4 = (2/3)((2S - 14) + 4) \quad | \cdot 3$

$3S + 12 = 2(2S - 10)$

$3S + 12 = 4S - 20$

$-S + 12 = -20$

$-S = -32$

$S = 32$

Then, D $= 2(32) - 14 = 50$. Denny is 50 and Sammy is 32 years old.

10. Let t be the tens digit and u be the ones digit of the two-digit number. Then:

$\begin{cases} t + u = 11 \\ 10t + u = 7u - 2 \end{cases}$

$\downarrow$

$\begin{cases} t + u = 11 \\ 10t - 6u = -2 \end{cases} \Big| \cdot 6$

$\downarrow$

$+ \begin{cases} 6t + 6u = 66 \\ 10t - 6u = -2 \end{cases}$

$16t = 64$

$t = 4$

Then, $u = 11 - 4 = 7$. The number is 47.

11.

	weight	protein content	protein *amount*
peanut butter	w_1	25%	$0.25w_1$
protein powder	w_2	90%	$0.9w_2$
Mixture	210 g	30%	$0.3(210) = 63$ g

Equations: $\begin{cases} w_1 + w_2 = 210 \\ 0.25w_1 + 0.9w_2 = 63 \end{cases}$

Substituting $210 - w_2$ for w_1 in the second equation:

$$0.25(210 - w_2) + 0.9w_2 = 63$$
$$52.5 - 0.25w_2 + 0.9w_2 = 63$$
$$0.65w_2 = 10.5$$
$$w_2 \approx 16.154 \text{ g}$$

Then, $w_1 = 210 - w_2 = 210 \text{ g} - 16.154 \text{ g} = 193.846 \text{ g}$. The mixture contains 194 grams of peanut butter and 16 grams of protein powder.

12. Let v_w be the speed of the wind and v_c the speed of the crow.

	distance	velocity	time
Flying with a tailwind	400	$v_c + v_w$	48 sec
Flying with a headwind	400	$v_c - v_w$	72 sec

Using $d = vt$, the two equations we get are:

$$\begin{cases} 400 = 48(v_c + v_w) \\ 400 = 72(v_c - v_w) \end{cases}$$
$$\downarrow$$
$$\begin{cases} 400 = 48v_c + 48v_w & \div 4 \\ 400 = 72v_c - 72v_w & \div 4 \end{cases}$$
$$\downarrow$$
$$\begin{cases} 100 = 12v_c + 12v_w & \cdot 3 \\ 100 = 18v_c - 18v_w & \cdot 2 \end{cases}$$
$$\downarrow$$

$$+\begin{cases} 300 = 36v_c + 36v_w \\ 200 = 36v_c - 36v_w \end{cases}$$
$$\overline{\qquad\qquad 500 = 72v_c \qquad}$$
$$v_c = 6.9\overline{4}$$

Then, let's substitute $6.9\overline{4}$ for v_c in the first equation:

$$400 = 48(6.9\overline{4} + v_w)$$
$$400 = 333.\overline{3} + 48v_w$$
$$48_w = 66.\overline{6}$$
$$v_w = 1.3\overline{8}$$

The speed of the crow is about 6.9 m/s and the wind speed is about 1.4 m/s.

Chapter 8: Bivariate Data

Scatter Plots, pp. 183-185

Page 183

1. a. It signifies a couple where the wife is 36 years old and the husband is 41 years old.

   b. Answers will vary. There are two dots where the wives are 20, and the husbands are about 24 and 26.
   Then, there are two dots where the wives are 39 and the husbands are about 37 and 41.
   Then, there are two dots where the wives are 50 and the husbands are about 45 and 48.

   c. The wife is 52, and the husband is 68.

   d. No. The youngest wives are 20 years old, married to husbands age 24 and 26. But the youngest husband is 21, married to a wife that is also 21.

   e. Yes. The dot that is the highest is also the furthest to the right.

   f. Yes, there is a relationship or association between the variables. As the wife's age increases, so does the husband's age. We see a pattern where the dots fall approximately on a line that is increasing.

Page 184

2. a. Final grade and the number of missed classes.
   b. One and two times.
   c. Answers will vary; check the student's answer. For example:

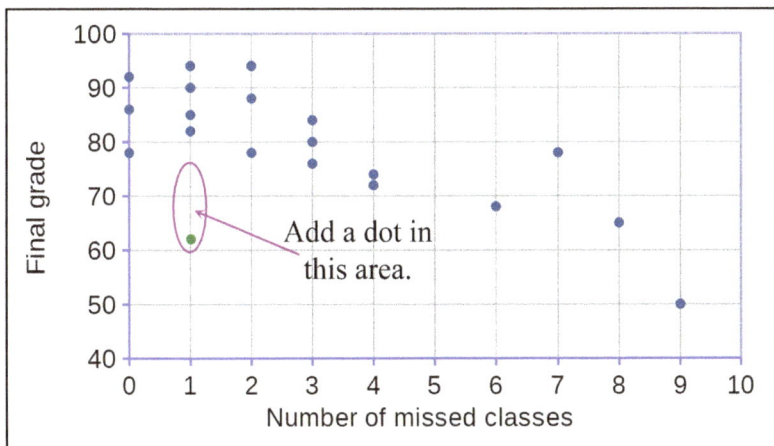

   d. Yes. I see that the more classes a student missed, the lower their grade, overall.

   e. No. The direct reason for the final grade is usually a combination of the final test and graded homework, or however the teacher has set up the requirements. Missing classes can indirectly cause a lower grade because the student is less likely to understand the material and less likely to be involved in the learning and the work, which then causes them to not do so well in the test(s) and also less likely to finish their homework.

Scatter Plots, cont.

Page 184

3. a. It signifies a bag of rice that weighs 2.5 kg and costs $6.
   b. Those bags are circled in the graph below. There are actually three of them: one costs $2 and weighs 1 kg, another costs $3 and weighs 1.5 kg, and the third one costs $4 and weighs 2 kg.
   c. Answers will vary. See the graph below for one example.

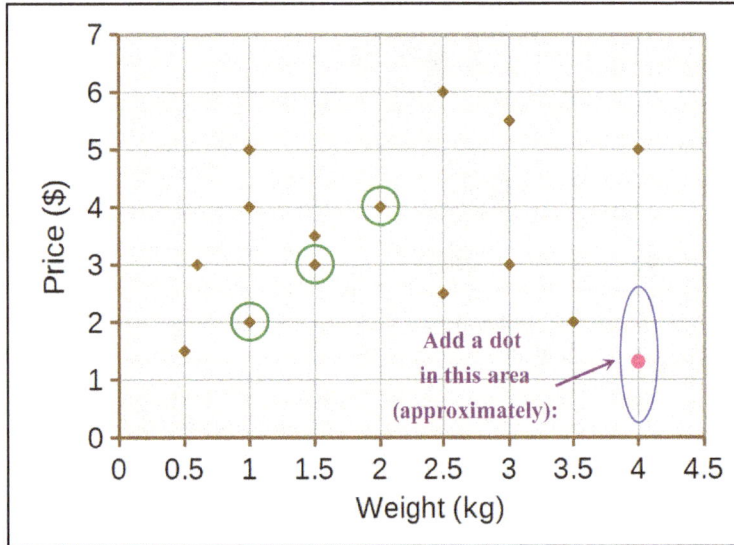

   d. The bag with the lowest price per kilogram is the one costing $2 for 3.5 kg of rice.
   e. Not really. The dots are kind of scattered all over the place, and not in any clear pattern.

Page 185

4. a. Answers will vary because the scaling on the axes can vary. See the graph on the right for one example. In it, the horizontal axis starts at 18:

   b. Yes, there is a relationship. The more shots a player made, the more goals they also tended to make (which makes sense!).

   c. This requires us to look at the goals/shots ratio for each player (or shots/goals). Imagine a diagonal line going through the graph from the bottom left corner to the upper right corner. The players that are above that line are better at making goals than the players below that line.

   The player with 6 goals for 28 shots is the best one. Their goals/shots ratio is 6/28 ≈ 0.214, which is quite a bit better than for any other player. They make a goal about 21% of the time when they shoot towards a goal.

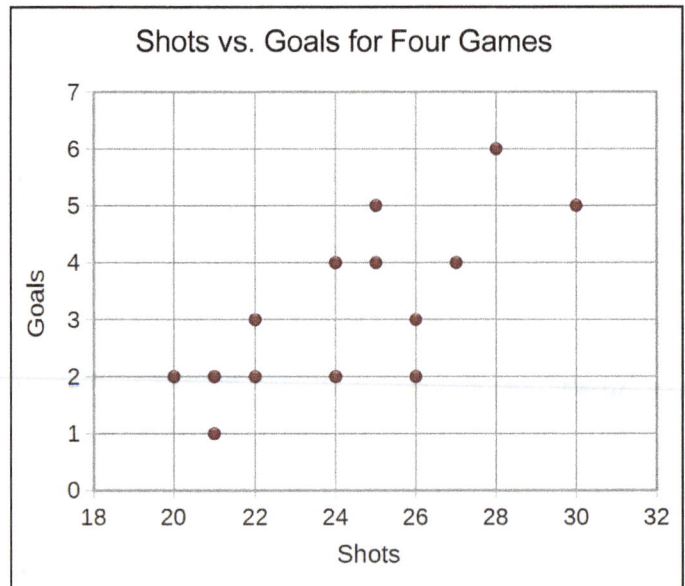

   Two other very good players are the ones with 4 goals for 24 shots, and the one with five goals for 30 shots. Their goals/shots ratios are 4/24 = 1/6 ≈ 16.7% and 5/30 = 1/6 ≈ 16.7%

186

Scatter Plot Features and Patterns, pp. 186-189

Page 187

1. a. nonlinear negative association
   b. positive linear association with one outlier and a cluster

2. a. There is a negative linear association. In general, heavier cars get a worse gas mileage than lighter cars. There is one definite outlier (about 2,350 kg and 36 mpg) and one other possible outlier at about 1,650 kg and 14 mpg. (There are no clear clusters.)
   b. A little over 1,600 kg (exactly 1,630 kg)
   c. About 2,350 kg and 36 mpg.
   d. 24 mpg

Page 188

3. The student graphs will vary because the scaling on the axes will vary; check the student's graph. For example:

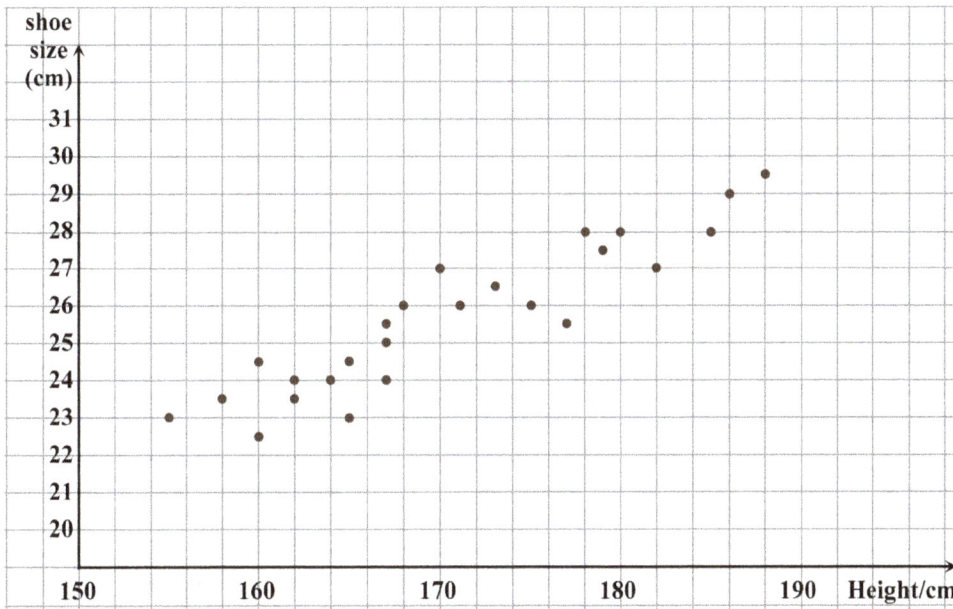

The scatter plot shows a positive linear association. (There are no clusters or outliers.)

Page 189

4. The student graphs will vary because the scaling on the axes will vary; check the student's graph. For example:

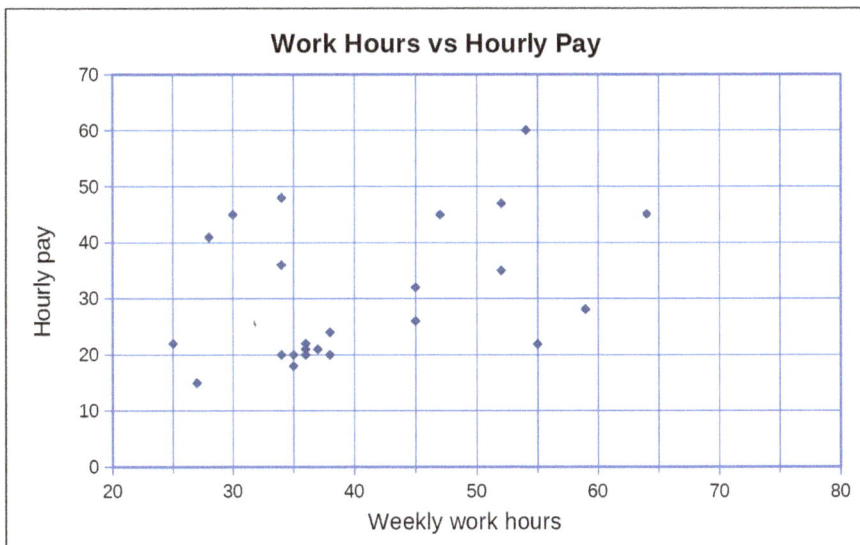

5. a. No, there is no association between the variables. The dots on the scatter plot are quite scattered.
   b. There is one cluster around 34 to 38 work hours and $18-$24 of pay.
   c. Three persons worked more hours than the person with the highest hourly pay. They worked 55 hours, 59 hours, and 64 hours.

Work Hours vs Hourly Pay

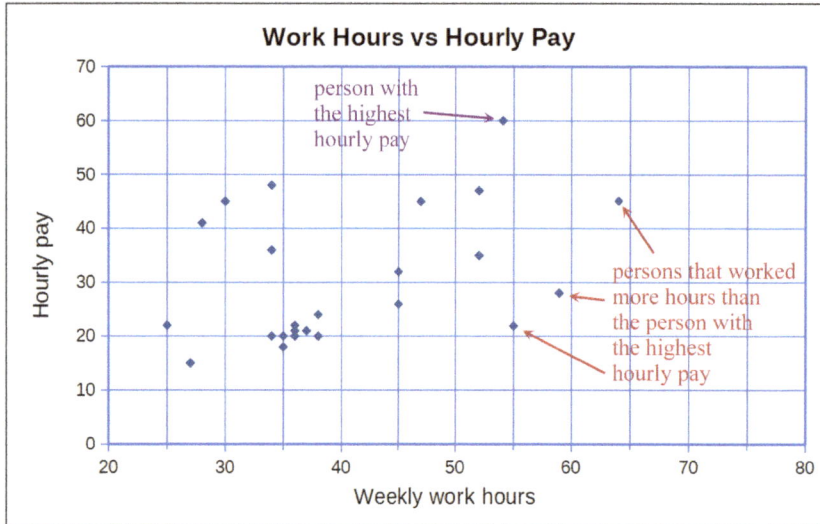

d. The person who worked the most hours worked 64 hours, and their pay is about $45 per hour, which means they earned 64 · $45 = $2,880. The person who worked the least hours worked 25 hours, and their pay is about $22 per hour, which means they earned 25 · $22 = $550.
   The person with the most work hours earned $2,880 − $550 = $2,330 more than the person with the least hours.

e. No. The person with the lowest hourly rate ($15 per hour) worked 27 hours.

Fitting a Line, pp. 190-193

1. a. Answers will vary. Check the line the student drew. For example:

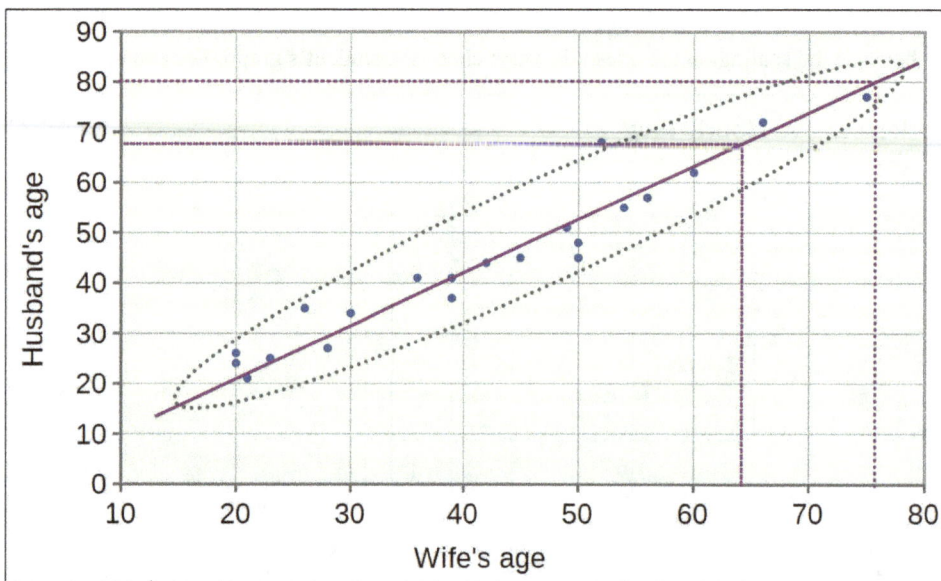

b. Answers will vary since the answer is based on the line the student drew. About 68 years.
c. Answers will vary since the answer is based on the line the student drew. About 76 years.

Fitting a Line, cont.

Page 191

2. Answers will vary. Check the student's line. For example:

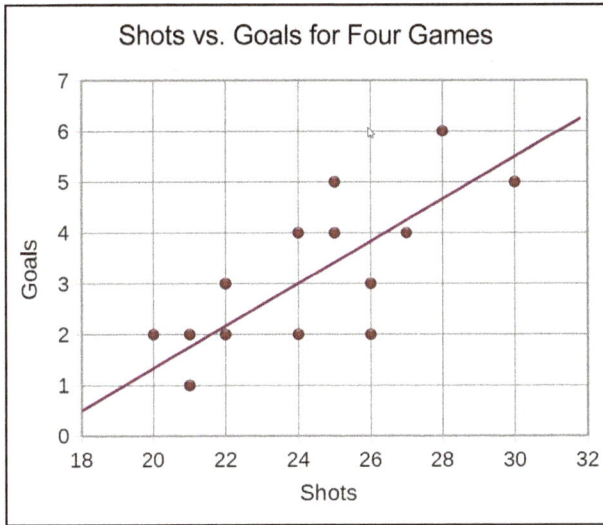

Page 192

3. a. Answers will vary; check the student's answer. Here are two possibilities:

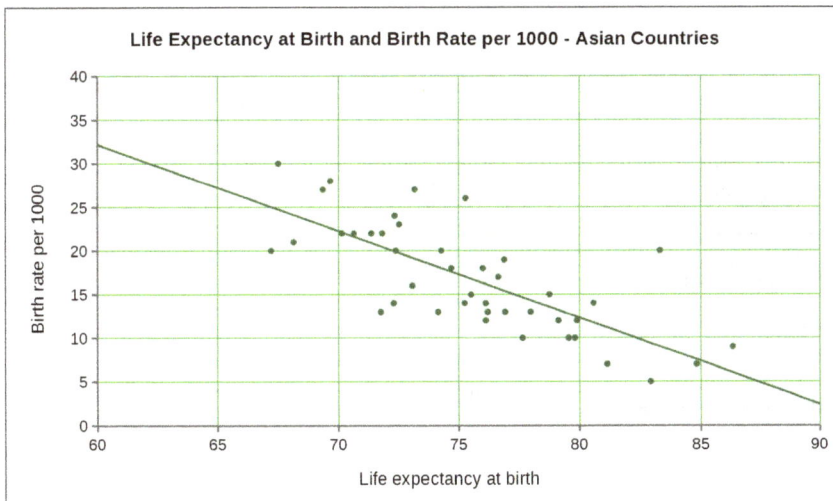

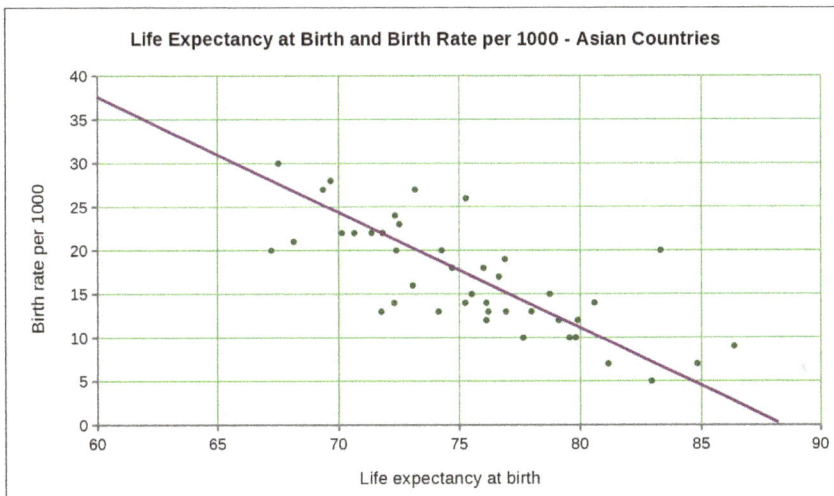

3. b. Answers will vary. Based on the top line above, it would be between 72 and 73 years.
   c. About 67.5 years. The predicted life expectancy is about 62 years.
   d. The lowest birth rate in this sample is 5 per 1000. That country has a life expectancy of about 83 years.
      The predicted life expectancy for a country with birth rate of 5 per 1000 is about 87 years, so the difference is
      <u>about 4 years</u>.
   e. The country with the highest life expectancy has a life expectancy of about 87 years, and a birth rate of 9 per 1000.
      The line would predict a birth rate of about 6 per thousand, so the difference is about 3 per 1000.

Page 193

4. a. Answers will vary; check the student's answer. Here is one possibility:

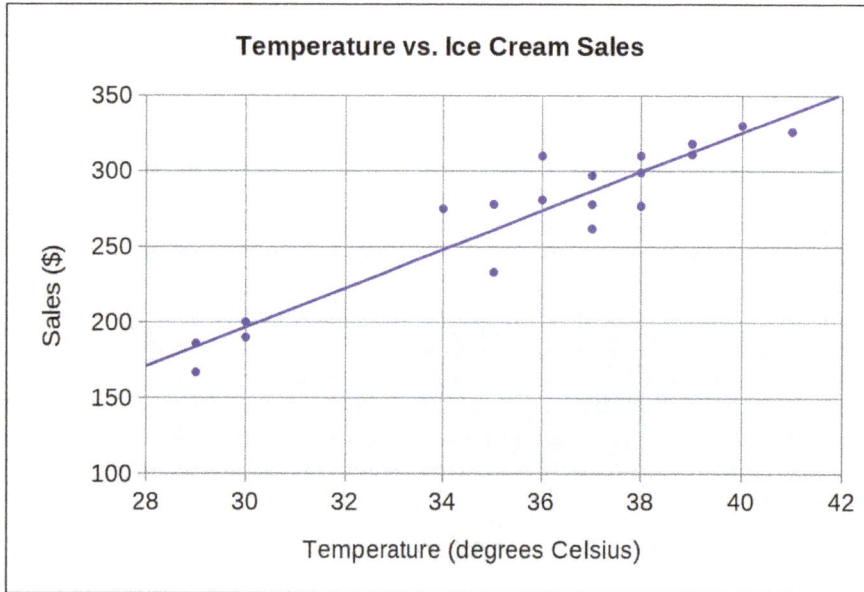

   b. Answers will vary; check the student's answer. Based on the line above, about $220.

5. a. Answers will vary; check the student's answer. Here is one possibility:

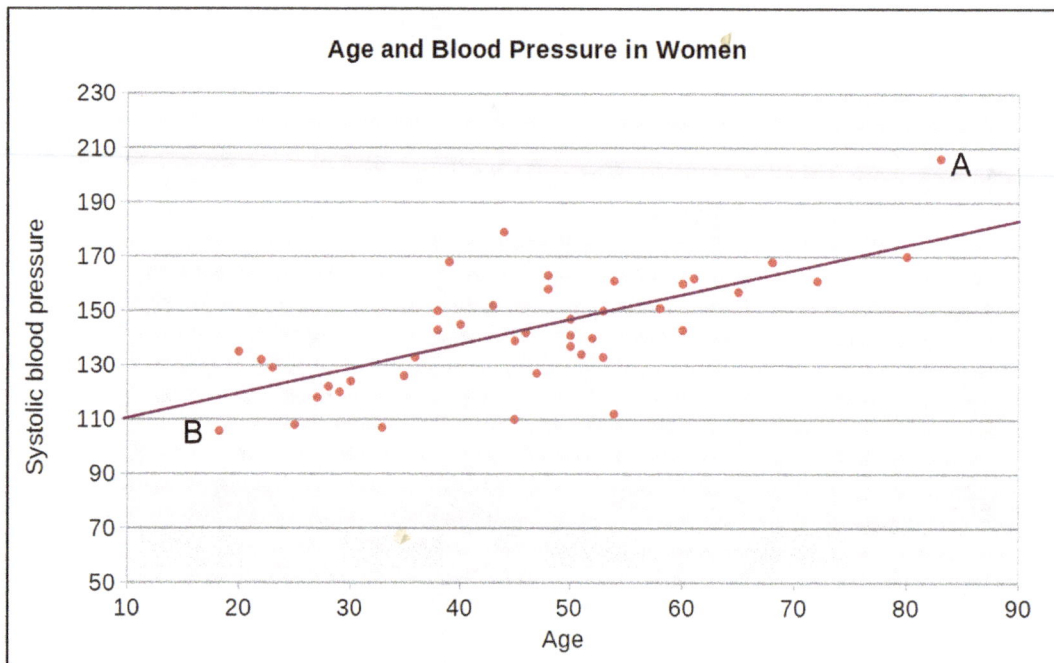

5. b. A line from A to B will leave many more points below the line than above it. Also, the total of the distances of the points to the line is much more for the points below the line than for the ones above it.

Age and Blood Pressure in Women

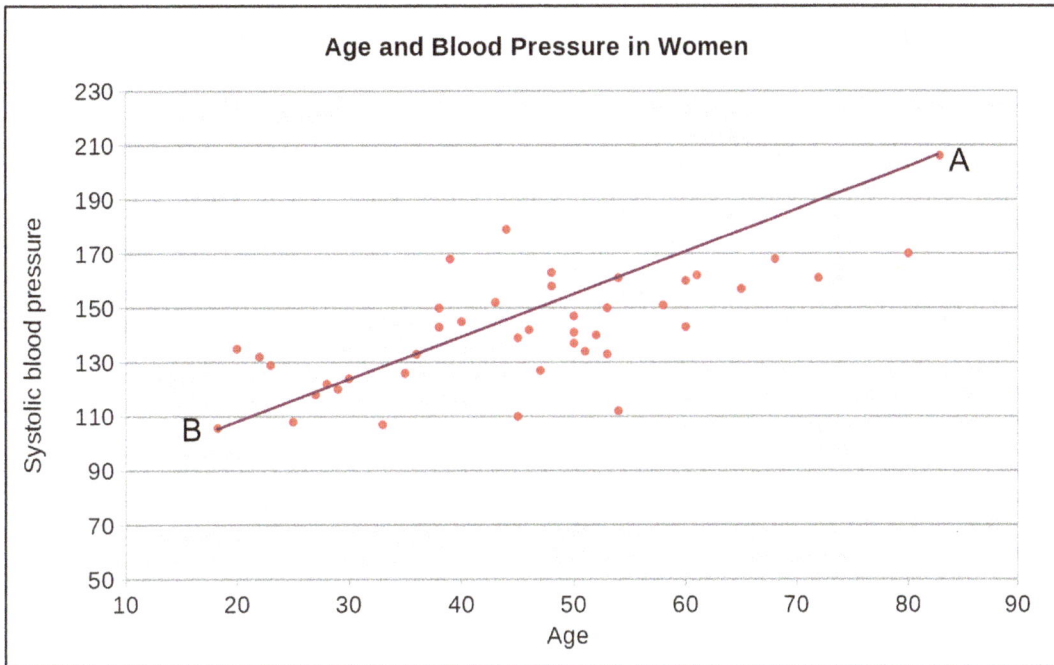

Equation of the Trend Line, pp. 194-198

Page 194

1. This is the only statement that is correct:
Each 1-minute increment in time is associated with a 0.07-mile increment in distance.

2. a. $d = 0.074(32) = \underline{2.368 \text{ miles}}$
   b. We set d to be 3.3 and solve the resulting equation: $3.3 = 0.074t$, from which $t = 3.3/0.074 \approx \underline{44.6 \text{ minutes}}$.
   c. Each five minutes added to the walk time is associated with a $\underline{0.37}$-mile increase in her walking distance.

Page 195

3. a. These are the correct statements:

   - Each 1-lb increase in weight is associated with a 0.16-inch increase in height.
   - Heavier dogs tend to be taller; and for each 5-lb increase in the weight, the dogs tend to be 0.8 inches taller.
   - The model predicts a height of 11.68 inches for a dog weighing zero pounds.
   - We should be careful in using this model to extrapolate the heights of dogs less than 55 pounds.

   b. We substitute 22.5 in place of h in the equation $h = 0.16w + 11.68$, and then solve for w (on the right):

   The equation predicts such a dog would weigh <u>about 68 pounds</u>.

$$22.5 = 0.16w + 11.68$$
$$22.5 - 11.68 = 0.16w$$
$$10.82 = 0.16w$$
$$w = 67.625$$

   c. The height of a 63-lb dog is predicted to be $h = 0.16(63) + 11.68 \approx 21.8$ inches or about 21 3/4 inches.
   d. No.
   e. The predicted height of a 75-lb dog is $h = 0.16(75) + 11.68 = 23.68$ inches.
   The difference is $24.25 - 23.68 = \underline{0.57 \text{ inches}}$.

4. a. It signifies that for each 1-kg increase in mass, the MPG decreases by -0.011115 miles per gallon.

   b. We would expect it to decrease by $100(0.011115) = 1.1115$ miles per gallon.

   c. MPG $= -0.011115(2500) + 44.156 = 16.3685$ or about 16.4 miles per gallon.

   d. We substitute 27.5 in place of MPG in the equation MPG $= -0.011115m + 44.156$, and then solve for m:

$$27.5 = -0.011115m + 44.156$$
$$27.5 - 44.156 = -0.011115m$$
$$-16.656 = -0.011115m$$
$$m = -16.656/(-0.011115)$$
$$m = 1498.5 \text{ kg}$$

The equation predicts such a car would weigh about 1,500 kg.

e.

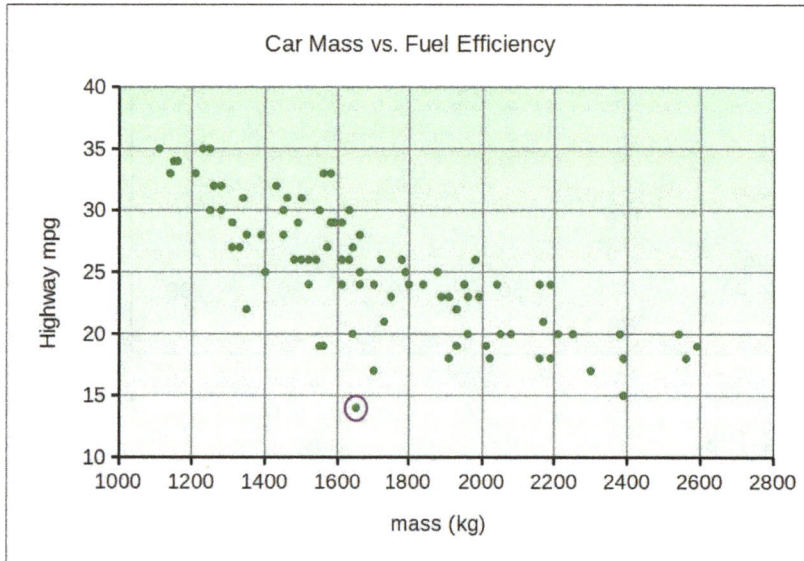

The equation would predict a fuel efficiency of $-0.011115(1650) + 44.156 \approx 25.8$ miles per gallon.

f. The image below has a horizontal line drawn at 27 mpg. The arrow points to the heaviest car among those that get 27 mpg or more. That car weighs about 1650 kg. So, that is the maximum that I would expect such a car to weigh.

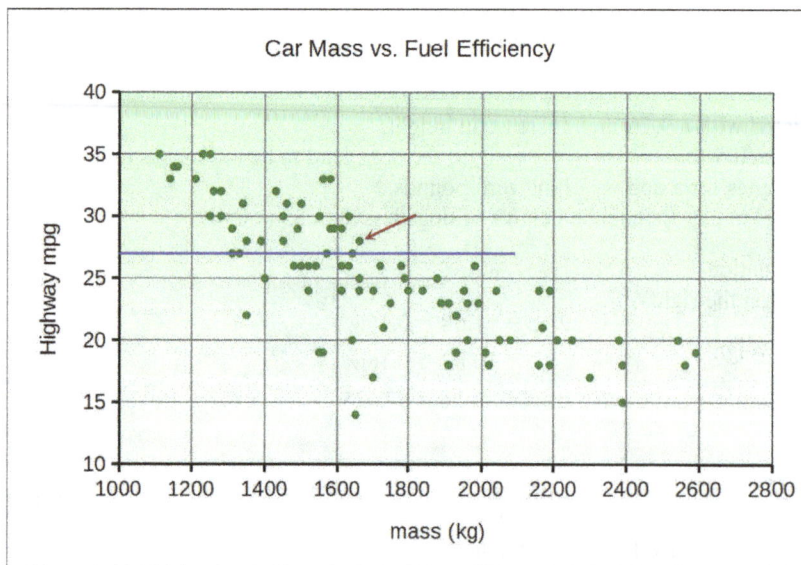

Equation of the Trend Line, cont.

5. a. Answers will vary; check the student's graph. Here is one possibility:

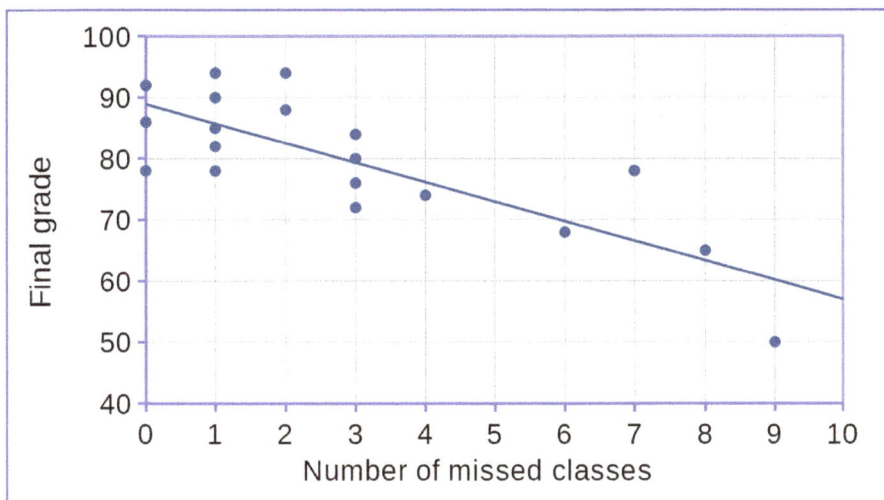

b. Answers will vary; check the student's answer. The equation for the line in the above graph is $G = -3.2m + 89$, where G is the grade and m is the number of the missed classes.

To find the equation, draw two points ON the line, preferably points that are on the gridlines so that it will be easier to read their coordinates. Then figure out the slope using those two points: it is the ratio of (change in y-values/change in x-values). Then read the y-intercept from where your line crosses the vertical axis.

Alternatively, you can use the coordinates of any one point on the line to figure out the y-intercept, once the slope is known. For example, if the line goes through the point (9, 60) and you have determined the slope is −3, then use the generic equation for a line $y = mx + b$. Substitute (9, 60) for x and y, and −3 for m, and solve for b:

$60 = -3(9) + b$, from which $b = 87$.

c. The slope −3.2 signifies that each missed class is associated with a 3.2-point decrease in the final grade.

d. The y-intercept (or G-intercept in this case) signifies that the equation predicts a grade of 89 if you miss zero classes.

e. If $m = 5$, then $G = -3.2(5) + 89 = 73$.

f. The slope (3) in her equation is positive, which would mean that for each missed class, the equation would predict a 3-point *increase* in the final grade, which clearly is not reasonable. The association between the variables is negative, not positive.

6. a. Answers will vary; check the student's line. For example:

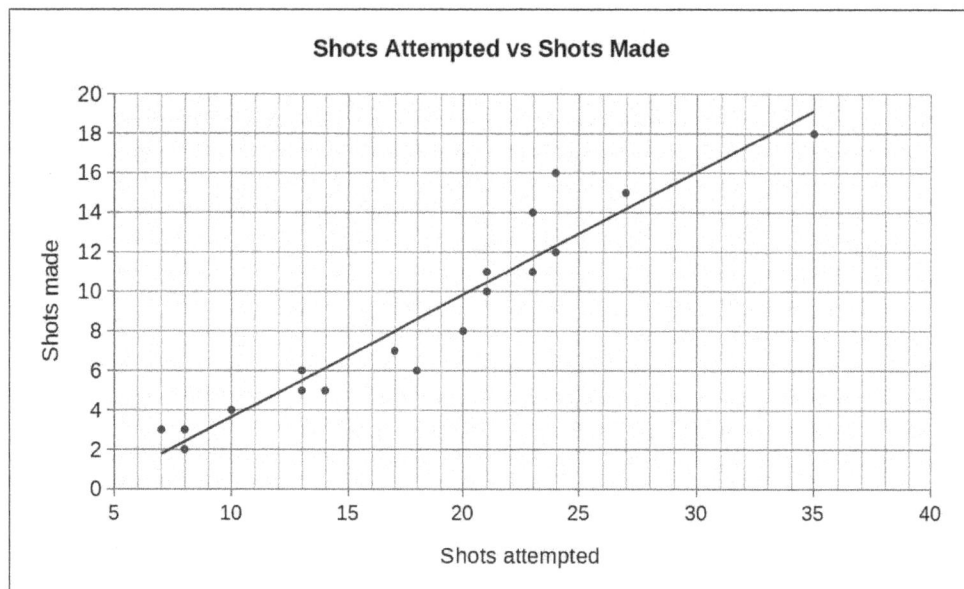

b. Answers will vary; check the student's equation. The equation for the line in the above graph is $y = 0.62x - 2.56$.

To find the equation, draw two points on the line, preferably points that are on the gridlines so that it will be easier to read their coordinates. Then figure out the slope using those two points: it is the ratio of (change in y-values/change in x-values). Then extend the line so it reaches the vertical axis, and read the y-intercept from there.

Alternatively, you can use the coordinates of any one point on the line to figure out the y-intercept, once the slope is known. For example, if the line goes through the point (30, 16) and you have determined the slope is 0.6, then use the generic equation for a line $y = mx + b$. Substitute (30, 16) for x and y, and 0.6 for m, and solve for b:

$16 = 0.6(30) + b$, from which $b = -2$.

c. The slope (in the example graph, 0.62) signifies that for each additional shot attempted, the number of shots made tends to increase by 0.62.

d. The y-intercept (in our example, -2.56) means that the equation predicts that for zero shots attempted, there would be -2.56 shots made. It does not make sense. We might need to be careful and not extrapolate below 7 attempted shots.

e. *shots made* $= 0.62(30) - 2.56 \approx \underline{16}$.

f. We set y to be 8 in our equation, and solve for x:

$$8 = 0.62x - 2.56$$
$$8 + 2.56 = 0.62x$$
$$10.56 = 0.62x$$
$$x \approx 17$$

The equation predicts that they attempted to make a shot 17 times.

Page 200

1. a. $23/48 \approx 47.9\%$ completed the course and 52.1% did not.
   b. $57/65 \approx 87.7\%$ completed the course and 12.3% did not.
   c. Yes, there is. Far greater proportion of the students who took the course in person completed it than those who took the online version.

2. a. See the table on the right.
   b. Most of the 20- to 30-year olds were against the plan. Most of the 31+ year-olds were for the plan.
   c. There are more people for it than against it, but the difference is not much.
   d. Yes, there is an association. Most of the younger people are against the plan, whereas most of the older are for it.
   e. Perhaps the 31+ crowd has a better financial situation overall, so they can afford the price increase better.

	for	against	Total
20-30	21	56	77
31+	58	14	72
TOTAL	79	70	149

Page 201

3. a. See the table on the right.
   b. There are more people in favor.
   c. Yes. Then there are 14 against and 12 in favor.
   d. 33.3% of the adults are in favor, 75% of the teens, 100% of the children, and 66.7% of everyone.
   e. Yes, there is an association. The older the person is, the less likely they are to support the thought of gathering at the amusement park and paying a fee.

	Fee	No fee	Total
Adults	6	12	18
Teens	6	2	8
Children	16	0	16
TOTAL	28	14	42

Puzzle corner. No, there isn't. Whether homeschooled or in public school, about half the students don't own a pet, about a third own one pet, and about a sixth own 2+ pets.

Relative Frequencies, pp. 202-206

Page 203

1. a. See the table on the right.
   b. The 60-to 79-year-olds.
   c. Yes. With advancing age, there are more people who do crossword puzzles frequently. In the youngest group, about 3/4 don't do them frequently, and in the oldest group, slightly over half do.

	Does crossword puzzles frequently		
	Yes	**No**	**Total**
20-39 years	23.6%	76.4%	100%
40-59 years	43.0%	57.0%	100%
60-79 years	53.3%	46.7%	100%

2. a.

	Owns a house or an apartment					
	Yes		**No**		**Total**	
	#	%	#	%	#	%
Has debt	274	42.5%	371	57.5%	645	100%
Has no debt	129	44.6%	160	55.4%	289	100%
TOTAL	**403**	(N/A)	**531**	(N/A)	**934**	

   b. No. The percentages for owning a house or an apartment or not owning one are quite similar for either group.
   c. It is more likely that they do have debt. This is because if we're randomly choosing a person from among the 403 that own a house/apartment, 274 of them have debt and only 129 don't. So, the probability for that is $274/403 \approx 68\%$ whereas the probability for choosing a person that doesn't have debt is only $129/403 \approx 32\%$.
   d. It is more likely that they do have debt ($371/531 \approx 70\%$ vs. $160/531 \approx 30\%$).
   e. It is more likely that they don't own a house/apartment ($371/645 \approx 57.5\%$, vs. $274/645 \approx 42.5\%$).

3. a. See the table below.

	Dessert or not?				Total
	Yes		No		
	#	%	#	%	
Expensive dish	8	57.1%	6	42.9%	14
Cheap dish	15	53.6%	13	46.4%	28
TOTAL	23	54.8%	19	45.2%	42

b. The row percentages are fairly similar, so there is no association.

Page 205

4. a.

	Had a summer job		
	Yes	No	Total
14-15 years	33	109	142
16-17 years	35	55	90
18-19 years	68	53	121
TOTALS	136	217	353

Relative frequencies:

	Had a summer job		
	Yes	No	Total
14-15 years	23.2%	76.8%	100%
16-17 years	38.9%	61.1%	100%
18-19 years	56.2%	43.8%	100%

b.

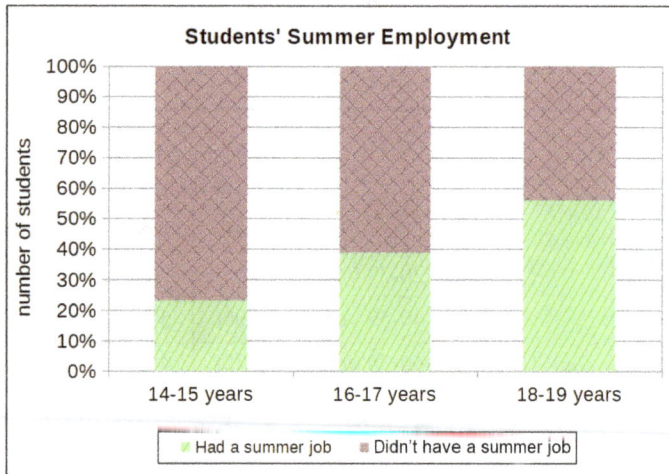

c. There *is* an association. As we go from the youngest age group to the oldest, the proportion of those who had a summer job keeps increasing. In the youngest group, only about 23% had had a summer job, whereas among the 18- to 19-year-olds, over half of them had had a summer job.

d. The true statements are:
   (iii) About 3/4 of the youngest age group had not had a summer job.
   (iv) About 2/5 of all the surveyed teens had had a summer job.

Relative Frequencies, cont.

5. a.

Favorite book genres by age

	30-50	**over 50**
Biographies & Memoirs	4.2%	28.2%
Comics	14.8%	5.2%
Mysteries	21.8%	41.1%
Poetry	2.6%	6.5%
Romance	25.2%	14.7%
Science Fiction & Fantasy	31.3%	4.3%
Total	99.9%	100%

b.

- 30- to 50-year-olds: Science fiction & fantasy, romance, and mysteries
- over 50 years old: mysteries, biographies & memoirs, and romance

c. Answers will vary. Check the student's answer. Here are some possibilities:

- With advancing age, comics and science fiction become far less popular among the readers.
- At the same time, the popularity of mysteries and biographies & memoirs increases a lot.
- Poetry is the least popular category overall, with only 4.5% of the respondents listing that as their favorite.
- Mysteries is the most popular favorite among the respondents, with 31.6% of them preferring that genre over others.

d. Yes, there is. There are noticeable differences between the two age groups in the genres of books that are most popular.

6. Answers will vary; check the student's results.

Page 207

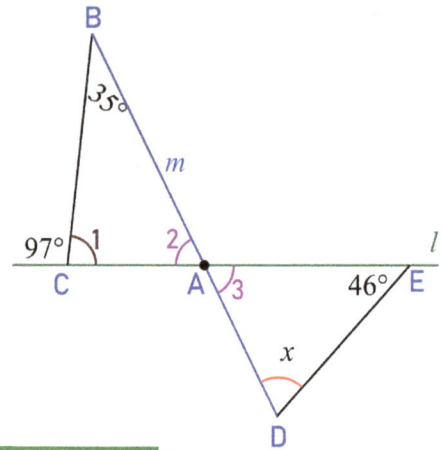

1. See the image on the right. Angle 1 is supplementary to the 97° angle, so it is 83°. In triangle CBA, angle 2 is $180° - 35° - 83° = 62°$. Angle 3 and angle 2 are vertical angles, so angle 3 is also 62°. In triangle AED, angle x is $180° - 62° - 46° = \underline{72°}$.

2. a. 8   b. 6   c. 9/4
   d. 0.4   e. 0.1   f. 1.1

3.

a.	$18 + t^3 = 134$	b.	$5x^2 = 220$	c.	$109 - v^3 = 11$
	$t^3 = 116$		$x^2 = 44$		$109 - 11 = v^3$
	$t = \sqrt[3]{116} \approx 4.877$		$x = \sqrt{44} \approx 6.633$		$98 = v^3$
			or $x = -\sqrt{44} \approx -6.633$		$v = \sqrt[3]{98} \approx 4.610$

4. a. Correct.
   b. Not correct. $\sqrt{121} = 11$ so it is a whole number.
   c. Correct.
   d. Not correct. Unending repeating decimals are rational; 0.19191919... is a rational number. As a fraction, it is 19/99.

Page 208

5. Let s be the unknown side. The diagonal divides the rectangle into two right triangles. Applying the Pythagorean Theorem to one of them (on the right), we get:

   The other side of the rectangle is $\underline{4.7 \text{ m long}}$.

$$2.80^2 + s^2 = 5.5^2$$
$$7.84 + s^2 = 30.25$$
$$s^2 = 22.41$$
$$x = \sqrt{22.41} \approx 4.7$$

6. We need to first find the length marked with y (see the image on the right). After that, we can use the Pythagorean Theorem to find x.

   To find y, we apply the Pythagorean Theorem in the triangle ABC:

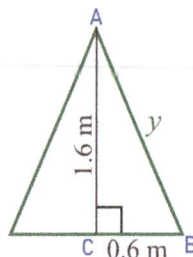

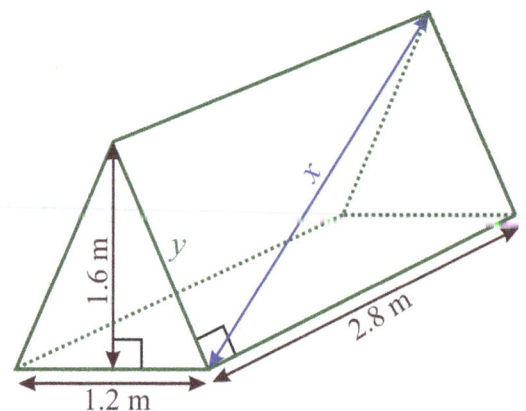

$$0.6^2 + 1.6^2 = y^2$$
$$0.36 + 2.56 = y^2$$
$$y^2 = 2.92$$
$$y = \sqrt{2.92}$$

Then, we use the Pythagorean Theorem in the triangle that is part of the tent's side, with sides y, x, and 2.8 meters:

$$y^2 + 2.8^2 = x^2$$
$$(\sqrt{2.92})^2 + 2.8^2 = x^2$$
$$2.92 + 7.84 = x^2$$
$$y^2 = 10.76$$
$$y = \sqrt{10.76} \approx 3.28$$

The diagonal marked with x is $\underline{3.3 \text{ m long}}$.

7. Answers will vary since the description did not give actual amounts of water.
   The first portion should be linear and decreasing. The portion for September should be increasing in a nonlinear manner, so that it rises fast at first and then slows down. The portions for October and January should be horizontal lines. The portion for November-December should be slowly increasing and linear. For example:

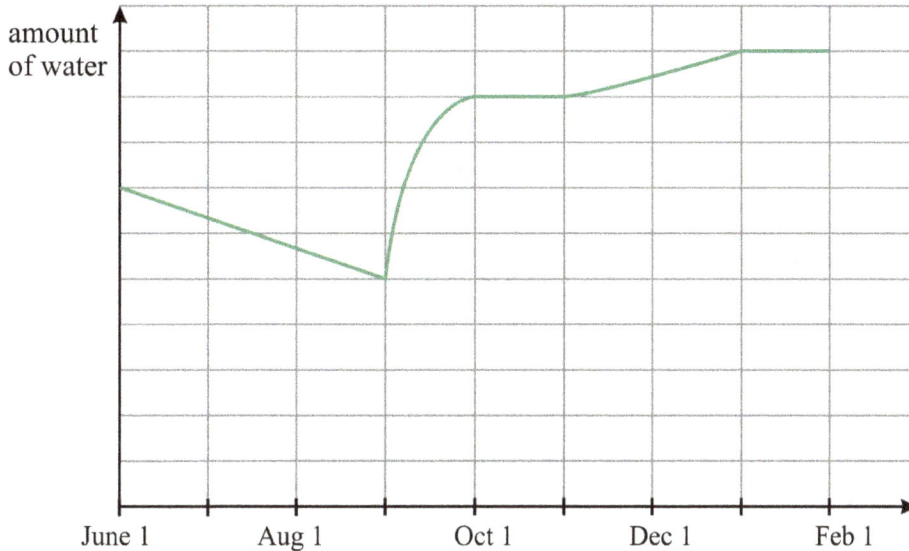

Page 209

8. a. $y = (2/3)x - 2$

   b. The slope of line N is $-3/2$. Its equation is of the form
   $y = -(3/2)x + b$. To find b, we will substitute $(3, 0)$ into the equation and solve for b:

   $0 = -(3/2)(3) + b$

   $0 = -9/2 + b$

   $b = 9/2$

   So, the equation is $\underline{y = -(3/2)x + 9/2}$.

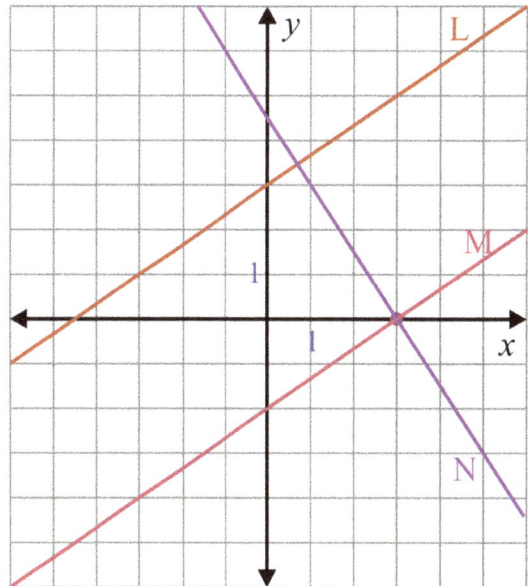

9.

a.
$$\begin{cases} 5x - 7y = 3 \\ -6x + 8y = 1 \end{cases} \quad \begin{matrix} \cdot 6 \\ \cdot 5 \end{matrix}$$

$$\downarrow$$

$$+ \begin{cases} 30x - 42y = 18 \\ -30x + 40y = 5 \end{cases}$$
$$\overline{\qquad -2y = 23}$$
$$y = -23/2$$

Substituting $y = -23/2$ in the second equation, we get:

$$-6x + 8(-23/2) = 1$$
$$-6x - 92 = 1$$
$$-6x = 93$$
$$x = -93/6 = -31/2$$

Solution: $(-31/2, -23/2)$ or $(-15.5, -11.5)$

b.
$$\begin{cases} 3x = -2(y + 1) \\ -x - 3y = 0 \end{cases}$$

$$\downarrow$$

$$\begin{cases} 3x = -2y - 2 \\ -x - 3y = 0 \end{cases}$$

$$\downarrow$$

$$\begin{cases} 3x + 2y = -2 \\ -x - 3y = 0 \end{cases} \quad \Big| \cdot 3$$

$$\downarrow$$

$$+ \begin{cases} 3x + 2y = -2 \\ -3x - 9y = 0 \end{cases}$$
$$\overline{\qquad -7y = -2}$$
$$y = 2/7$$

Substituting $y = 2/7$ in the second equation, we get:

$$-x - 3(2/7) = 0$$
$$-x - 6/7 = 0$$
$$x = -6/7$$

Solution: $(-6/7, 2/7)$

10. Let x be the number of chickens and y be the number of cows. Then:

$$\begin{cases} x + y = 42 \\ 2x + 4y = 100 \end{cases}$$

Solving for y in the first equation, we get $y = 42 - x$. Substituting that in place of y in the second, we get:

$$2x + 4(42 - x) = 100$$
$$2x + 168 - 4x = 100$$
$$168 - 2x = 100$$
$$-2x = -68$$
$$x = 34$$

Then, $y = 42 - 34 = 8$. There are 34 chickens and 8 cows.

Page 210

11. a. There is a positive linear association. We do not see any outliers or clusters.
    b. Answers will vary; check the student's line. For example:

Age and Resting Heart Rate

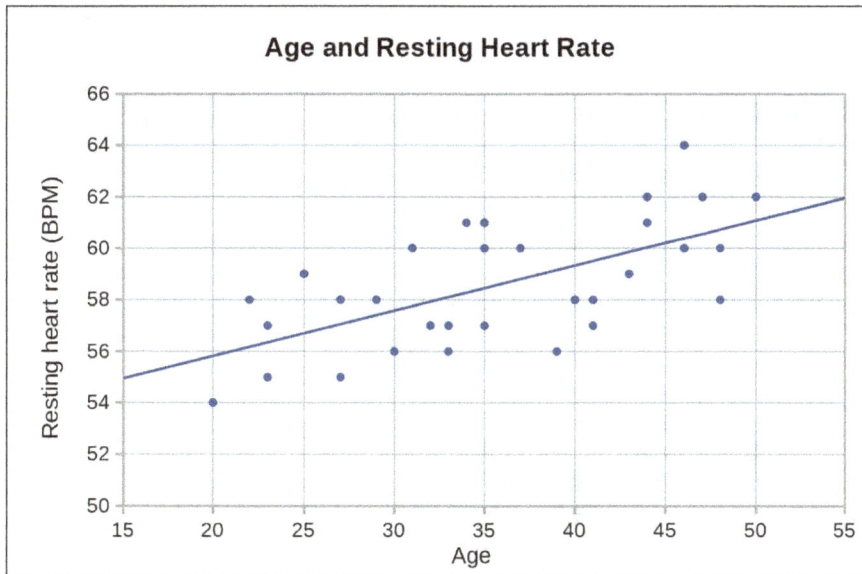

c. Answers will vary; check the student's equation. The equation for the line in the graph above is approximately $y = 0.18x + 52$.

d. Answers will vary; check the student's answer. Using the equation above, it increases by 0.18 beats per minute.

e. Answers will vary; check the student's answer. Using the equation above, if $x = 30$, we get the heart rate as $y = 0.18(30) + 52 = 57.4$ beats per minute.

f. For a newborn, the equation predicts a heart rate of $0.18(0) + 52 = 52$ beats per minute.
   Since in reality, a newborn heart rate is 120 to 160 bpm, we cannot use this equation to extrapolate much (if any) past the range of the original data.

Page 211

12. a. (i) and (iii)
    b. See the chart on the right.
    c. Yes. When you compare the percentages in the two columns, they are definitely different. The majority of female clients have haircuts either two or three months apart, whereas half of the male clients have a haircut every month. The males have haircuts more often than females, and that makes sense because shorter hairstyles need a haircut more often.

Relative frequencies:

	Male	Female
every month	50%	12%
every two months	28%	30%
every three months	15%	37%
every four months	7%	20%
TOTALS	100%	99%

1. a. It is a negative linear association.

    b. There is at least one outlier: the country with about 89 years of life expectancy and a birth rate of 25 per 1000. Some other dots in the lower end of the life expectancy could perhaps be considered outliers, also, but without further statistical tools, we cannot say for sure.

    c. Answers will vary; check the student's trend line. For example:

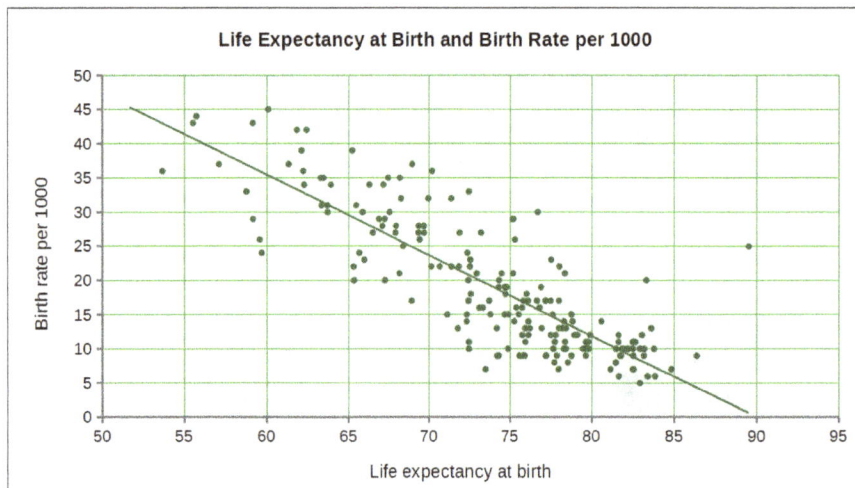

Life Expectancy at Birth and Birth Rate per 1000

    d. Answers will vary; check the student's trend line equation.

    To find the equation, draw two points on the line, preferably points that are on the gridlines so that it will be easier to read their coordinates. Then figure out the slope using those two points: it is the ratio of (change in y-values/change in x-values).

    In the case of the above graph, we could use for example (80, 12) and (70, 23.5) (estimated from the graph). The slope becomes $(12 - 23.5)/(80 - 70) = -11.5/10 = -1.15$.

    Then, we figure out the y-intercept. The generic equation for a line is $y = mx + b$. We substitute into this the slope of -1.15 and one of the points, say (80, 12) to get the equation $12 = -1.15(80) + b$, from which $b = 104$.

    So, the equation is $y = -1.15x + 104$.

    For comparison, the equation, as calculated by a spreadsheet program, is $y = -1.181x + 106.3$. The equation we got, $y = -1.15x + 104$, is not exactly the same, but that is to be expected when we are estimating the coordinates of the points from the graph.

    e. The slope of -1.15 means that each 1-year increase in life expectancy is associated with a decrease of 1.15 per 1000 in the birth rate.

    f. We solve the equation $25 = -1.15x + 104$:

    $$25 = -1.15x + 104$$
    $$25 - 104 = -1.15x$$
    $$-79 = -1.15x$$
    $$x = 79/1.15 \approx 69 \text{ years}$$

    g. Birth rate $= -1.15(55) + 104 \approx 41$ per 1000

    h. We calculated the predicted life expectancy at birth when the birth rate is 25 per 1000 in question (f). It was 69 years, or more precisely, 68.7 years. The difference between 89.5 and that is 20.8 years. That's quite a bit!

2. a. We use the given equation of the line to calculate the coordinates for two points, and then we can draw the line.

The d-values for these two points can be any. Here, I chose 0 and 3,000 as being the beginning and end points of the horizontal axis.

First, if $d = 0$, then C $= 0.111(0) + 76.9 = 76.9$. Then, if $d = 3,000$, then C $= 0.111(3,000) + 76.9 = 409.9$. So, the two points to draw the line between are approximately (0, 77) and (3,000, 410).

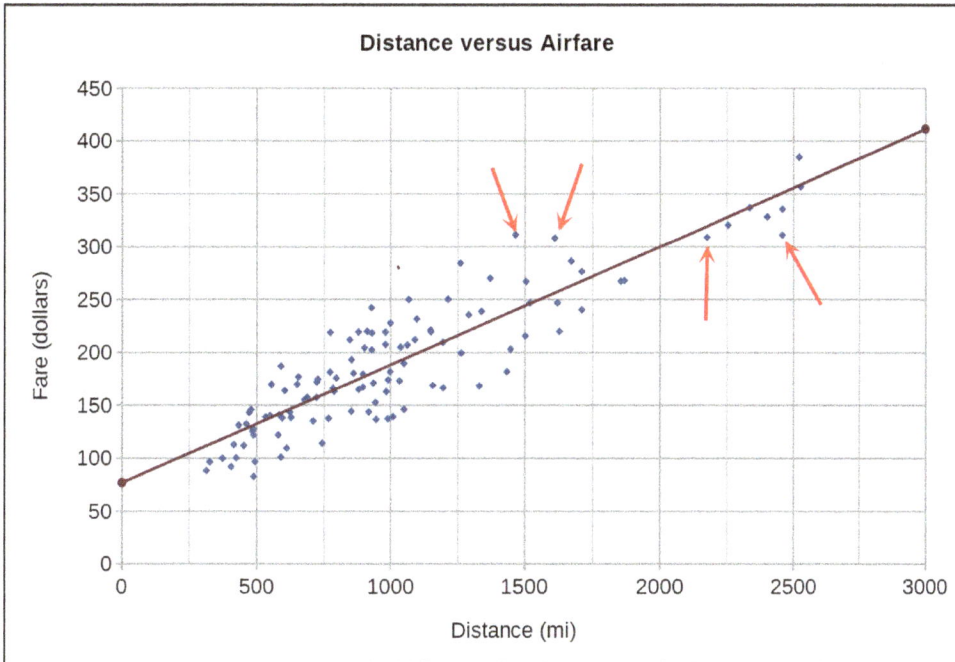

b. Those four points are signaled by the arrows in the image above. The distances are about 1,450, 1,600, 2,200, and 2,450 miles.

c. When the cost is $310, we set C to be 310 in the equation C $= 0.111d + 76.9$ and solve for d:

$$310 = 0.111d + 76.9$$
$$310 - 76.9 = 0.111d$$
$$233.1 = 0.111d$$
$$d = 233.1/0.111 = 2,100$$

The equation predicts a distance of <u>2,100 miles</u>.

d. When d is 800 miles, C $= 0.111(800) + 76.9 = $ <u>$165.70</u>.

e. We use the slope. The slope of 0.111 signifies that each 1-mile increment in distance is associated with a $0.111-increase in the airfare. This means that a 100-mile increase in distance is associated with a $100(0.111) = 11.10 increase in airfare.

f. When the cost is $300, the distance is 2,009.91 miles (solve the equation $300 = 0.111d + 76.9$ for d).

When the cost is $250, the distance is 1,559.46 miles. The difference is 450.45 miles, or <u>about 450 miles</u>.

3. a.

	soccer	baseball	basketball	Total
Asia	17	1	5	23
Europe	85	7	60	152
Middle East	40	3	10	53
North America	11	45	26	82
Total	153	56	101	310

	soccer	baseball	basketball	Total
Asia	74%	4%	22%	100%
Europe	56%	5%	39%	100%
Middle East	75%	6%	19%	100%
North America	13%	55%	32%	100%

b. Answers will vary; check the student's answer. For example: Soccer is the most popular favorite sport everywhere except among the employees in North America, who prefer baseball. Baseball is not very popular outside of North American employees. Basketball is in the middle; about 1/3 of the North American respondents, about 2/5 of the European ones, and about 1/5 of the Middle Eastern and Asian ones consider it their favorite.

c. Yes, there is an association. The relative frequencies vary a lot based on the location. For example, soccer is the favorite of 75% of the respondents in Middle East, but only of 13% of those in North America. And baseball is not the preferred sport of many outside of North America.

d. $85/153 = 55.\overline{5}$. There is a 55.6% chance that that person is from Europe.

e. $10/101 \approx 0.0990$. There is a 9.9% chance that that person is from the Middle East.

f. $5/23 \approx 0.2174$. There is a 21.7% chance that that person prefers basketball.

4. a. 9
   b. 2
   c. 24

Test Answer Keys

Math Mammoth Grade 8
Tests Answer Key

Chapter 1 Test

Grading

My suggestion for grading the chapter 1 test is below. The total is 33 points. Divide the student's score by the total of 33 to get a decimal number, and change that decimal to percent to get the student's percentage score.

Question #	Max. points	Student score
1	8 points	
2	8 points	
3	3 points	
4	3 points	
5a	1 point	
5b	1 point	

Question #	Max. points	Student score
6	2 points	
7	2 points	
8	3 points	
9	2 points	
TOTAL	**33 points**	/ 33

1. a. $81, -81$ b. $5/8, 1,000$ c. $25, 1/144$ d. $1/64, 1/25$

2.

a. $24v^8$	b. $-10b^8 a^{15}$	c. $25x^2$	d. $\dfrac{1}{8x^3}$
e. $\dfrac{2}{s^2}$	f. $\dfrac{1}{s^6}$	g. $\dfrac{x^{10}}{32}$	h. $\dfrac{5x^5}{3}$

3. a, d, e

4. $0.4 \cdot 10^6$ $9 \cdot 10^5$ $5.2 \cdot 10^7$ $64 \cdot 10^6$ $2 \cdot 10^8$

5. a. $7 \cdot 10^9$ or $7,000,000,000$ times bigger
   b. $2.5 \cdot 10^{10}$ or $25,000,000,000$ times bigger

6. The area is 341 square feet (given to three significant digits, as the numbers in the problem).

7. $12,000 \cdot \$2.65 \cdot 5 \cdot 36 \approx \$5,720,000$. We give the answer to three significant digits because $2.65 and 12,000 are accurate to three significant digits. The quantities 5 days and 36 weeks, being perfectly exact, are accurate to any number of significant digits, so they do not affect how we round the final answer.

8. 1.5 km/s $\cdot$ 60 s/min $\cdot$ 60 min/hr $\cdot$ 24 hr/day $\cdot$ 365 day/year $= 47,304,000$ km/year $\approx 4.7 \cdot 10^7$ km/year
   Sounds travels $4.7 \cdot 10^7$ km in one year, in salt water.

9. We divide 10 grams by the mass of one copper atom to find the number of atoms in 10 grams of copper:

$$\frac{10 \text{ g}}{1.055 \cdot 10^{-22} \text{ g}} = 0.948 \cdot 10^{23} \text{ atoms} = 9.48 \cdot 10^{22} \text{ atoms}$$

Chapter 2 Test

Grading

My suggestion for grading the chapter 2 test is below. The total is 30 points. Divide the student's score by the total of 30 to get a decimal number, and change that decimal to percent to get the student's percentage score.

Question #	Max. points	Student score
1	2 points	
2	3 points	
3	2 points	
4	2 points	
5a	1 point	
5b	1 point	
5c	1 point	
5d	1 point	

Question #	Max. points	Student score
6	3 points	
7	3 points	
8	3 points	
9a	1 point	
9b	2 points	
10	2 points	
11	3 points	
TOTAL	**30 points**	/ 30

1.

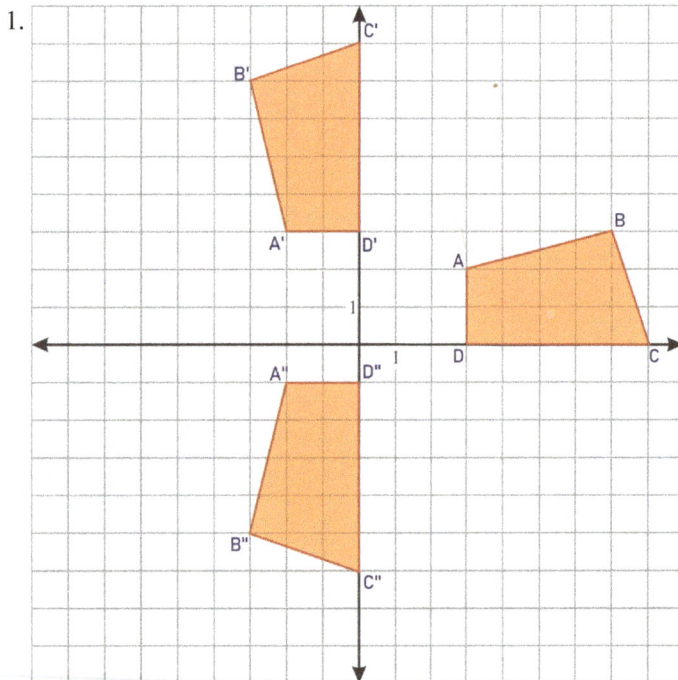

2. Student answers may vary; check the student's answer. At least three transformations are necessary: a dilation with a scale factor of 1/2 (can be from a variety of points), a rotation, and a translation. For example:

First, dilate the bigger arrow from point A and with scale factor 1/2. Then, rotate the resulting image 90 degrees clockwise around the origin. Lastly, translate the figure four units down.

Or, first rotate the large arrow 90 degrees clockwise around the origin. Then dilate it from point A' and with scale factor 1/2. Lastly translate the figure four units down.

3. (4, 0), (3, 2), and (1, 0)

4. The points are (2, −2), (2, −1), (0, −2) and (0, −1).

5. a. rotation

   b. and c.

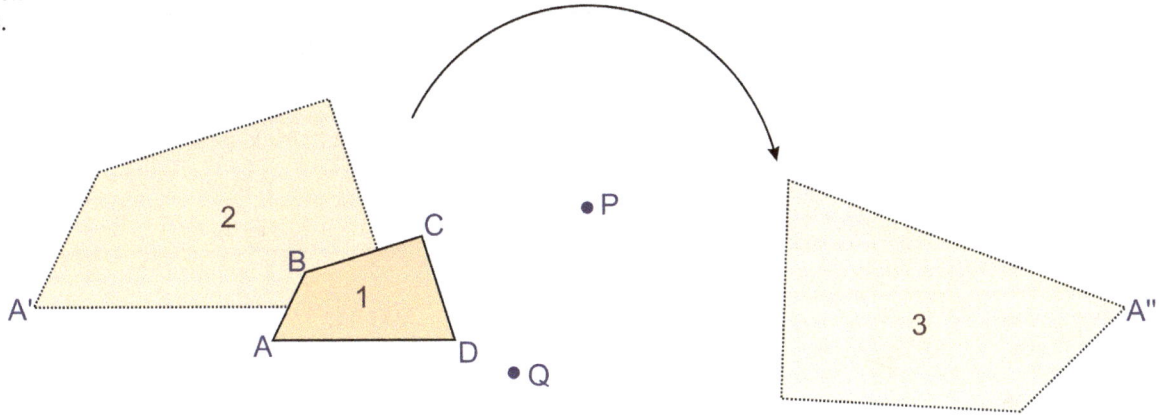

   d. Only (iv) and (v) are preserved.

6. Student reasoning may vary; please check the student's reasoning. For example:

   Angle BCD is supplementary to the 115° angle, so it is 65°. Angle CBD is 42° because it is an alternate interior angle with the given 42° angle. Angle x is the third angle in the triangle BCD, so it is $180° - 65° - 42° = \underline{73°}$.

7. Angle BAC is supplementary to the 150° so it measures 30°. Angle ABC is the third angle in triangle ABC, so it is $180° - 30° - 77° = 73°$. Lastly, x is a vertical angle with angle ABC, so, it also measures $\underline{73°}$.

8. Yes, they are. The third angle in the smaller triangle measures $180° - 25° - 60° = 95°$. And the third angle in the larger triangle measures $180° - 25° - 95° = 60°$. So, both triangles have angles that measure 25°, 60°, and 95°. This means they are similar triangles.

9. a. $V = (4/3)\pi \, (4.0 \text{ cm})^3 \approx 270 \text{ cm}^3$

   b. $V = \pi \, (4.0 \text{ cm})^2 \cdot 16 \text{ cm} \approx 800 \text{ cm}^3$.

10. The volume of the cone is $\underline{\text{one-third}}$ of the volume of the cylinder. You don't need to calculate the volumes to find that out, because the formulas for the volumes show this fact easily. The volume of a cylinder is $A_b h$ where A_b is the area of the bottom face and h is its height. The volume of a cone is $(1/3)A_b h$ or $1/3$ of the volume of the cylinder with the same bottom face and the same height.

11. Yes, you should believe her. The volume of one lime is $(4/3)\pi \, (3 \text{ cm})^3 \approx 113 \text{ cm}^3$ or 113 ml. The volume of the four limes is 452 ml, which is more than 1 1/2 cups. Now, the lime is not 100% juice; it also has rind and other material inside the fruit. Still, it is reasonable to expect to get half a cup of juice from these four limes. (On average, four limes do give you half a cup of lime juice.)

Chapter 3 Test

My suggestion for grading the chapter 3 test is below. The total is 24 points. Divide the student's score by the total of 24 to get a decimal number, and change that decimal to percent to get the student's percentage score.

Question #	Max. points	Student score
1	8 points	
2	2 points	
3	3 points	
4	3 points	
5a	1 point	
5b	1 point	

Question #	Max. points	Student score
6a	1 point	
6b	1 point	
7	4 points	
TOTAL	**24 points**	/ 24

1.

a.
$$6 - 3y - 2 + 8y = 4y + 1 - 9y - 5$$
$$5y + 4 = -5y - 4$$
$$10y + 4 = -4$$
$$10y = -8$$
$$y = -4/5$$

b.
$$12 - (x - 2) = 10 - 2x$$
$$12 - x + 2 = 10 - 2x$$
$$14 - x = 10 - 2x$$
$$14 + x = 10$$
$$x = -4$$

c.
$$4x + 20 - x = 2(x - 5) - 6x$$
$$4x + 20 - x = 2x - 10 - 6x$$
$$3x + 20 = -4x - 10$$
$$7x + 20 = -10$$
$$7x = -30$$
$$x = -30/7 = -4\,2/7$$

d.
$$-35 - 3(x - 4) = 10x + 40 - x + 13x$$
$$-35 - 3x + 12 = 10x + 40 - x + 13x$$
$$-3x - 23 = 22x + 40$$
$$-25x - 23 = 40$$
$$-25x = 63$$
$$x = -63/25 = -2\,13/25$$

2. $R = V/C$

3. Let L be Lucas's age now. The equation is $L + 6 = (2/3)(66 + 6)$, or $L + 6 = (2/3)\,72$, from which $L = 42$.

4. Let x be the increase in price. One way to write an equation from this situation is $0.8(288 + x) = 249$, from which $230.4 + 0.8x = 249$, and $x = 23.25$. He should increase the price by \$23.25.

5. a. It has one solution.
   b. Answers will vary; check the student's answer. For example, the equation could be changed to $-8y + 12 = 4(3 - 2y)$ or to $8y + 4 = 4(1 + 2y)$, just anything where everything cancels out, so that the equation reduces to $0 = 0$.

6. a. It has an infinite number of solutions.
   b. Answers will vary; check the student's answer. For example, the equation could be changed to $-2(9y - 3) + 8y = 8 - 10y$ or $-2(9y - 4) + 8y = 2 - 10y$. The terms with y need to cancel out, but the constant terms should not.

7.

a.
$$\frac{3x - 5}{4} - 1 = 2x$$
$$3x - 5 - 4 = 8x$$
$$3x - 9 = 8x$$
$$-5x - 9 = 0$$
$$-5x = 9$$
$$x = -9/5 = -1\,4/5$$

b.
$$\frac{y - 2}{3} = 3y + \frac{5 - y}{2}$$
$$2y - 4 = 18y + 15 - 3y$$
$$2y - 4 = 15y + 15$$
$$-13y - 4 = 15$$
$$-13y = 19$$
$$y = -19/13 = -1\,6/13$$

Chapter 4 Test

My suggestion for grading the chapter 4 test is below. The total is 32 points. Divide the student's score by the total of 32 to get a decimal number, and change that decimal to percent to get the student's percentage score.

Question #	Max. points	Student score
1a	2 points	
1b	2 points	
2a	2 points	
2b	1 point	
2c	3 point	
3	4 points	
4a	1 point	
4b	1 point	
4c	1 point	
4d	1 point	

Question #	Max. points	Student score
5a	4 points	
5b	1 points	
6a	2 points	
6b	2 points	
6c	1 point	
6d	2 points	
6e	1 point	
6f	1 point	
TOTAL	**32 points**	/ 32

1. a. Two things need changed: (1) The output for Joe should not be left as a question mark but as a number.
   (2) Currently, Jane maps to two outputs: 10 and 8. One or the other row for Jane needs eliminated.

   b. No, because 5 would be mapped to both Sally and John. In a function, there needs to be exactly one output for each input, so there cannot be two outputs for 5.

2. a. Function 3. The initial value is 16.
   b. Function 2 is a linear function.
   c. Function 1: $-8/3$. Function 2: -1.2. Function 3: $-5/3$.

3. Interval -3 to -1: increasing and linear
   Interval -1 to 1.1: constant and linear
   Interval 1.1 to 1.8: decreasing and nonlinear
   Interval 1.8 to 5: increasing and nonlinear

4. a. Yes.
   b. 150 dollars per week.
   c. 200 dollars
   d. We solve the equation $150w + 200 = 1{,}200$ for w, to get $w = 100/15 = 20/3 = 6.\overline{6}$.
      After 7 weeks Leo will have saved at least $1,200.

5. a.

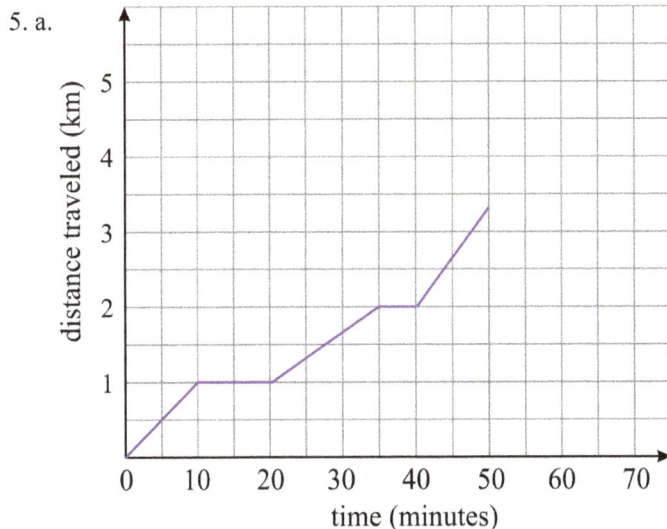

   b. 3 1/3 km

6. a. 4 gal/min. It means the water is flowing to the tub at the rate of 4 gallons per minute.
  b. 12 gallons. It means there were 12 gallons of water in the tub when Janet started filling it.
  c. $V = 12 + 4t$ where V is the volume of water, and t is time in minutes.
  d. The student graphs will vary since the scaling for the axes is not given in the problem. Check that the student's graph will show the correct line for the equation $V = 12 + 4t$. The line should go through the points (2, 20), (4, 28), and others that are on that line. For example:

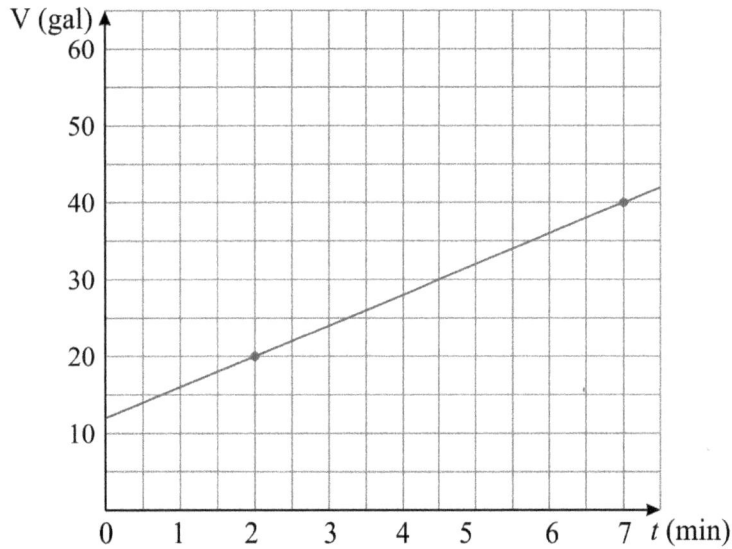

  e. 52 gallons
  f. In 5 minutes.

Grading
My suggestion for grading the chapter 5 test is below. The total is 35 points. Divide the student's score by the total of 35 to get a decimal number, and change that decimal to percent to get the student's percentage score.

Question #	Max. points	Student score
1a	1 point	
1b	1 point	
1c	2 points	
1d	2 points	
1e	1 point	
2a	2 points	
2b	2 points	
2c	2 points	
2d	2 points	

Question #	Max. points	Student score
3a	1 point	
3b	2 points	
3c	1 point	
4a	2 points	
4b	2 points	
5	3 points	
6a	4 points	
6b	2 points	
7	3 points	
TOTAL	**35 points**	/ 35

1. a. The slope is 400 m^2/(25 L) = 16 m^2/L.

   b. Paint 2. Paint 2 covers 16 m^2 per liter, whereas Paint 1 only covers 12 m^2 per liter.

   c. Paint 1: 30 m^2/(12 m^2/L) = 2.5 liters.
      Paint 2: 30 m^2/(16 m^2/L) = 1.875 liters.

   d. See the graph on the right.

   e. For one liter, Paint 2 covers 4 m^2 more than Paint 1.
      So, for 40 liters, Paint 2 covers 40 L · 4 m^2 = 160 m^2 more than Paint 1.

2. a. $y = (3/4)x - 8$
   b. $y = (-5/2)x - 2$
   c. $x = -11$
   d. $y = 6x - 22$

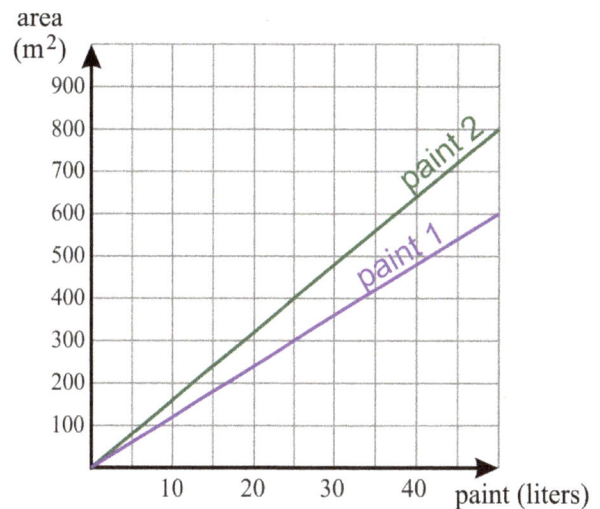

area (m^2) vs. paint (liters) graph showing paint 1 and paint 2 lines.

3. a. $W = 7.6 - 0.4t$, where W is the weight in kilograms, and t is the time in months.
   b. The answers will vary since the scaling on the vertical axis will vary. Check the student's graph. For example:

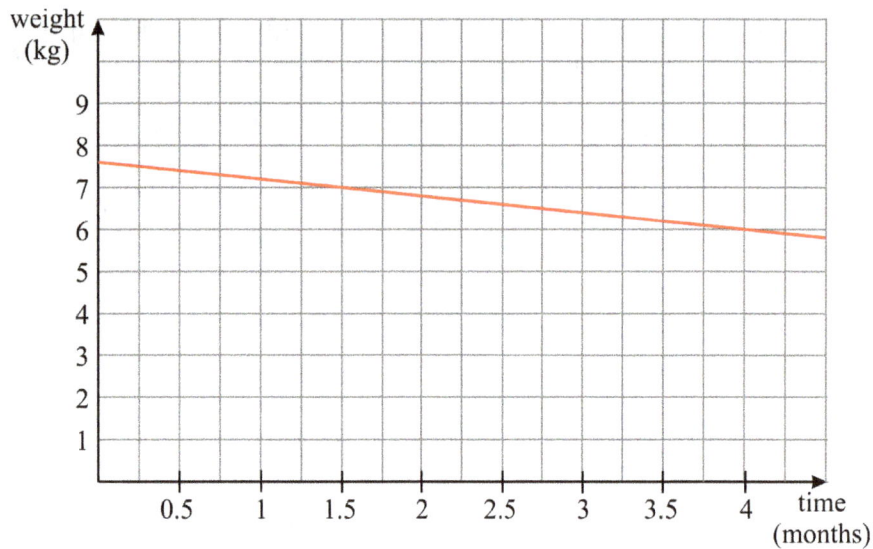

Here is another possibility, where the vertical axis does not start from zero:

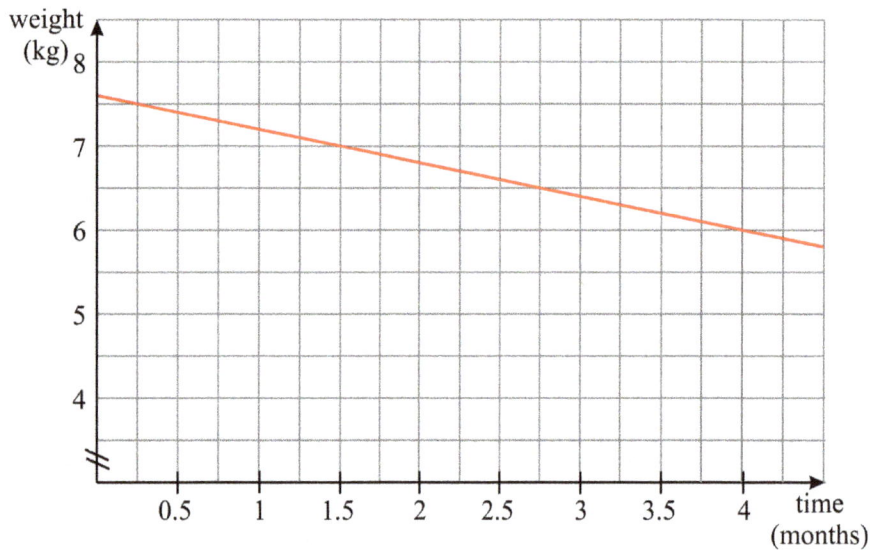

c. Solving the equation $5 = 7.6 - 0.4t$, we get $t = 2.6/0.4 = $ <u>6.5 months</u>.

4. a. This line goes through the points (0, 10) and (15, 30), from which we can calculate the slope as $20/15 = 4/3$. It crosses the y-axis at $y = 10$, so the equation is $y = (4/3)x + 10$.

   b. Here we cannot easily see the y-intercept. But we can find two points the line goes through: $(-5, 0)$ and $(2.5, -20)$. From those, we can calculate the slope as $-20/7.5 = -20/(15/2) = -40/15 = -8/3$.

   So, the equation is of the form $y = (-8/3)x + b$. Now we will substitute one of the points on the line, such as $(-5, 0)$, into that equation, to find the value of b. We get $0 = (-8/3)(-5) + b$, from which $b = -40/3$. So, the equation is $\underline{y = (-8/3)x - 40/3}$. If the student gave it in standard form, it is $8x + 3y = -40$.

5. The equation is $y = 3x - 2$. See the graph on the right.

6. a. The slope of Line L is $7/(-5) = -7/5$. Its equation is $y = (-7/5)x + b$. Substituting $(2, -1)$ into that equation, we get $-1 = (-7/5)(2) + b$, from which $b = 9/5$. So, the equation is $y = (-7/5)x + 9/5$.

   Since M is perpendicular to L, its slope is $5/7$. The equation of M is of the form $y = (5/7)x + b$. We substitute $(-2, -3)$ into that, and get: $-3 = (5/7)(-2) + b$. From that, $b = -3 + 10/7 = -11/7$. So, the equation of line M is $y = (5/7)x - 11/7$.

   b. The equation of L in standard form is $7x + 5y = 9$, and of line M, $5x - 7y = 11$.

7. The slope of this line is $(-28 - (-1))/(10 - 1) = -27/9 = -3$. For $(6, s)$ to be on the same line, the slope calculated using the points $(1, -1)$ and $(6, s)$ needs to be -3. Writing that as an equation, we get $(s - (-1))/(6 - 1) = -3$. See its solution below:

$$\frac{s - (-1)}{6 - 1} = -3$$

$$\frac{s + 1}{5} = -3 \quad \Big| \cdot 5$$

$$s + 1 = -15$$

$$s = -16$$

So, the solution is that $s = -16$.

You can also solve this in this manner. After finding out that the slope is -3, we can advance from the point $(1, -1)$, adding one to the x-value and subtracting 3 from the y-value each time, until we come to the x-value of 6: $(1, -1) \rightarrow (2, -4) \rightarrow (3, -7) \rightarrow (4, -10) \rightarrow (5, -13) \rightarrow (6, -16)$.

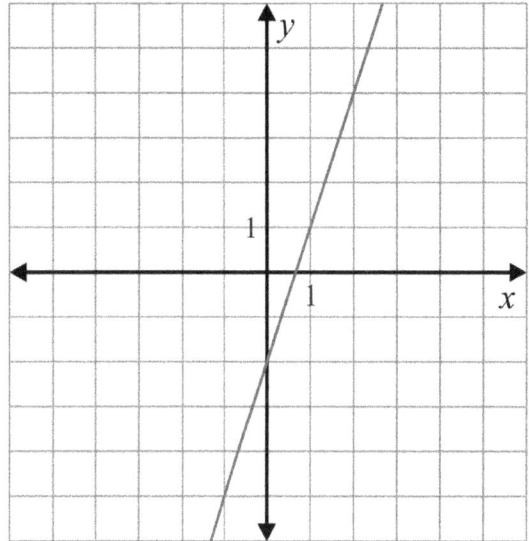

Chapter 6 Test

Grading
My suggestion for grading the chapter 6 test is below. The total is 31 points. Divide the student's score by the total of 31 to get a decimal number, and change that decimal to percent to get the student's percentage score.

Question #	Max. points	Student score
1	4 points	
2	3 points	
3	6 points	
4a	2 points	
4b	2 points	
4c	1 point	

Question #	Max. points	Student score
5	3 points	
6	3 points	
7	3 points	
8	4 points	
TOTAL	**31 points**	/ 31

1. (b), (c), and (d) are correct. The statement (a) is almost correct. It is true that 0.141414 is rational, but not because its decimal expansion repeats. Its decimal expansion does not actually repeat indefinitely anyway; its decimal expansion terminates. The reason it is rational is because we can write it as a ratio of two whole numbers: 141,414/1,000,000.

2. a. $6 < \sqrt{44} < 7$ b. $-3 < -\sqrt{5} < -2$ c. $2 < \sqrt[3]{21} < 3$

3. $-\sqrt{25}$ $-2\sqrt{2}$ $\sqrt[3]{27}$ $\sqrt{80} - 5$

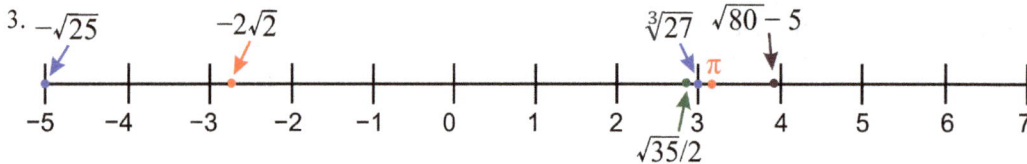

4.

a.	b.	c.
$x^2 = 37$ $x = \sqrt{37}$ or $x = -\sqrt{37}$	$5a^2 = 90$ $a^2 = 18$ $a = \sqrt{18}$ or $a = -\sqrt{18}$	$w^3 = 125$ $w = 5$

5.
$$100x = 23.232323\ldots$$
$$- \ x = \ \ \ 0.232323\ldots$$
$$99x = 23$$
$$x = \underline{23/99}$$

6. Let x be the hypotenuse. Using the Pythagorean Theorem, we get:

$$(\sqrt{21})^2 + (\sqrt{15})^2 = x^2$$
$$21 + 15 = x^2$$
$$x^2 = 36$$
$$x = 6$$
$$\text{or } x = -6$$

Since this is a triangle, we omit the negative solution. The hypotenuse is <u>6 units long</u>.

7. The horizontal distance between the two points comes from their x-coordinates and is $|10 - (-2)| = 12$ units. The vertical distance comes from the y-coordinates and is $|8 - 5| = 3$ units long.

Let x be the desired distance between the two points. Using the Pythagorean Theorem:

$$12^2 + 3^2 = x^2$$

$$144 + 9 = x^2$$

$$153 = x^2$$

$$x = \sqrt{153} \approx 12.37 \text{ units}$$

8. In the image, there is a right triangle with 0.3 km and 0.5 km legs. Its hypotenuse is marked with x. Applying the Pythagorean Theorem to that triangle, we get:

$$x^2 = 0.3^2 + 0.5^2$$

$$x^2 = 0.09 + 0.25$$

$$x^2 = 0.34$$

$$x = \sqrt{0.34} \approx 0.6 \text{ km}$$

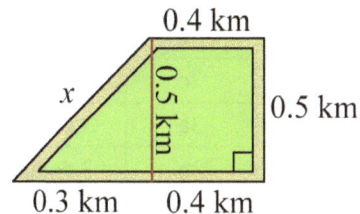

Now we can calculate the perimeter of the track: it is 0.4 km + 0.5 km + 0.7 km + 0.6 km = 2.2 km. When she runs around it three times, she runs a total distance of 6.6 km.

Chapter 7 Test

Grading

My suggestion for grading the chapter 7 test is below. The total is 27 points. Divide the student's score by the total of 27 to get a decimal number, and change that decimal to percent to get the student's percentage score.

Question #	Max. points	Student score
1	3 points	
2a	1 point	
2b	1 point	
2c	1 point	
3a	3 points	
3b	3 points	

Question #	Max. points	Student score
4	2 points	
5	3 points	
6	3 points	
7	3 points	
8	4 points	
TOTAL	**27 points**	/ 27

1. a. One solution.
   b. An infinite number of solutions.
   c. No solutions.

2. a. s = 3
   b. See the graph on the right.
   c. The lines are parallel and never intersect.

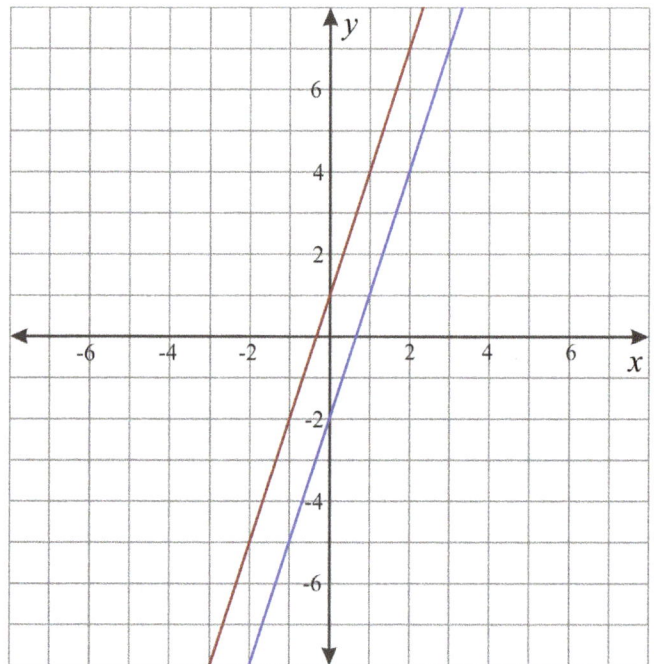

3.

a. $\begin{cases} x = 10 - y \\ 5y = 7(x - 2) \end{cases}$

↓

$\begin{cases} x + y = 10 \\ 5y = 7x - 14 \end{cases}$

↓

$\begin{cases} x + y = 10 \quad \big| \cdot 7 \\ -7x + 5y = -14 \end{cases}$

↓

$\begin{aligned} 7x + 7y &= 70 \\ + \quad -7x + 5y &= -14 \\ \hline 12y &= 56 \\ y &= 56/12 = 14/3 \end{aligned}$

Substituting $y = 14/3$ in the first equation, we get:

$x = 10 - 14/3$

$x = 30/3 - 14/3 = 16/3$

Solution: $(16/3, 14/3)$

b. $\begin{cases} 3x - y = 14 \quad \big| \cdot 3 \\ 6x + 3y = -12 \end{cases}$

↓

$\begin{aligned} \quad 9x - 3y &= 42 \\ + \quad 6x + 3y &= -12 \\ \hline 15x &= 30 \\ x &= 2 \end{aligned}$

Substituting $x = 2$ in the first equation, we get:

$\begin{aligned} 3(2) - y &= 14 \\ 6 - y &= 14 \\ y &= -8 \end{aligned}$

Solution: $(2, -8)$

4.

$\begin{cases} x + 3y = -2 \\ 2x - 5y = -15 \end{cases}$

Solving for x in the top equation, we get $x = -2 - 3y$.
Now substituting that in place of x in the second, we get:

$\begin{aligned} 2(-2 - 3y) - 5y &= -15 \\ -4 - 6y - 5y &= -15 \\ -4 - 11y &= -15 \\ -11y &= -11 \\ y &= 1 \end{aligned}$

Then, substituting 1 for y in the top equation, we get:

$\begin{aligned} x + 3(1) &= -2 \\ x + 3 &= -2 \\ x &= -5 \end{aligned}$

Solution: $(-5, 1)$

5.

$$\begin{cases} x + 6y = -1 \\ y = 2.2x + 0.8 \end{cases}$$

Substituting $2.2x + 0.8$ in place of y in the first equation, we get:

$$\begin{aligned} x + 6(2.2x + 0.8) &= -1 \\ x + 13.2x + 4.8 &= -1 \\ 14.2x + 4.8 &= -1 \\ 14.2x &= -5.8 \\ x &\approx -0.4084507 \end{aligned}$$

Then, substituting -0.4084507 for x in the top equation, we get:

$$\begin{aligned} -0.4084507 + 6y &= -1 \\ 6y &= -0.5915493 \\ y &\approx -0.09859155 \end{aligned}$$

Solution: $(-0.408, -0.099)$

6. Let A be Ann's age and D be Adam's age. Then:

$$\begin{cases} A + 3 = 2(D + 3) \\ A - 7 = 2.5(D - 7) \end{cases}$$

From the top equation, we get $A = 2(D + 3) - 3 = 2D + 3$. Substituting that for A in the bottom equation, we get:

$$\begin{aligned} 2D + 3 - 7 &= 2.5(D - 7) \\ 2D - 4 &= 2.5D - 17.5 \\ -4 &= 0.5D - 17.5 \\ 13.5 &= 0.5D \\ D &= 27 \end{aligned}$$

Then, $A = 2D + 3 = 57$.

Solution: Ann is 57 and Adam is 27.

7. Let x be the number of tables that seat 4 people, and y be the number of tables that seat 6. Then:

$$\begin{cases} x + y = 28 \\ 4x + 6y = 132 \end{cases}$$

We substitute $28 - x$ in place of y in the second equation, and get:

$$\begin{aligned} 4x + 6y &= 132 \\ 4x + 6(28 - x) &= 132 \\ 4x + 168 - 6x &= 132 \\ 168 - 2x &= 132 \\ -2x &= -36 \\ x &= 18 \end{aligned}$$

Then, $y = 28 - 18 = 10$. The restaurant has 18 tables that seat 4, and 10 tables that seat 6.

8. Let d be the distance that Train 1 travels until they meet and t be the amount of time since 3 PM when they meet. Then:

	distance	velocity	time
Train 1	d	90	t
Train 2	$70 - d$	105	t

Using the basic formula of $d = vt$ for both trains, we get the system of equations:

$$\begin{cases} d = 90t \\ 70 - d = 105t \end{cases}$$

We substitute $90t$ for d in the bottom equation:

$$\begin{aligned} 70 - 90t &= 105t \\ 70 &= 195t \\ t &= 70/195 = 14/39 \approx 0.358974 \end{aligned}$$

Then, $d = 90t = 90(14/39) \approx 32$ km. For Train 2, the distance is 70 km $-$ 32 km = 38 km. The time 0.358974 hours is about 22 minutes.

So, the trains meet at 3:22 PM, and the second train has traveled about 38 km by that point.

Chapter 8 Test

Grading
My suggestion for grading the chapter 8 test is below. The total is 25 points. Divide the student's score by the total of 25 to get a decimal number, and change that decimal to percent to get the student's percentage score.

Question #	Max. points	Student score
1a	2 points	
1b	1 point	
1c	1 point	
1d	2 points	
1e	1 point	
1f	2 points	
1g	2 points	

Question #	Max. points	Student score
2a	2 points	
2b	1 point	
2c	1 point	
2d	1 point	
3a	3 points	
3b	1 point	
3c	1 point	
4a	1 point	
4b	3 points	
TOTAL	**25 points**	/ 25

1. a. Yes. There is a negative linear association between the variables.
   b. (24, 31.3)
   c. Student graphs will vary; check the student's graph. For one example, see the graph below. We're not looking for anything super accurate; the student's line should go through somewhere in the middle, without leaving many more points below the line than above it, or vice versa.

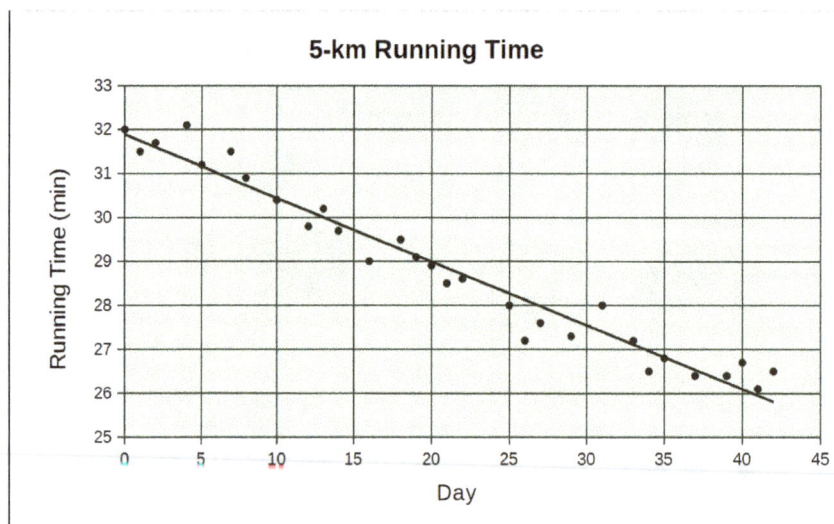

   d. Answers will vary; check the student's equation. For the line here, we can see it goes through (0, 31.9) approximately and (20, 29). From those two points, we can calculate the slope as $(29 - 31.9)/20 = -0.145$. The y-intercept is 31.9. So, the equation is $y = -0.145x + 31.9$, or, if using d for day and t for running time, $t = -0.145d + 31.9$. For comparison, the equation calculated by a spreadsheet program (for the line in this graph) is approximately $y = -0.144x + 31.887$.

   e. For day 24, the equation predicts a running time of $t = -0.145(24) + 31.9 = 28.42$ minutes.

   f. The slope of -0.145 signifies that for each additional day, the running time tends to decrease by 0.145 minutes.

   g. The y-intercept (or t-intercept) is 31.9 minutes. It signifies that at the start of the program, at zero days, the model predicts a running time of 31.9 minutes.

2. a. The slope of 1.34 signifies that each additional inch in diameter is associated with a $1.34-price increase.

b. No, we cannot. The equation predicts a negative price for a pizza with 1-inch diameter. (In fact, a pizza with 4.44-inch diameter would have a price of $0, and smaller ones would have a negative price.)

c. Price = 1.34(15) − 5.95 = $14.15.

d. We solve the equation 15 = 1.34x − 5.95 for x:

$$15 = 1.34x - 5.95$$
$$20.95 = 1.34x$$
$$x \approx 15.63$$

The equation predicts a size of 15.6 inches for a pizza that costs $15.

3. a. Yes. Since the total number of males and females is similar, if there wasn't an association, the numbers of students taking any of the languages, or no language, would also be similar. But instead, we see that twice as many females as males took Spanish, and more than twice as many females as males took French.
   b. $9/13 \approx 69\%$
   c. $117/145 \approx 81\%$

4. a. The table "Has two or more siblings" does not have any association between the variables.

b. This can be seen by checking the relative frequencies. They could be calculated as percentages, however, in this case it is not even necessary to do that. It is easy to notice that in the sibling situation, for each age group, those who said "Yes" are about half of those who said "No" (which means the relative frequencies for "Yes" would be close to 33% in each age group, and for "No", about 67%). This table is the only one with that kind of situation.

End-of-Year Test

Use your judgment in grading. You can give points or partial points for partial answers.

Question #	Max. points	Student score
Exponents and Scientific Notation		
1	8 points	
2	9 points	
3	4 points	
4	2 points	
5	2 points	
subtotal	/ 25	
Irrational Numbers		
6	5 points	
7	5 points	
8	3 points	
9	2 points	
subtotal	/ 15	
Geometry		
10	3 points	
11	2 points	
12	3 points	
13	2 points	
14a	3 points	
14b	3 points	
15	3 points	
16	3 points	
subtotal	/ 22	
Linear Equations		
17	4 points	
18	4 points	
19	6 points	
20	2 points	
21	2 points	
22	3 points	
subtotal	/21	
Functions		
23	2 points	
24a	1 point	
24b	2 points	
24c	2 points	

Question #	Max. points	Student score
Functions		
25a	1 point	
25b	1 point	
25c	1 point	
25d	1 point	
25e	1 point	
26a	2 points	
26b	1 point	
26c	1 point	
26d	1 point	
subtotal	/17	
Graphing Linear Equations		
27a	1 point	
27b	1 point	
27c	2 points	
28	3 points	
29	3 points	
30	3 points	
subtotal	/13	
The Pythagorean Theorem		
31	4 points	
32	3 points	
33	3 points	
subtotal	/10	
Systems of Linear Equations		
34	6 points	
35	3 points	
36	3 points	
37	3 points	
subtotal	/15	
Bivariate Data		
38	3 points	
39	3 points	
40	3 points	
41	5 points	
subtotal	/14	
TOTAL	/152	

Exponents and Scientific Notation

1. a. −16   b. 16   c. 1/49   d. 36
   e. 0.031   f. 110,000   g. −8/27   h. 64

2.

a. $-8s^3$	b. $144x^2$	c. y^{15}
d. $-6x^8$	e. $\dfrac{1}{y^6}$	f. $\dfrac{1}{4v^2}$
g. $\dfrac{49x^2}{9y^2}$	h. $\dfrac{-x^3}{125}$	i. $\dfrac{81b^4}{c^{20}}$

3. a. $1.93 \cdot 10^8$ b. $3.0805 \cdot 10^{12}$
   c. $4.6 \cdot 10^{-4}$ d. $9 \cdot 10^{-7}$

4. $\dfrac{6.0 \cdot 10^{24} \text{ kg}}{1.0 \cdot 10^{26} \text{ kg}} = \dfrac{6.0}{10^2} = 6/100 = 3/50$. The earth's mass is (about) 3/50 of Neptune's mass.

5. We need to divide to find out how many gold atoms "fit" into 99 grams of gold:

$\dfrac{9.9 \cdot 10^1 \text{ g}}{3.3 \cdot 10^{-22} \text{ g}} = 3 \cdot 10^{23}$. There are about $3 \cdot 10^{23}$ gold atoms in 99 grams of gold.

Irrational Numbers

6.

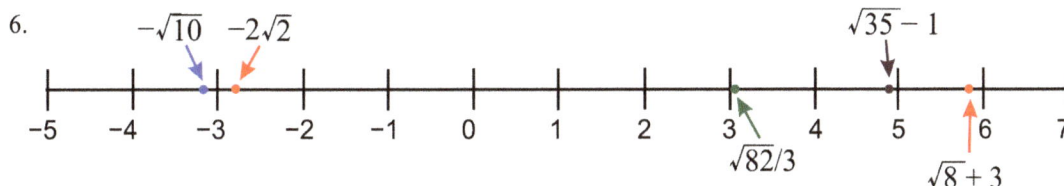

7.

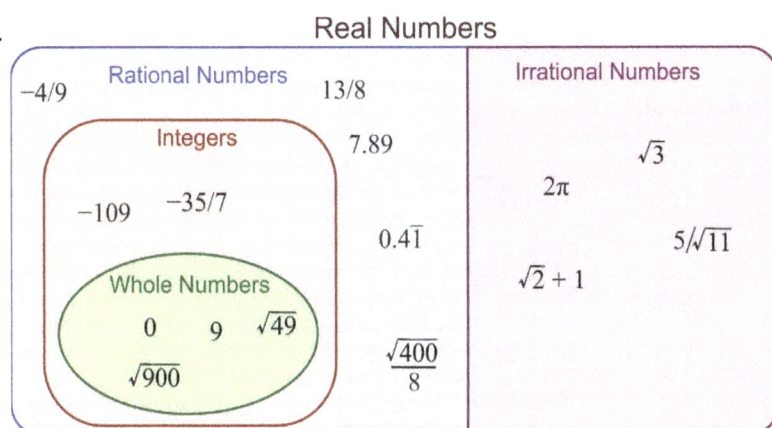

8. a. $x = \sqrt{54}$ or $x = -\sqrt{54}$ b. $n = 7$ or $n = -7$ c. $z = 4$

9. Let $x = 0.\overline{71}$. Then $100x = 71.\overline{71}$. Subtracting those, we get:

$$\begin{array}{rl} 100x = & 71.717171\ldots \\ -\ \ x = & 0.717171\ldots \\ \hline 99x = & 71 \\ \end{array}$$

$$x = \underline{71/99}$$

Geometry

10. Answers will vary. Check the student's answer. For example:

First, dilate triangle ABC from point A with scale factor 2/3.
Then translate it 8 units to the right.
Lastly, reflect it in the horizontal line $y = -1.5$.
(See the image on the right.)

But there are many possible answers. Here is another one.

First, reflect the triangle ABC in the horizontal line $y = -1.5$.
Then, translate it 8 units to the right.
Lastly, dilate it from point A" with scale factor 2/3.

Another one:
First, translate the triangle ABC 8 units to the right.
Then dilate it from point A' with scale factor 2/3.
Lastly, reflect it in the horizontal line $y = -1.5$.

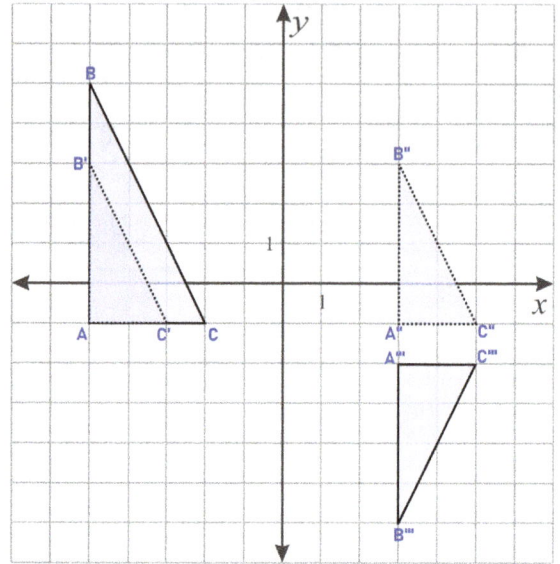

11. Answers will vary. Check the student's answer. For example:

First, rotate trapezoid ABCD 90° counterclockwise
around the origin.
Then, dilate it from point C' with scale factor 1/2.
(See the image on the right.)

Another way:

First, dilate trapezoid ABCD from point C with scale factor 1/2.
Then rotate it 90° counterclockwise around the origin.

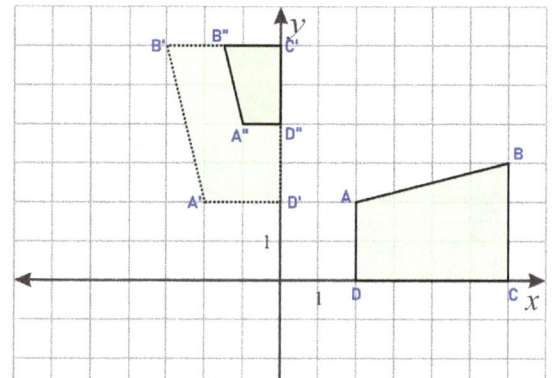

12. The bottom row of the table gives the coordinates after all
the transformations.

Position	Vertices		
Original	A(−2, 5)	B(−5, 4)	C(−3, 0)
After reflection in y-axis	A'(2, 5)	B'(5, 4)	C'(3, 0)
After translation (2 units down, 1 to the left)	A"(1, 3)	B"(4, 2)	C"(2, −2)
After rotation (90° clockwise around the origin)	A'''(3, −1)	B'''(2, −4)	C'''(−2, −2)

13. Since the sum of the angles in a triangle is 180°, in triangle CDE,
angle $a = 180° - 52° - 51° = 77°$.
Similarly, in triangle ABC, angle $b = 180° - 52° - 59° = 69°$.

Angles a, x, and b form a straight angle, so, $x = 180° - a - b$
$= 180° - 77° - 69° = \underline{34°}$.

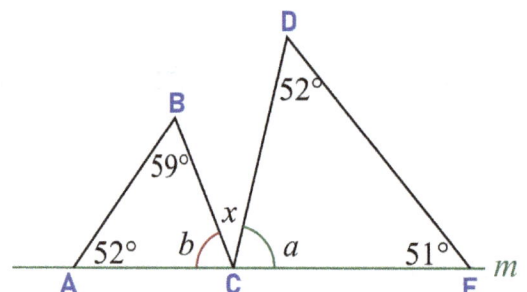

14. a. Angle BAC = $180 - (x + 26) = 154 - x$. Now, the angle sum of triangle ABC is $180°$, so, we can write an equation using that fact, and then solve for x:

$$\begin{aligned} \angle BAC + \angle ABC + \angle BCA &= 180 \\ 154 - x + x - 48 + x - 7 &= 180 \\ x + 99 &= 180 \\ x &= 81 \end{aligned}$$

b. Since m and n are parallel, angles y and BAC are corresponding angles, thus congruent. So, $y = 154° - x = 154° - 81° = \underline{73°}$.

15. $V = (4/3) \cdot \pi \, (3.0 \text{ in})^3 \cdot (2/3) \approx 75.398$ cubic inches. In cups, this is 5.2 cups or about 5 1/4 cups.

16. Let h be the height of the cup. Then, the volume is given by $V = \pi \, (3.1 \text{ cm})^2 \cdot h = 340$ ml. This is an equation that we can use to solve for h. Since 1 ml = 1 cubic centimeter, the equation becomes $\pi \, (3.1 \text{ cm})^2 \cdot h = 340 \text{ cm}^3$, from which $h = (340 \text{ cm}^3)/(\pi \cdot 3.1^2 \text{ cm}^2) \approx \underline{11.3 \text{ cm}}$.

Linear Equations

17.

a.			b.		
$10s + 8$	$=$	$7s - 2(s - 5)$		$20 - 3(x + 4)$	$= 14 - 5x$
$10s + 8$	$=$	$7s - 2s + 10$		$20 - 3x - 12$	$= 14 - 5x$
$10s + 8$	$=$	$5s + 10$		$8 - 3x$	$= 14 - 5x$
$5s$	$=$	2		$2x$	$= 6$
s	$=$	$2/5$		x	$= 3$

18.

a.	$\dfrac{2x - 3}{5} - x = 2$ $\quad \Big| \cdot 5$	b.	$\dfrac{y - 3}{4} = \dfrac{1 - y}{5}$ $\quad \Big| \cdot 20$ or cross-multiply
	$2x - 3 - 5x = 10$		$4(1 - y) = 5(y - 3)$
	$-3x = 13$		$4 - 4y = 5y - 15$
	$x = -13/3$		$4 - 9y = -15$
			$-9y = -19$
			$y = 19/9$

19.

a.	b.	c.
$6x - 1 = 6(x - 1)$	$-5x + 1 = 6(x - 1) - 5$	$6x - 12 = 6(x - 2)$
$6x - 1 = 6x - 6$	$-5x + 1 = 6x - 6 - 5$	$6x - 12 = 6x - 12$
$-1 = -6$	$-5x + 1 = 6x - 11$	$0 = 0$
No solutions.	$-11x = -12$	An infinite number of solutions.
	$x = 12/11$	Any value of x is a solution.
	One solution.	

20. Let d be the amount of discount. The non-discounted blocks cost 3000($1.35) = $4,050.
The discounted blocks cost 1500(1.35 − d). The total of these equals $5,775.

$$
\begin{aligned}
4050 + 1500(1.35 - d) &= 5775 \\
4050 + 2025 - 1500d &= 5775 \\
6075 - 1500d &= 5775 \\
-1500d &= -300 \\
d &= 3/15 = 1/5 = 0.2
\end{aligned}
$$

The discount was $0.20 per block. In other words, he paid $1.15 each for the 1500 blocks.

21. Let x be the first one of the four consecutive numbers. Then:

$$
\begin{aligned}
x + (x + 1) + (x + 2) + (x + 3) &= 2342 \\
4x + 6 &= 2342 \\
4x &= 2336 \\
x &= 584
\end{aligned}
$$

The numbers are <u>584, 585, 586, and 587</u>.

22. Let p be the original price of the item. Then:

$$
\begin{aligned}
1.06(0.73p) &= 34.82 \\
0.7738p &= 34.82 \\
p &\approx 44.9987
\end{aligned}
$$

The item cost $45.00 originally.

Functions

23. a. Because 3 is mapped to two different outputs: to 0 and to 3.
    b. The number 6 works. If you place either 3 or 9 there, then you will have the same input mapping to two distinct outputs, which would make it not a function.

24. a. Farm B's pricing system is a linear function. $C = 6.25w$.
    b. For Farm A, the rate of change is $(21.5 - 15)/(3 - 2) = 6.5$, or $6.50 per kg.
       For Farm B, the change of rate is 6.25, or $6.25 per kg.
    c. At Farm A, 4 kg will cost about $27, and at Farm B, $25. So, Farm B has the better deal.
       For 7 kg, Farm B charges you $40 and Farm B $43.75, so, Farm A has the better deal.

25. a. $10.
    b. That there is an initial fee of $10 just to get to go riding.
    c. $1 per minute.
    d. Horse riding will cost you $1 per minute, on top of the $10 initial fee.
    e. cost = $10 + t$, where t is the number of minutes you will go riding.

26. a. From $x = 0$ to $x = 4$:   linear and increasing
   From $x = 4$ to $x = 8$:   nonlinear and decreasing
   From $x = 8$ to $x = 11$: linear and decreasing
   From $x = 11$ to $x = 15$: nonlinear and increasing

b.

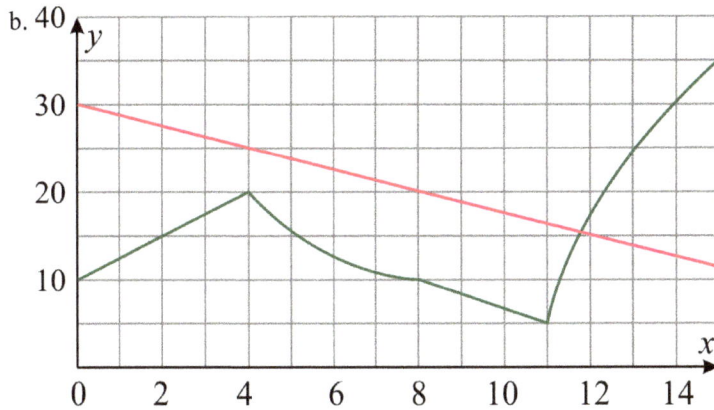

c. $y = -(5/4)x + 30$
d. For the function in green: the rate of change is $-5/3$. For the function in red, it is $-5/4$.

Graphing Linear Equations

27. a. $y = (-2/3)x + 4$
  b. $y = -3$
  c. $y = 5x - 25$

28. a. Fridge 1 (120 kWh versus 100 kWh). It consumes 20 kWh more than Fridge 2.
  b. Fridge 1: $E = 40t$. Fridge 2: $E = (100/3)t$. It is also acceptable to write it with a rounded decimal, as $E = 33.3t$.

c.

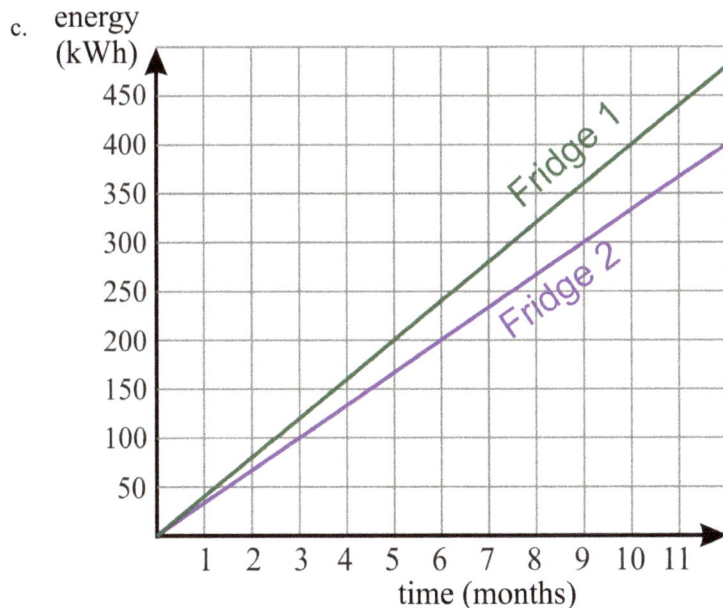

29.

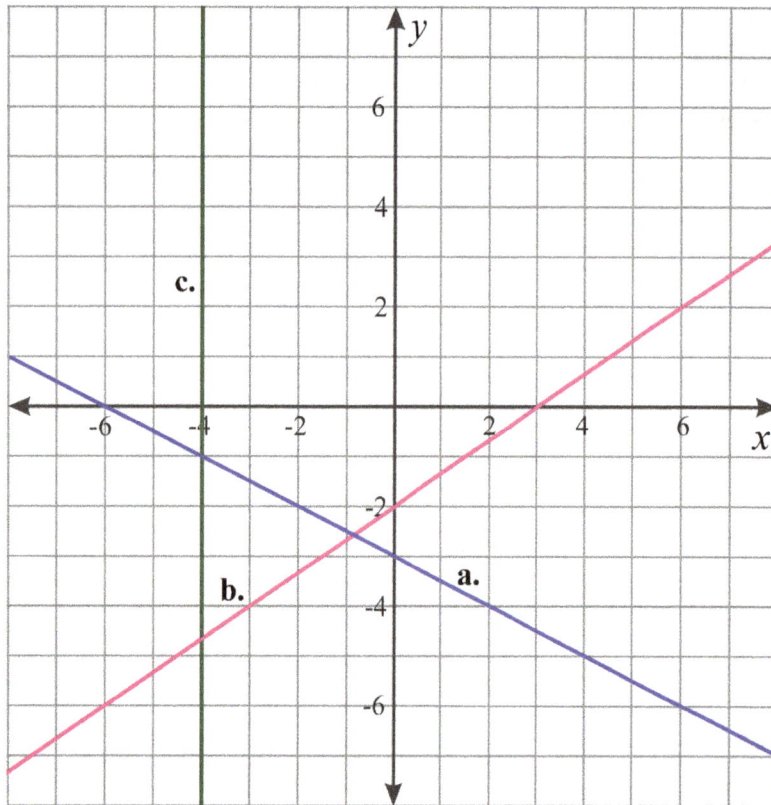

30. The slope of this line can be calculated using the two given points. It is $(-6 - 14)/(-7 - 3) = -20/(-10) = 2$. The equation of this line is therefore of the form $y = 2x + b$. Substituting $(3, 14)$ into it, we get $14 = 2(3) + b$, from which $b = 8$. So, the equation is $y = 2x + 8$. Since point $(a, 2)$ is on this line, let's substitute those values into the equation of the line:

$$2 = 2a + 8$$
$$-6 = 2a$$
$$a = -3$$

So, $\underline{a = -3}$. There are also other ways to arrive to the final answer, such as using the formula for the slope.

The Pythagorean Theorem

31. Using the Pythagorean Theorem, we get:

a. $r^2 + 17.5^2 = 26.6^2$	b. $x^2 + x^2 = (\sqrt{70})^2$
$r^2 = 26.6^2 - 17.5^2$	$2x^2 = 70$
$r^2 = 401.31$	$x^2 = 35$
$r = \sqrt{401.31} \approx 20.0$	$x = \sqrt{35}$
We ignore the negative root since this is a length of a side. The unknown side measures 20.0 units.	We ignore the negative root since this is a length of a side. The unknown side measures $\sqrt{35}$ units.

32. The rafter, the height of 1 ft 10 in, and half of the 6 ft 8 in span form a right triangle. In this triangle, using inches instead of feet and inches, the two legs measure 22 in and 40 in. Now, let r be the length of the rafter. According to the Pythagorean Theorem:

$$r^2 = 22^2 + 40^2$$
$$r^2 = 2{,}084$$
$$r = \sqrt{2{,}084} \approx 45.651$$

The decimal portion, 0.651 inches, can be converted into 16th parts of an inch this way. Let x be the number of 16th parts of an inch that equals 0.651. Then, $x/16 = 0.651$, from which $x = 0.651(16) = 10.416$. So, the rafter measures 3 ft 9 10/16 in.

33. a. To find the height, we will use the right triangle ABC. First, we need to find the length of the diagonal of the bottom square (d). From the Pythagorean Theorem:

$$d^2 = 36.0^2 + 36.0^2$$
$$d^2 = 2592$$
$$d = \sqrt{2592}$$

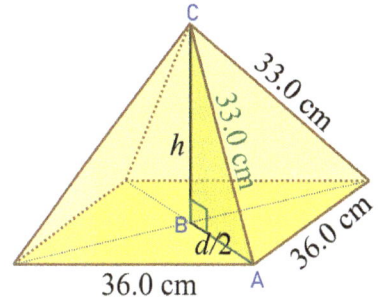

Next, we apply the Pythagorean Theorem to triangle ABC. Note that one of its legs is $d/2$ (half of the diagonal), the other leg is the height of the pyramid (h), and the hypotenuse is the 33-cm edge of the pyramid.

$$h^2 + (\sqrt{2592}/2)^2 = 33.0^2$$
$$h^2 + 2592/4 = 1{,}089$$
$$h^2 = 1{,}089 - 648$$
$$h = \sqrt{441} = 21$$

The height of the pyramid is 21.0 cm.

b. The volume is V = 36.0 cm · 36.0 cm · 21.0 cm / 3 = 9,072 cm^3.

34.

a. $\begin{cases} 2x - 3y = 8 \\ 3x + 4y = -5 \end{cases}$ $\begin{array}{l} \cdot\, 3 \\ \cdot\, (-2) \end{array}$

$\downarrow$

$\begin{array}{r} \begin{cases} 6x - 9y = 24 \\ -6x - 8y = 10 \end{cases} \\ \hline \end{array}$

$$-17y = 34$$
$$y = -2$$

Substituting $y = -2$ in the first equation, we get:

$$2x - 3(-2) = 8$$
$$2x + 6 = 8$$
$$2x = 2$$
$$x = 1$$

Solution: $(1, -2)$

b. $\begin{cases} -x = 4(y + 5) \\ 2x = -12y - 10 \end{cases}$

Solving for x from the top equation, we get that $x = -4(y + 5)$ which simplifies to $-4y - 20$.
Now, substituting that for x in the bottom equation, we get:

$$2(-4y - 20) = -12y - 10$$
$$-8y - 40 = -12y - 10$$
$$4y - 40 = -10$$
$$4y = 30$$
$$y = 30/4 = 15/2$$

Substituting $y = 15/2$ in the first equation, we get:

$$-x = 4(15/2 + 5)$$
$$-x = 4(25/2)$$
$$-x = 50$$
$$x = -50$$

Solution: $(-50, 15/2)$

35. a. No solutions.   b. One solution.   c. An infinite number of solutions.

36. Let x be the number of tables that seat 4, and y be the number of tables that seat 6.

We can write this system of equations: $\begin{cases} x + y = 106 \\ 4x + 6y = 500 \end{cases}$

Solving for y from the top equation, we get $y = 106 - x$. Substituting that in the bottom equation, we get:

$$4x + 6(106 - x) = 500$$
$$4x + 636 - 6x = 500$$
$$636 - 2x = 500$$
$$-2x = -136$$
$$x = 68$$

Then, $y = 106 - x = 106 - 68 = 38$. The restaurant has <u>68 tables that seat 4, and 38 tables that seat 6</u>.

37. Let G be Greta's age and S be Susan's age. Then: $\begin{cases} G + 10 = (3/4)(S + 10) \\ G + S = 127 \end{cases}$

From the bottom equation, we can solve that $G = 127 - S$. Substituting that in the top equation, we get:

$$\begin{array}{rl} 127 - S + 10 = (3/4)(S + 10) & \\ 137 - S = (3/4)S + 7.5 & \quad \cdot\, 4 \\ 548 - 4S = 3S + 30 & \quad +\, 4S \\ 548 = 7S + 30 & \quad -\, 30 \\ 518 = 7S & \quad \div\, 7 \\ S = 74 & \end{array}$$

Then, $G = 127 - 74 = 53$. <u>Greta is 53 and Susan is 74</u>.

38. a. Nonlinear and decreasing association.   b. No association.   c. Linear and increasing association.

39. There is no association between the variables. For each age group, there is about an equal number of people who exercise and who do not exercise. (In other words, in each age group the relative frequencies for "Exercises" and "Does not exercise" would be close to 50%.).

40. a. 6   b. 24   c. 3

41. a. Answers will vary. Check the student's answer. For example:

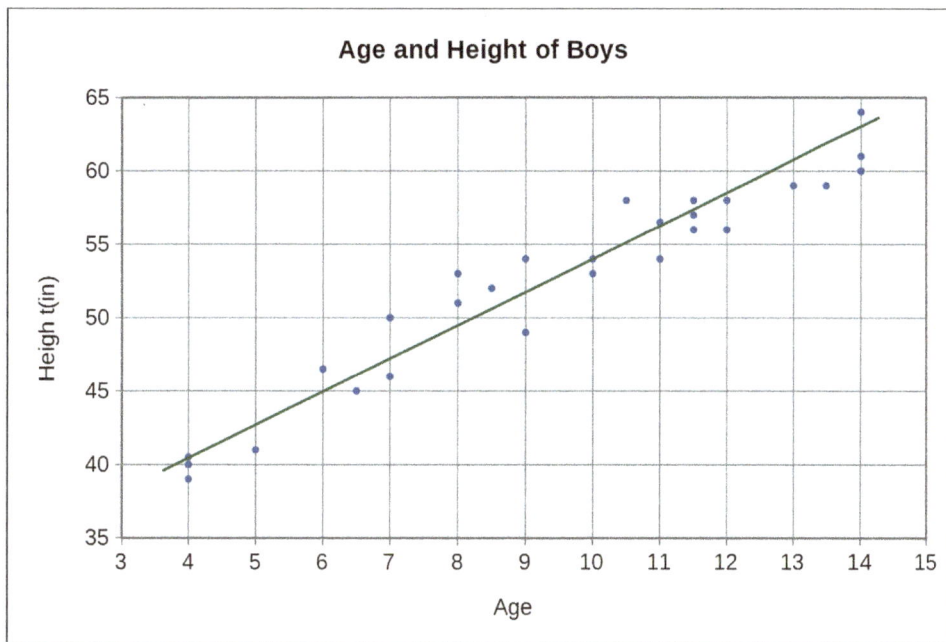

Age and Height of Boys

b. Answers will vary. Check the student's answer. The line above goes through (6, 45), and (10, 54). Therefore, its slope is 9/4 = 2.25, and its equation is of the form $y = 2.25x + b$. Substituting (10, 54) into this allows us to solve for b: $54 = 2.25(10) + b$, from which $b = 31.5$. So, the equation is $y = 2.25x + 31.5$.

c. It means that each 1-year increment of age is associated with a 2.1-inch increment in height. In other words, boys tend to grow 2.1 inches per year.

d. The y-intercept of 32.6 inches means that this equation predicts a newborn baby to be 32.6 inches tall. However, we know newborns are not that tall; they are typically between 18-22 inches tall. They grow very fast during the first year. Then from about age 2 onward, the growth follows fairly closely a linear pattern. This shows us that we cannot extrapolate backwards all the way to zero years using this data and this equation.

e. We solve the equation $50.5 = 2.1x + 32.6$ for x:

$$50.5 = 2.1x + 32.6$$
$$50.5 = 2.1x + 32.6$$
$$17.9 = 2.1x$$
$$x = 8.52$$

The equation predicts the age of about 8.5 years for a boy that is 50.5 inches tall.

Cumulative Reviews
Answer Key

Math Mammoth Grade 8
Cumulative Reviews Answer Key

Cumulative Review, Chapters 1-2

1. $(-3, 8)$, $(1, 6)$, and $(-2, 4)$. See the image on the right.

2. $(14, 9)$, $(14, 13)$, $(11, 9)$, and $(11, 13)$

   We start with $(12, -6)$, $(12, -10)$, $(9, -6)$, and $(9, -10)$, and do the transformations in the backwards order. First, after the reflection, the coordinates are $(12, 6)$, $(12, 10)$, $(9, 6)$, and $(9, 10)$.

   Then we move it 3 units up and 2 to the right, and get $(14, 9)$, $(14, 13)$, $(11, 9)$, and $(11, 13)$.

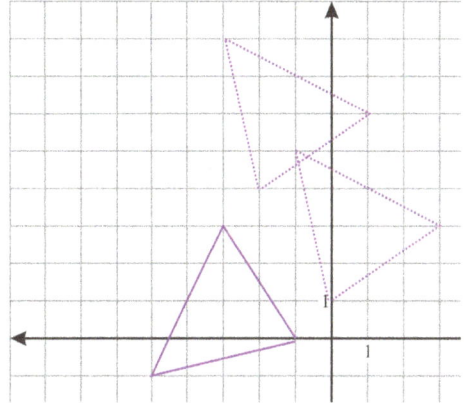

3.

a. Draw a dilation of triangle ABC using origin as center, and the scale factor of 1/2.	b. Draw a dilation of triangle ABC from point B, again using the scale factor of 1/2.

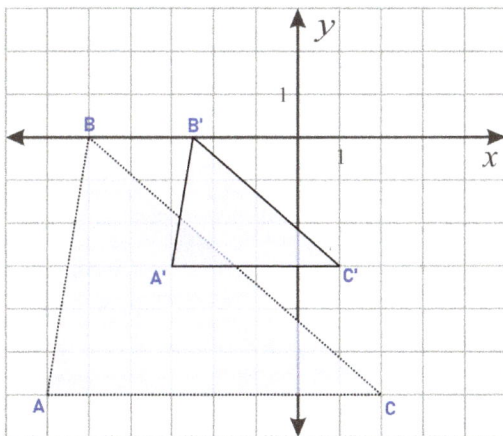

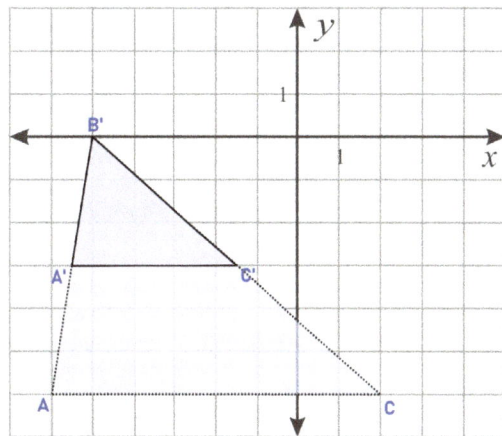

4. a. -16 b. $1/36$ c. $3/1000$ d. $1/5$
   e. $1/27{,}000$ f. 117 g. $25/36$ h. $1/121$

5.

a. $\dfrac{1}{5s^2}$	d. $\dfrac{1}{x^5y^{15}}$	g. $8a$
b. $\dfrac{t^6}{16}$	e. $\dfrac{1}{9x^2y^6}$	h. $\dfrac{16y^2}{-1000y^3} = -\dfrac{2}{125y}$
c. $\dfrac{1}{x^2y^6}$	f. $\dfrac{16}{w^8}$	i. $\dfrac{27x^3}{-3x} = -9x^2$

6. a. $x = 4$ b. $y = 3$ c. $x = 4$

7.

$0.07 \cdot 10^9$	$\boxed{7 \cdot 10^4}$	$\boxed{7 \cdot 10^9}$
70,000,000	$70 \cdot 10^8$	$0.7 \cdot 10^5$
$0.7 \cdot 10^8$	7,000,000,000	70,000

8. a. $9.3 \cdot 10^6$ b. $2.3 \cdot 10^7$ c. $1.45 \cdot 10^9$

9. a. It means $2.1 \cdot 10^{25}$.

   b. Answers will vary; check the student's answer. For example: $7 \cdot 10^{12}$ and $3 \cdot 10^{12}$ or $7 \cdot 10^6$ and $3 \cdot 10^{18}$.

10. The population of Lydia's hometown is $\dfrac{1.5 \cdot 10^9}{2 \cdot 10^4} = 0.75 \cdot 10^5 = 75{,}000$.

11. a. 20 m^2 b. 120 kg
    c. \$24.20/hr   d. 36.4 mi/gal

Cumulative Review, Chapters 1-3

1. Answers will vary; check the student's answer. For example:

First, reflect the pentagon in the *x*-axis. Then, translate it four units to the right. Lastly, dilate it from point E" and with a scale factor of 2. (See the image on the right.)

Or, first, rotate the figure 180 degrees around the origin. Then dilate it from the point E' with scale factor 2.

There are many other possible ways to do it. Each sequence should include a reflection in the *x*-axis or a rotation 180 degrees around the origin, a dilation with a scale factor of 2, and a possible translation. These can be in different orders.

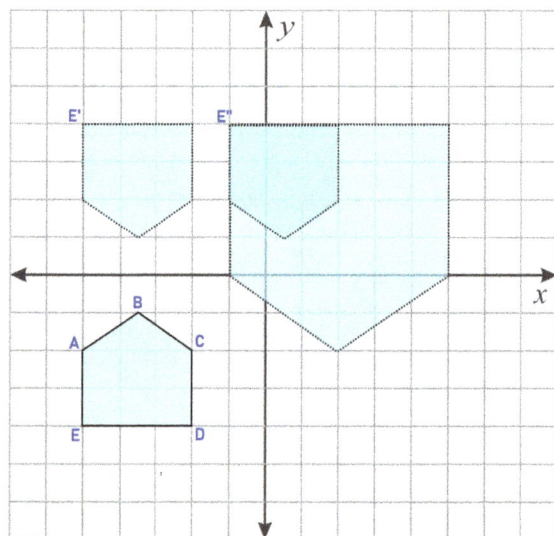

2.

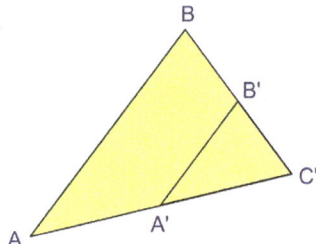

3. Please refer to the image on the right. The student will also need to name angles, points, and/or lines in the figure to be able to reference them in the explanation.

Student explanations will vary; please check the student's explanation.

For example:

Points A, B, and C mark the triangle ABC. In triangle ABC, $\angle \alpha = 180° - 72° - 65° = 43°$. Since lines *n* and *m* are parallel and line *s* is a transversal to them, α and α' are corresponding angles, and thus congruent. The angle in question is supplementary to α, so it is $180° - 43° = \underline{137°}$.

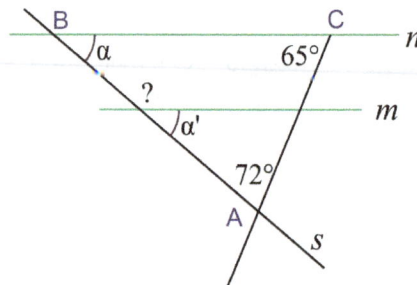

4. a. The first one is a rotation. The second is a reflection. (The reflection is not in the *x* or *y*-axis. It is a reflection in a line — a line that goes through the points where $\overline{B'C'}$ and $\overline{B"C"}$ intersect).

   b. All the attributes stay the same except position.

5.

a. $2 \cdot 10^8 + 8 \cdot 10^7$	b. $1.3 \cdot 10^6 + 2 \cdot 10^7$	c. $3 \cdot 10^5 - 9 \cdot 10^4$	d. $2.6 \cdot 10^8 - 5 \cdot 10^7$
$= 20 \cdot 10^7 + 8 \cdot 10^7$	$= 1.3 \cdot 10^6 + 20 \cdot 10^6$	$= 30 \cdot 10^4 - 9 \cdot 10^4$	$= 26 \cdot 10^7 - 5 \cdot 10^7$
$= 28 \cdot 10^7$	$= 21.3 \cdot 10^6$	$= 21 \cdot 10^4$	$= 21 \cdot 10^7$
$= 2.8 \cdot 10^8$	$= 2.13 \cdot 10^7$	$= 2.1 \cdot 10^5$	$= 2.1 \cdot 10^8$

6. (b) is in error. It should be $(2 \cdot 10)^{-3} = \dfrac{1}{8 \cdot 10^3} = \dfrac{1}{8000}$.

7. Let h be the height of the cone. The volume of the cone is $(1/3) \cdot \pi \cdot (6 \text{ ft})^2 \cdot h$, and on the other hand, the volume is 11.0 cubic yards, or $11 \cdot 27 = 297$ cubic feet. So, we can write an equation:

$$\frac{\pi \cdot (6 \text{ ft})^2 h}{3} = 297 \text{ ft}^3$$

$$\pi \cdot 36 \text{ ft}^2 \cdot h = 3 \cdot 297 \text{ ft}^3$$

$$h = \frac{3 \cdot 297 \text{ ft}^3}{\pi \cdot 36 \text{ ft}^2}$$

$$h = 7.88 \text{ ft}$$

8. a. 14.8 km (to the accuracy of the least accurate measurement, which was to the tenth of a kilometer)

   b. 448 mi (to the accuracy of the least accurate measurement, which was to the mile)

9. 170 meters (to two significant digits, just as 4.2 cm has two significant digits)

10. $2.2 \cdot 10^9$ times

11. The time it takes is $t = \dfrac{d}{v} = \dfrac{4.0208 \cdot 10^{13} \text{ km}}{3.00 \cdot 10^5 \text{ km/s}} \approx 1.340267 \cdot 10^8$ seconds. To convert this to years, we need to

divide it by $60 \cdot 60 \cdot 24 \cdot 365$. Doing that, we get $\dfrac{1.340267 \cdot 10^8 \text{ seconds}}{60 \cdot 60 \cdot 24 \cdot 365} = \underline{4.25 \text{ years}}$.

Note that this is given to <u>three</u> significant digits, since the speed of light was given to three significant digits (and the distance was given to five.)

1.

Original figure	Dilation	Reflection
D(−5, −4)	D'(−2.5 , −2)	D"(−2.5, 2)
E(−6, −2)	E'(−3, −1)	E"(−3 , 1)
F(−1, −2)	F'(−0.5 , −1)	F"(−0.5, 1)
G(−2, −4)	G'(−1, −2)	G"(−1, 2)

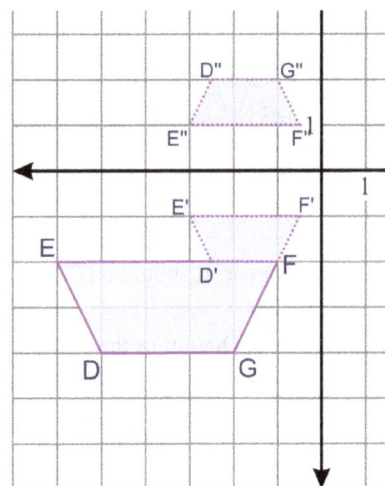

2. a. See the image on the right. The vertices were
originally at (−3, 4), (−3, 6), (−1, 6), and (1, 2).

   Working backwards from the given, final coordinates, we get:
   1. After a reflection in the x-axis: (6, −4), (6, −6), (4, −6), and (2, −2).
   2. After a translation 3 units to the left: (3, −4), (3, −6), (1, −6), and (−1, −2).
   3. After a rotation 180° around the origin: (−3, 4), (−3, 6), (−1, 6), and (1, 2).

   b. A reflection in the vertical line $x = 1.5$.

3. Let p be the price before the increases. Then we can write the equation
   $1.04 \cdot 1.075 \cdot 1.05p = \$136.45$, from which
   $p = \$136.45/(1.04 \cdot 1.075 \cdot 1.05) = \$116.24$.
   Its price had been $\underline{\$116.24}$ before the increases.

4. We cannot know whether they are similar or not.

   If $\overline{AB}$ and $\overline{QR}$ were parallel, then triangles ABP and PQR would be
   similar, and otherwise not. But it is not stated whether $\overline{AB}$ and $\overline{QR}$ are
   parallel, so we cannot know whether the two triangles are similar or not.

   (If $\overline{AB}$ and $\overline{QR}$ were parallel, then angle BAP and PQR would be alternate interior angles and thus congruent.
   Similarly, angles QRP and PBA would be alternate interior angles and congruent. Angles BPA and QPR are
   congruent, being vertical angles. The triangles ABP and PRQ would have three congruent angles, and thus would
   be similar.)

5. Angle ABC is supplementary to the 137° angle, so it measures 43°. In triangle ABC, $y = 180° − 82° − 43° = 55°$.
   Angle x is an alternate interior angle with angle y, so it is congruent with y and also measures 55°.

6. The volume of the bucket, in cubic inches, is $V − \pi (5.75 \text{ in})^2 \cdot 11 \text{ in} \approx 1142.558$ cubic inches.
   In gallons, this is $1142.558 \text{ in}^3 \cdot 1 \text{ gal}/(231 \text{ in}^3) \approx 4.9$ gallons. Yes, four gallons of water fit in the
   bucket and you should not believe your friend.

7. a. 230 cm = _91_ in   b. 54 L = _14_ gal

   c. 24.5 ft = _7.47_ m   d. 437 in = _11.1_ m

8. $\dfrac{1 \cdot 10^2 \text{ g}}{2.99 \cdot 10^{-23} \text{ g}} \approx 0.334448 \cdot 10^{25}$ atoms $\approx 3.34 \cdot 10^{24}$ atoms.

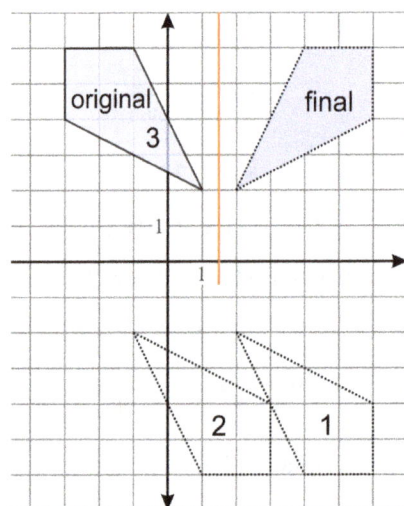

9. a. After multiplying both sides by 2, the term x gets forgotten.

b. When the term $-(x - 1)/5$ is multiplied by 5, since there is a minus sign in front of it, it should become $-x + 1$.

a.
$$5 + x + \frac{x-3}{2} = 6 \qquad \Big| \cdot 2$$
$$10 + 2x + x - 3 = 12$$
$$7 + 3x = 12 \qquad \Big| -7$$
$$3x = 5 \qquad \Big| \div 3$$
$$x = 5/3 = 1 \; 2/3$$

b.
$$x - 2 = 2x - \frac{x-1}{5} \qquad \Big| \cdot 5$$
$$5x - 10 = 10x - x + 1$$
$$5x - 10 = 9x + 1 \qquad \Big| -9x$$
$$-4x - 10 = 1 \qquad \Big| +10$$
$$-4x = 11 \qquad \Big| \div (-4)$$
$$x = -11/4 = -2 \; 3/4$$

10. a. $a_2 = 4m - a_1 - a_3 - a_4$

b. At least 92 points.

11. Let x be the first such even whole number. Then the other two are $x + 2$ and $x + 4$.
We get the equation $x + x + 2 + x + 4 = 13{,}788$, from which $3x = 13{,}782$ and $x = 4{,}594$.
The three numbers are 4,594, 4,596, and 4,598.

Cumulative Review, Chapters 1-5

1. It will have no solutions if the coefficient of the x-term is the same on both sides of the equation, and if the constant terms will not cancel out. We can simplify both sides of the equation (separately) to get:

$4x + 28 = 10 - 2ax + 10$
$4x + 28 = -2ax + 20$

The constant terms on the two sides are different (28 and 20). So, when $-2a = 4$, or $\underline{a = -2}$, the equation will have no solutions.

2. a.
$$A = \frac{pq}{2} \qquad \Big| \cdot 2$$
$$2A = pq \qquad \Big| \div q$$
$$\frac{2A}{q} = p$$

Solution: $p = (2A)/q$

b. $p = (2 \cdot 0.6 \text{ m}^2)/(0.9 \text{ m}) = (1.2 \text{ m}^2)/(0.9 \text{ m}) = 12/9 \text{ m} = 4/3 \text{ m}$, or about $\underline{133 \text{ cm}}$.

3. a. $V = \pi(4.5 \text{ in})^2 \cdot 4 \text{ in} \div 12 \approx 21 \text{ in}^3$

b. We first need to calculate the full volume of the 9-inch diameter circular cake: $\pi(4.5 \text{ in})^2 \cdot 4 \text{ in} \approx 254 \text{ in}^3$.
Let h be the height of the square cake. We can write the equation 8 in $\cdot$ 8 in $\cdot$ $h = 254 \text{ in}^3$, from which $h = 254/64$ inches, or about 4 inches. So both cakes need to be about the same height to have the same volume.

4.

a.
$$2x = \frac{x-5}{6} - 2 \qquad \Big| \cdot 6$$
$$12x = x - 5 - 12$$
$$11x = -17 \qquad \Big| \div 11$$
$$x = -17/11$$

b.
$$\frac{x-1}{4} + \frac{2x+7}{3} = 0 \qquad \Big| \cdot 12$$
$$3(x - 1) + 4(2x + 7) = 0$$
$$3x - 3 + 8x + 28 = 0$$
$$11x = -25 \qquad \Big| \div 11$$
$$x = -25/11$$

5. a. $d = 600t$
   b. 600 mph
   c. Yes, because it is of the form $y = mx$ (or, dependent variable = constant times the independent variable.) Here, m is 600.
   d. Student graphs will vary because the scaling on the d-axis may vary. For example:

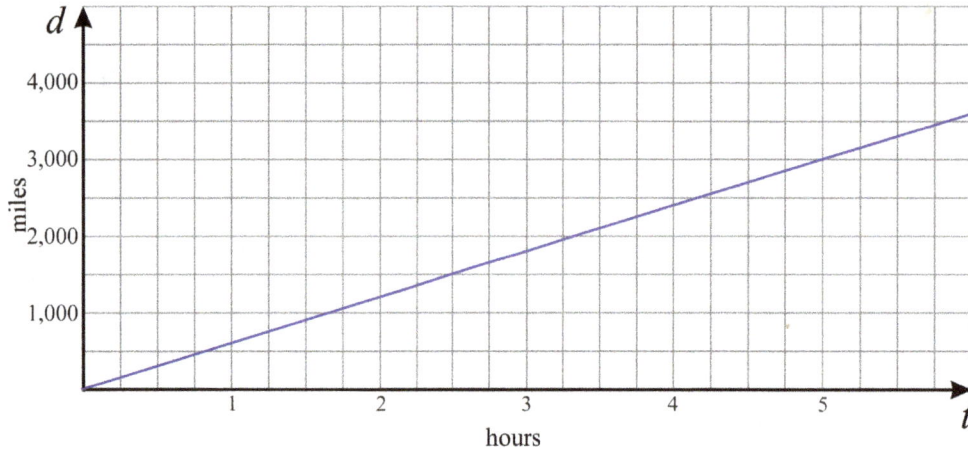

   e. The time 1 hr 40 minutes needs to be in hours before we can use the equation. 1 hr 40 min = 1 2/3 hr = 5/3 hr. Then, $d = 600(5/3) = \underline{1,000 \text{ miles}}$.

6. a. No. While the time increases steadily by half an hour each time, the distance does not increase by the same amount each time. For example, from 0.5 to 1 hour, the distance increases by 300 miles, but from 1 to 1.5 hours, it increases by 350 miles.

   b. The function in question #5 has a constant rate of change of 600 mph at all times.
   The function in question #6 has a rate of change of (2,400 mi − 1,750 mi)/1 hr = 650 mph.
   <u>The function in #6 has a larger rate of change from 3 to 4 hours.</u>

7.

a. $(2x^3)^{-1} = \dfrac{1}{2x^3}$	c. $(ab^3)^{-2} = \dfrac{1}{a^2b^6}$	e. $(-2x)^5 \cdot (5x)^{-1} = \dfrac{-32x^4}{5}$
b. $(8a^{-5})^2 = \dfrac{64}{a^{10}}$	d. $(-3s^{-2}t)^3 = \dfrac{-27t^3}{s^6}$	f. $z^4 \cdot (-3z^3)^{-2} = \dfrac{1}{9z^2}$

8. a. From $x = -25$ to $x = -20$
   b. From $x = -10$ to $x = -2.5$
   c. From $x = -2.5$ to $x = 7.5$

9. Answers will vary. Check the student's graph. For example:

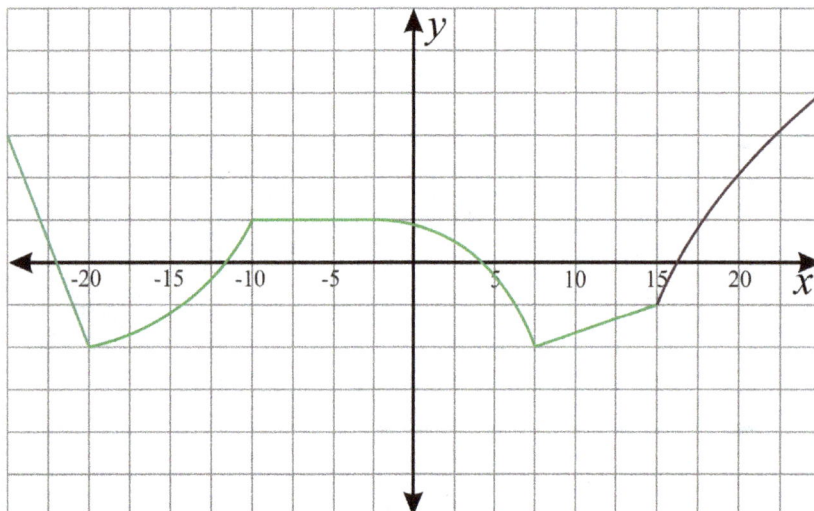

1.

<table>
<tr><td>a.</td><td>

$2 - \dfrac{x+1}{5} = x$ | $\cdot\, 5$

$10 - (x+1) = 5x$

$10 - x - 1 = 5x$

$9 - x = 5x$

$9 = 6x$ | $\div 6$

$x = 3/2$

</td><td>b.</td><td>

$4y + \dfrac{1-2y}{10} = 3y$ | $\cdot\, 10$

$40y + (1-2y) = 30y$

$38y + 1 = 30y$

$8y = -1$ | $\div 8$

$y = -1/8$

</td></tr>
<tr><td>c.</td><td>

$\dfrac{x-10}{3} = \dfrac{3x+4}{5}$ | $\cdot\, 15$ (or cross-multiply)

$5(x-10) = 3(3x+4)$

$5x - 50 = 9x + 12$

$-4x = 62$

$x = -31/2$

</td><td>d.</td><td>

$-x + \dfrac{7x-3}{4} = \dfrac{x}{2} - 2$ | $\cdot\, 8$

$-8x + 2(7x-3) = 4x - 16$

$-8x + 14x - 6 = 4x - 16$

$6x - 6 = 4x - 16$

$2x = -10$

$x = -5$

</td></tr>
</table>

2. a. $7.2 \cdot 10^{-5}$ b. $2.33 \cdot 10^{8}$ c. $4 \cdot 10^{-4}$ d. $3.09 \cdot 10^{8}$

3. a. $>$ b. $>$ c. $<$
   d. $=$ e. $<$ f. $<$

4. a. No solutions.
   b. One solution.
   c. One solution.
   d. An infinite number of solutions.

5. Of these, (2) and (3) are functions. In (1) the input 131 has two outputs. In (4), each input (each date) has millions of outputs, if you consider the population of the entire world.

6. a. Function 1 (90).
   b. Functions 2 and 3 are linear.
   c. Function 1. Its rate of change in that interval is $-50/4 = -25/2$.
   d. Function 1 is constant. Function 2 is increasing. Function 3 is decreasing.

7. No, it is not. It is not of the form $y = mx + b$ (or, using A and t, $A = ma + b$). Another way to see this is to calculate some values, and check whether the rate of change stays the same.

We can easily see that as the t-values increase by ones, the A-values do not increase by the same amount each time. So, the function is not linear. In fact, this is an example of an *exponential function*.

t	A
0	5,000
1	5,300
2	5,618
3	5,955.08

8. a. has only one answer (see below), but the answer for (b) will vary. Check the student's answer. The output values should NOT always decrease or increase by the same amount.

a.

Input (x)	−5	−4	−3	−2	−1	0	1	2	3	4	5
Output (y)	−12	−9	−6	−3	0	3	6	9	12	15	18

b. For example:

Input (x)	−5	−4	−3	−2	−1	0	1	2	3	4	5
Output (y)	63	56	49	42	35	−2	5	6	−9	23	0

9. $V = 30 \text{ cm} \cdot 30 \text{ cm} \cdot 24 \text{ cm} / 3 = 7{,}200 \text{ cm}^3$

10. a. Company 1: $C = 42h + 1800$. Company 2: $C = 36h + 2400$.
   b. Company 1 will charge \$2,430, whereas Company 2 will charge \$2,940. <u>Company 1 is the better deal</u>.
   c. We will set the expressions for the costs equal, and solve that equation for h:

$$
\begin{aligned}
42h + 1800 &= 36h + 2400 \\
6h &= 600 \\
h &= 100
\end{aligned}
$$

For 100 hours of work, both companies will cost the same (\$6,000).

Cumulative Review, Chapters 1-7

1. $-2 \cdot 10^5 \ < \ -2 \cdot 10^{-5} \ < \ 2 \cdot 10^{-5} \ < 0.0005 < 2 \cdot 10^5$

2. Yes, they are. In triangle ABC, the third angle is $180° - 86° - 36° = 58°$. This means its angles measure 86°, 58°, and 36°. Triangle DEF also has angles of 58° and 86°. So, based on the AA criterion for similar triangles, the triangles are similar.

   Alternatively, you could calculate the third angles for both triangles and note that both triangles have angles of 86°, 58°, and 36°, which means they are similar triangles.

3. a. Let C denote the cost and d the number of days. Then, $C = 55d + 10$.
   b. The rate of change is \$55/day. It means that each additional day carries an additional cost of \$55.
   c. The initial value is \$10. It means that before you can even rent the snowboard, you have to pay a fixed \$10 fee.

4. distance from shore

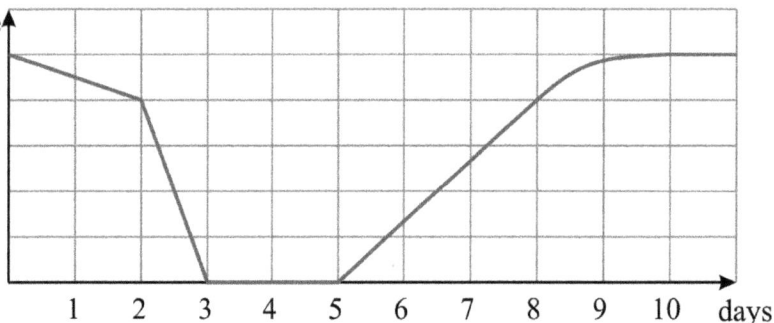

5. Let l denote each of the legs of the isosceles triangle and b the base. Since the perimeter is 65, we can write the equation $2l + b = 65$. If the base would be 7 units longer, the triangle would be equilateral, which means $b + 7 = l$. So, we have this system of equations:

$$
\begin{cases}
2l + b = 65 \\
b + 7 = l
\end{cases}
$$

To solve it, we will substitute $b + 7$ for l in the top equation:

$$
\begin{aligned}
2(b + 7) + b &= 65 \\
2b + 14 + b &= 65 \\
3b + 14 &= 65 \\
3b &= 51 \\
b &= 17
\end{aligned}
$$

Then, from the equation $b + 7 = l$ we can solve the $l = 17 + 7 = 24$.

The sides of the isosceles triangle measure <u>17, 24, and 24 units</u>.

6. Angle BAD and the 52° angle are corresponding angles, which means ∠BAD = 52°. Then, ∠BAC = ∠BAD − ∠CAD = 52° − 24° = 28°. Then, ∠ACD and ∠BAC are alternate interior angles, thus congruent. So, x = ∠BAC = <u>28°</u>.

7. See the image on the right.

   a. $y = (3/2)x - 2$, or in standard form, $3x - 2y = 4$.

   b. The equation is of the form $y = -3x + b$. Substituting the point $(-4, 5)$ into this, we get $5 = -3(-4) + b$, from which $b = -7$. So, the equation of the line is $\underline{y = -3x - 7}$, or in standard form, $\underline{3x + y = -7}$.

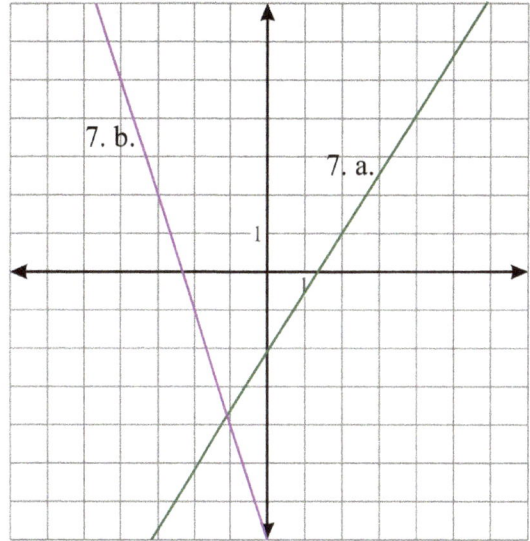
7. b.   7. a.

8.

  a. Passes through $(4, 1)$ and has slope $1/4$ **(i)** $y = (1/4)x - 2$

  b. Passes through $(-1/4, 1/4)$ and has slope 0 **(ii)** $x = 1/4$

  c. Passes through $(1/4, 1/4)$ and has no slope   **(iii)** $y = (1/4)x$

  d. Passes through $(4, -1)$ and $(-8, -4)$ **(iv)** $y = 1/4$

9. See the graphs on the right.
   a. The x-intercept is 2 and the y-intercept is -4.
   b. The x-intercept is -2 and the y-intercept is -6.

10. Let p be its price before the increases. Then,
   $1.03 \cdot 1.04 \cdot 1.09p = 268.55$, from which
   $p = 268.55/(1.03 \cdot 1.04 \cdot 1.09) = 230.00$.
   The item cost \$230 before the price increases.

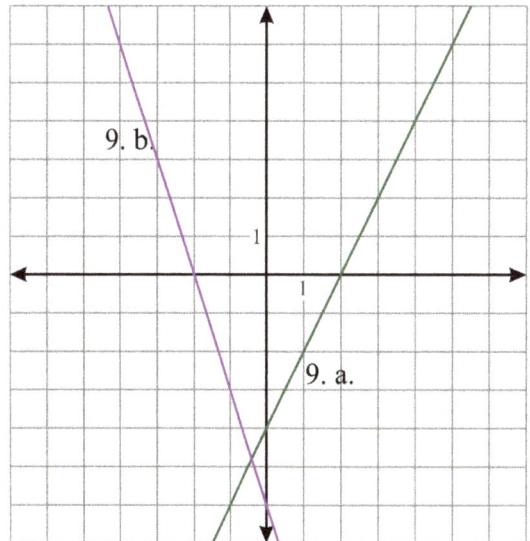
9. b.   9. a.

11. (1) $V_{total} = \pi \cdot (3.25 \text{ cm})^2 \cdot 25 \text{ cm} / 3 \approx 276.5256 \text{ cm}^3$.

    (2) $V_{cut} = \pi \cdot (1.9 \text{ cm})^2 \cdot 14.5 \text{ cm} / 3 \approx 54.8156 \text{ cm}^3$.

    (3) $V_{glass} = 276.5256 \text{ cm}^3 - 54.8156 \text{ cm}^3 \approx 220 \text{ cm}^3$.

    Note that since most of the measurements were given to two significant digits, we give the final answer also to two significant digits.

Cumulative Review, Chapters 1-8

1. The bottom row of the table gives the coordinates after all the transformations.

Original	A(5, −1)	B(3, −4)	C(1, −3)
After reflection	A'(5, 1)	B'(3, 4)	C'(1, 3)
After dilation	A"(10, 2)	B"(6, 8)	C"(2, 6)
After rotation	A'''(2, −10)	B'''(8, −6)	C'''(6, −2)

2. a. One solution.
   b. $10s - 6 = 10s$ or $2s - 6 = 2s$.

3.

a. $\begin{cases} 3x - y = 14 \\ 6x + 3y = -12 \end{cases}$

We can solve for y from the top equation: $y = 3x - 14$. Substituting that for y in the second equation, we get:

$$6x + 3(3x - 14) = -12$$
$$6x + 9x - 42 = -12$$
$$15x - 42 = -12$$
$$15x = 30$$
$$x = 2$$

Then, substituting 2 for x in the top equation, we get:

$$3(2) - y = 14$$
$$-y = 14 - 6 = 8$$
$$y = -8$$

Solution: $(2, -8)$

b. $\begin{cases} 3x - 8(y - 2) = 0 \\ 2x = 8y + 1 \end{cases}$

$\downarrow$

$\begin{cases} 3x - 8y + 16 = 0 \\ 2x = 8y + 1 \end{cases}$

$\downarrow$

$\begin{cases} 3x - 8y = -16 \\ 2x - 8y = 1 \quad | \cdot (-1) \end{cases}$

$\downarrow$

$\begin{array}{c} \begin{cases} 3x - 8y = -16 \\ -2x + 8y = -1 \end{cases} \\ \hline + \qquad x \qquad = -17 \end{array}$

Substituting $x = -17$ in the second equation, we get:

$$2(-17) = 8y + 1$$
$$-34 = 8y + 1$$
$$8y = -35$$
$$y = -35/8$$

Solution: $(-17, -35/8)$

4. a. Not correct. Corrected form: $\sqrt{121}$ is rational because it is a square root of a perfect square (it equals 11).
   b. Not correct. Corrected form: 0.831831831 is rational because it is a terminating decimal. Or, because it is the fraction 831,831,831/1,000,000,000.
   c. Correct.
   d. Correct.
   e. Not correct. Corrected form: $\sqrt{63}/4$ is irrational because $\sqrt{63}$ is irrational, 4 is rational, and an irrational number divided by a rational number is irrational.

5. Let V be the volume of the 3% milk in this mixture. The chart looks like this:

	volume (ml)	butterfat percentage	butterfat amount
3% milk	V	3	0.03V
15% cream	1000 − V	15	0.15(1000 − V)
Mixture	1000	12	120

From the last column, we can write the equation $0.03V + 0.15(1000 - V) = 120$. Here is its solution:

$$0.03V + 0.15(1000 - V) = 120$$
$$0.03V + 150 - 0.15V = 120$$
$$150 - 0.12V = 120$$
$$-0.12V = -30$$
$$V = 250$$

You should mix 250 mL of the 3% milk and 750 mL of the 15% cream in order to get 1 liter of a mixture that has 12% butterfat.

6. In the spot marked below the variable x gets switched to y. The corrected solution is on the right.

(1) $\quad \begin{cases} 3x - 7y = 4 \\ -5x + y = -4 \end{cases} \cdot 7$

$\downarrow$

$\begin{cases} 3x - 7y = 4 \\ -35x + 7y = -28 \end{cases}$

$\overline{\quad\quad\quad\quad\quad\quad}$

$-32y = -24 \quad \leftarrow$ Here is where the x and y got switched.

$y = 3/4$

$\downarrow$

(1) $\quad 3x - 7(3/4) = 4$

$3x - 21/4 = 4$

$3x = 4 + 21/4 = 37/4$

$3x = 37/4$

$x = 37/12$

---

(1) $\quad \begin{cases} 3x - 7y = 4 \\ -5x + y = -4 \end{cases} \cdot 7$

$\downarrow$

$\begin{cases} 3x - 7y = 4 \\ -35x + 7y = -28 \end{cases}$

$\overline{\quad\quad\quad\quad\quad\quad}$

$-32x = -24$

$x = 3/4$

$\downarrow$

(1) $\quad 3(3/4) - 7y = 4$

$9/4 - 7y = 4$

$-7y = 4 - 9/4$

$-7y = 7/4$

$y = -1/4$

Solution: $(3/4, -1/4)$

7. We ignore the negative roots since these are lengths of sides of triangles.

a.	b.
$11.2^2 + 7.6^2 = s^2$	$(\sqrt{26})^2 + x^2 = 7^2$
$11.2^2 + 7.6^2 = s^2$	$26 + x^2 = 49$
$183.2 = s^2$	$x^2 = 23$
$s = \sqrt{183.2} \approx 13.5$	$x = \sqrt{23}$

8. Let h be the height of the triangle. We apply the Pythagorean Theorem in the triangle BCD, in order to find the height of the triangle:

$8.25^2 + h^2 = 16.5^2$

$h^2 = 16.5^2 - 8.25^2$

$h^2 = 204.1875$

$h = \sqrt{204.1875} \approx 14.289$ in

Then, the area is A = 16.5 in (14.289 in) / 2 ≈ <u>118 square inches</u>.

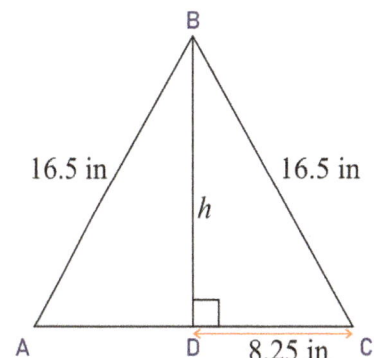

9. a. See the table on the right.

b. Yes. If there wasn't, we would see similar numbers for each hobby for boys and girls. But for example for arts and crafts, there are over twice as many girls than boys. Or, for reading, there are about 1.5 times as many girls as boys.

10. $\quad V = \pi r^2 h$

$\dfrac{V}{\pi h} = r^2$

$r = \sqrt{\dfrac{V}{\pi h}}$

Since this has to do with the radius of a circle, we discard the negative root.

Favorite hobbies of 2nd graders

	Boys	Girls	Total
Sports	26	21	47
Music	12	13	25
Reading	13	20	33
Arts & Crafts	5	13	18
Video games	16	7	23
Cooking	2	5	7
Photography	6	5	11
Total	80	84	164

11. a. b. Answers will vary. Check the student's answer. Here are two examples:

(1)

Tree Diameter vs. Height

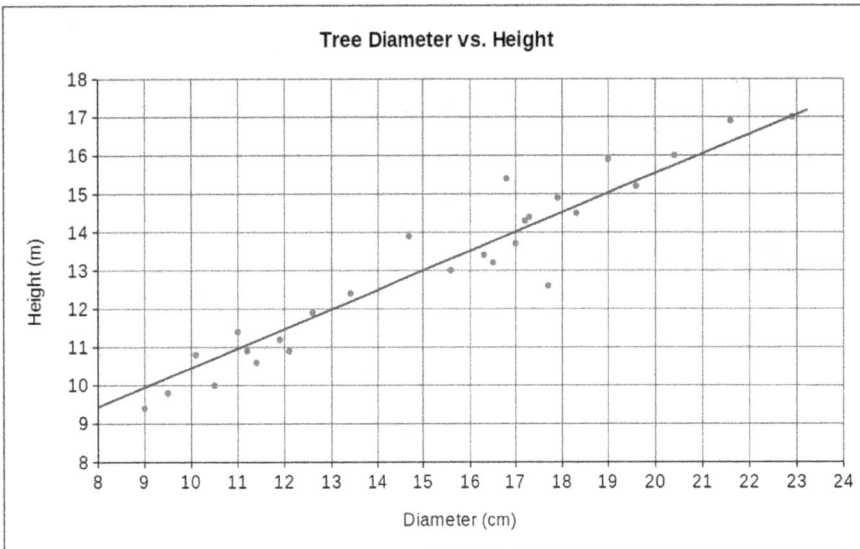

Approximate equation: The line goes through the points (8, 9.5) and (15, 13), from which the slope is 3.5/7 = 0.5. The equation is $y = 0.5x + b$. Substituting (15, 13), we get $13 = 0.5(15) + b$, from which $b = 5.5$.
So, the equation is $y = 0.5x + 5.5$.

(2)

Tree Diameter vs. Height

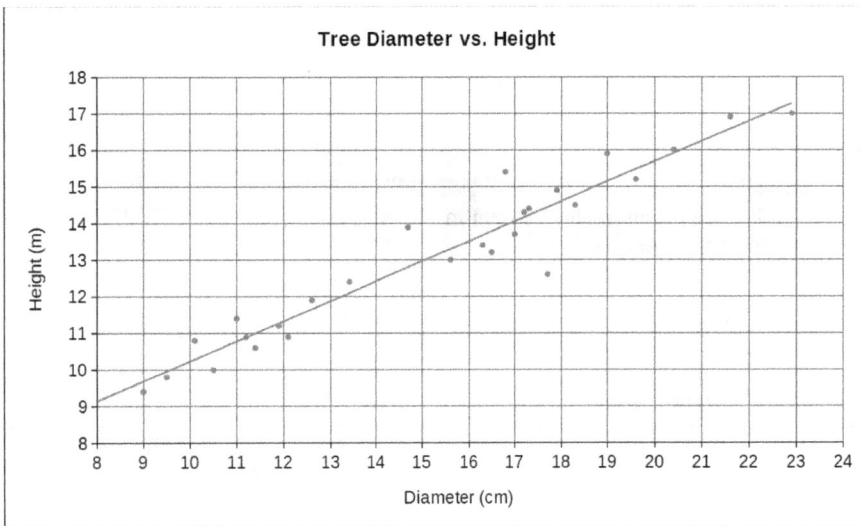

Approximate equation: $y = 0.55x + 4.77$ (calculated by a spreadsheet program).

c. The slope means that each additional centimeter in tree diameter is associated with a 0.5-meter (or 0.55-meter) increase in height.

d. It signifies that the model predicts that a tree with a diameter of zero centimeters is 5.5 m (4.77 m) high. This clearly doesn't make sense, so, we cannot extrapolate this model that much backwards.

e. Student answers will vary since the equations will vary. If the tree is 16.5 m tall, we will calculate the diameter using the equation $y = 0.5x + 5.5$:

$$16.5 = 0.5x + 5.5$$
$$0.5x = 11$$
$$x = 22$$

The equation predicts that the diameter of the tree is 22 cm .

www.ingramcontent.com/pod-product-compliance
Lightning Source LLC
Chambersburg PA
CBHW080528220326
41599CB00032B/6235